Resistance in Weeds from Herbicide Metabolism

Resistance in Weeds from Herbicide Metabolism

Edited by

Vijay K. Nandula
USDA-NIFA, Kansas City
Missouri, USA

Roland Beffa
Senior Scientist Consultant, Liederbach
Hesse, Germany

WILEY

Published by John Wiley & Sons, Inc., Hoboken, New Jersey.
Published simultaneously in Canada.

Library of Congress Cataloging-in-Publication Data applied for

Hardback ISBN: 9781119686644

Cover Design: Wiley
Cover Image: Courtesy of Roland Beffa

Set in 9.5/12.5pt STIXTwoText by Straive, Chennai, India

My wife Aparna and my children Ajay and Indu

Vijay K. Nandula

My wife Luigina and my children Florence and Mathias, and Elodie

Roland Beffa

Contents

List of Contributors

Hugh J. Beckie
Australian Herbicide Resistance Initiative
School of Agriculture and Environment
The University of Western Australia
Perth, Western Australia, Australia

Roland Beffa
Senior Scientist Consultant
Liederbach, Hesse, Germany

Stewart Brown
School of Natural and Environmental Sciences
Newcastle University
Newcastle Upon Tyne, UK

Alina Goldberg Cavallieri
School of Natural and Environmental Sciences
Newcastle University
Newcastle Upon Tyne, UK

Robert Cicchillo
Crop Protection Discovery and Development
Corteva Agriscience
Indianapolis, IN, USA

Jeanaflor Crystal T. Concepcion
Department of Crop Sciences
University of Illinois
Urbana, IL, USA

Nina Gracel Dimaano
College of Agriculture and Food Science
University of the Philippines Los Banos
Los Banos, Philippines

Stephen O. Duke
National Center for Natural Product Research
School of Pharmacy
University of Mississippi
Oxford, Mississippi, USA

Razvan Dumitru
Biotechnology
Corteva Agriscience
Johnston, IA, USA

Robert Edwards
School of Natural and Environmental Sciences
Newcastle University
Newcastle Upon Tyne, UK

Todd Gaines
Department of Agricultural Biology
Colorado State University
Fort Collins, CO, USA

Samantha Griffin
Crop Protection Discovery and Development
Corteva Agriscience
Indianapolis, IN, USA

Heping Han
Australian Herbicide Resistance Initiative
School of Agriculture and Environment
The University of Western Australia
Perth, Western Australia, Australia

Satoshi Iwakami
Graduate School of Agriculture
Tokyo University of Agriculture and Technology
Fuchu, Tokyo, Japan

Mithila Jugulam
Texas A&M AgriLife
Research and Extension Center
Beaumont

Shiv S. Kaundun
Herbicide Bioscience
Syngenta
Jealott's Hill International Research Centre
Bracknell, Berkshire, UK

Eshagh Keshtkar
Department of Agronomy
Tarbiat Modares University
Tehran, Iran

Per Kudsk
Department of Agroecology
Aarhus University
Slagelse, Denmark

Vijay K. Nandula
USDA-NIFA
Kansas City, Missouri, USA

Nawaporn Onkokesung
School of Natural and Environmental Sciences
Newcastle University
Newcastle Upon Tyne, UK

Sara Franco Ortega
School of Natural and Environmental Sciences
Newcastle University
Newcastle Upon Tyne, UK

Balaji Aravindhan Pandian
Enko Chem Inc. Mystic
USA

P. V. Vara Prasad
Department of Agronomy
Kansas State University
USA

Christopher Preston
School of Agriculture, Food & Wine
University of Adelaide
Glen Osmond South Australia
Australia

Dean E. Riechers
Department of Crop Sciences
University of Illinois
Urbana, IL, USA

Dilpreet Riar
Crop Protection Discovery and Development
Corteva Agriscience
Indianapolis, IN, USA

Carlos Alberto Gonsiorkiewicz Rigon
Department of Agricultural Biology
Colorado State University
Fort Collins, CO, USA

Alice A. Wright
USDA-ARS
Houma, LA, USA

Terry Wright
Biotechnology
Corteva Agriscience
Indianapolis, IN, USA

Carla Yerkes
Crop Protection Discovery and Development
Corteva Agriscience
Indianapolis, IN, USA

Qin Yu
Australian Herbicide Resistance Initiative
School of Agriculture and Environment
The University of Western Australia
Perth, Western Australia, Australia

About the Editors

Dr. Vijay K. Nandula, provides US national leadership in Weed Science and Integrated Pest Management in his current role as a National Program Leader in the Division of Plant Systems-Protection in the Institute of Food Production and Sustainability at the National Institute of Food and Agriculture (NIFA) of USDA. Prior to joining USDA-NIFA, Vijay worked as a Research Plant Physiologist in the Crop Production Systems Research Unit, Agricultural Research Service, USDA, where he established a recognized research program on biology, physiology, and management of herbicide-resistant weeds. As an Assistant Research Professor at the Mississippi State University, Vijay initiated an aggressive and comprehensive research program to address herbicide-resistant weed management issues across the southeastern United States. He also edited the landmark book *Glyphosate Resistance in Crops and Weeds: History, Development, and Management*. Vijay has published 76 peer-reviewed journal articles, delivered more than 150 presentations at regional, national, international professional society meetings, and served as an Associate Editor for the journal *Weed Science* of the Weed Science Society of America (WSSA) for several years.

Dr. Roland Beffa, after receiving his PhD from the University of Lausanne (Switzerland) in Plant Physiology and Biochemistry, did research and development for 23 years in academia and 22 years in industry across several countries in Europe and in the United States. Dr. Beffa has developed knowledge and experience in plant and fungal physiology, biochemistry, genetic and genomics, in addition to strong experience in agronomy related to disease and weed control. During the last 15 years, he dedicated his research and applied work to the evolution of herbicide resistance, especially on the molecular mechanisms involved in herbicide detoxification causing metabolic resistance in weeds. Dr. Beffa is currently working independently as a senior scientist consultant with projects supporting stakeholders involved in crop disease and weed control including farmers, advisors, retailors, authorities, industry, and academic scientists. One recent project is dedicated to using genomics knowledge to develop models aiming to help farmers to decide the best strategies for mitigating the evolution of weed resistance to herbicides in the fields. Dr. Beffa is a Member of the Weed Science Society of America (WSSA) and a Board Member of the European Weed Research Society (EWRS).

Preface

Among other mechanisms, metabolism, i.e., herbicide detoxification, is a widespread mode of resistance to herbicides in many weeds that infest crops in agronomic systems. It is mediated by complex enzyme systems such as cytochrome P450 monooxygenases, glutathione *S*-transferases, and glycosyltransferases. These enzymes as well as cofactors, such as glutathione, can be activated or catalyzed in certain crop species (so far, only monocotyledonous) by certain chemicals called safeners. Safeners are applied in combination with herbicides to provide tolerance in grass crops such as wheat, rice, corn, and sorghum against certain thiocarbamate, chloroacetamide, sulfonylurea, and aryoxyphenoxypropionate herbicides applied preemergence or postemergence. Also certain genetically modified herbicide-tolerant crops are able to highly detoxify the applied herbicides. Metabolism of herbicides usually occurs in three phases: a conversion of the herbicide molecule into a more hydrophilic metabolite (phase 1); followed by conjugation to biomolecules such as glutathione or sugar (phase 2); and further conjugation, breakup, or oxidation reactions, followed by transport into vacuoles or cell walls where additional breakdown occurs (phase 3). This results in the transformation of the herbicide's active ingredient into non-phytotoxic metabolites. Phase 2 can occur directly, depending on the chemical structure of the herbicide being detoxified. The next and most important phase after the confirmation of herbicide resistance in a weed population is deciphering of the underlying resistance mechanism(s), which can greatly determine the effectiveness of resistance management strategies. Nontarget site resistance, in particular metabolic resistance, is of high concern since it is not related to a given herbicide target (mode of action) but to chemical structures. Therefore, it can confer resistance to a broad spectrum of herbicides representing several modes of action. HRAC classification of herbicides according to its mode of action is useful to manage target site resistance (TSR) but would need to be updated to manage non-target site resistance (NTSR) mechanisms, in particular metabolic resistance. Therefore, an immediate and urgent challenge for weed scientists is to understand and characterize the mechanisms involved in the evolution of metabolic resistance and their regulation. This will contribute to find novel solutions for resistance management to sustain the limited herbicide portfolio (low innovation rate with few herbicides with novel modes of action introduced in the past 30 years and constant increase in the costs to commercialize new active ingredient, up to about US $300 million per new active ingredient). This will provide new insights for the development of integrated weed management strategies. Metabolic resistance research in weeds has mostly been limited to grass species such as rigid ryegrass, blackgrass, and *Echinochloa* spp. However, in recent years there are increasing reports of dicot species, such as Palmer amaranth and tall waterhemp, which have developed resistance to multiple herbicide mechanisms of action by enhancing metabolic degradation. Thus, both grass- and dicotyledonous-species can develop metabolic herbicide resistance. The main objective of this book is to present an up-to-date single source of information on current research and understanding of metabolic resistance to herbicides in weeds by revisiting history of related research including crop tolerance, reporting recent advances brought about by technologies, and identifying future research opportunities. The book will primarily be used as a reference source. There are currently no books available that are similar or comparable to this book. This book will be useful across disciplines as varied as weed science, agronomy, crop science, crop protection, soil science, botany, chemistry, biochemistry, genomics, genetics, molecular and cell biology, ecology, plant physiology, and biotechnology. This book provides a global perspective and addresses local issues.

April 30, 2025

Vijay K. Nandula
Kansas City, Missouri, USA

Roland Beffa
Liederbach, Hesse, Germany

USDA Disclaimer

The views and opinions provided herein are my own and do not represent the views of the National Institute of Food and Agriculture, or the United States Department of Agriculture, or the United States.

Vijay K. Nandula

Acknowledgments

We sincerely acknowledge the patience, support, and encouragement of the editorial team at John Wiley and Sons, Inc. despite the progress of this manuscript project being severely affected by the COVID-19 pandemic. We are very grateful to the corresponding authors and their co-authors for coming through every time and fulfilling our requests for revisions, updates, and clarifications on the respective chapters, and for sticking with us. We wholeheartedly recognize the work of reviewers for providing comprehensive reviews on time always. We are very thankful for the love, support, and backing from our families during this book project.

Vijay K. Nandula
Roland Beffa

1

Introduction to Herbicide Metabolism, Resistance, and Related Concepts

Vijay K. Nandula

USDA-NIFA, Kansas City, Missouri, USA

1.1 Introduction

Weeds were in existence even before humans took up the cultivation of plants for food, feed, fuel, and fiber. Before the advent of synthetic organic-based herbicides in the 1940s, weeds were controlled for thousands of years by mechanical, cultural, and biological means. 2,4-Dichlorophenoxyacetic acid (2,4-D) was the first herbicide to be used selectively to control weeds. Since then, several herbicides belonging to different chemical classes and possessing diverse modes of action have been synthesized and commercialized around the world. Herbicides have widely contributed to increasing world food production in an efficient, economical, and environmentally sustainable manner. However, repeated application(s) of the same herbicide or a different herbicide with a similar mode of action on the same field, growing season after growing season, has contributed to the widespread occurrence of resistance to herbicides in several weed species. Currently, there are 534 unique cases (species × site of action) of herbicide-resistant weeds globally, with 273 species (156 dicots and 117 monocots); weeds have evolved resistance to 21 of the 31 known herbicide sites of action and to 168 different herbicides; and herbicide-resistant weeds have been reported in 102 crops in 75 countries.[1]

One of the most important phases after the confirmation of herbicide resistance in a weed population is the deciphering of the underlying resistance mechanism(s). The mechanism of resistance to herbicide(s) in a weed population can greatly determine the effectiveness of resistance management strategies. For example, a target site mutation could endow cross-resistance to herbicides with similar mechanism of action, or metabolic resistance can bestow

Resistance in Weeds from Herbicide Metabolism, First Edition.
Edited by Vijay K. Nandula and Roland Beffa.

the ability to survive herbicides across more than one mechanism of action. In general, mechanisms of resistance to herbicides are broadly categorized into target-site- and non-target-site-based mechanisms. The major category in the non-target-site-based mechanisms is resistance via the metabolism/breakdown of herbicide molecules to non-phytotoxic metabolites that are further subject to conversion to secondary compounds that are also found in plants.

1.2 Herbicide Resistance and Tolerance

The Weed Science Society of America (WSSA)[2] defines herbicide resistance and herbicide tolerance as follows. Herbicide resistance is the inherited ability of a plant to survive and reproduce following exposure to a dose of herbicide normally lethal to the wild type. In a plant, resistance may be naturally occurring or induced by techniques such as genetic engineering or the selection of variants produced by tissue culture or mutagenesis. Herbicide tolerance is the inherent ability of a species to survive and reproduce after herbicide treatment. This implies that there is no selection or genetic manipulation to make the plant tolerant as it is naturally tolerant.

The natural tolerance of many crops to herbicides is thus exploited through the concept of herbicide selectivity, which is the ability of an herbicide to kill or slow the growth of some plants (weeds) without affecting others (crops). Herbicide selectivity or tolerance is mostly determined by the ability of a crop plant to metabolize an herbicide to non-phytotoxic metabolites. Metabolism of herbicides usually occurs in three phases: a conversion of the herbicide molecule into a more hydrophilic metabolite (phase 1), followed by conjugation to biomolecules such as glutathione/sugar (phase 2), and further conjugation/breakup/oxidation reactions followed by transport to vacuoles or cell walls where additional breakdown occurs (phase 3).[3–5]

Safeners. In general, safeners are chemicals applied in combination with herbicides to provide tolerance to grass crops such as wheat, rice, corn, and sorghum against certain thiocarbamate, chloroacetamide, sulfonylurea, and aryloxyphenoxypropionate herbicides that are applied preemergence or postemergence. Safeners enhance herbicide detoxification in "safened" plants. Safening agents activate/catalyze cofactors such as glutathione and enzyme systems such as cytochrome P450 (CYP) monooxygenases, glutathione-*S*-transferases (GSTs), and glycosyl transferases.[6] The safener-mediated induction of herbicide-detoxifying enzymes appears to be part of a general stress response.[4,6] These enzymes deactivate herbicide molecules by modifying side chains, which are then conjugated to biochemical moieties such as sugar and amino acid residues. Some of these conjugates are further deposited in vacuoles and cell walls.

1.3 Layout of Content in the Following Chapters

1.3.1 Chapter 2: Synthetic Auxin Herbicide-Tolerant Crops

This chapter reviews the overarching processes and approaches utilized for the discovery of 2,4-D- and dicamba-tolerant trait technologies. Additionally, this chapter discusses deregulation, adoption, potential future challenges, and opportunities with these two synthetic auxin herbicide-tolerant trait technologies.

1.3.2 Chapter 3: Metabolic Degradation of Glyphosate in Soil Microbes, Endophytes, Crops, and Weeds

Glyphosate is the most used herbicide worldwide. Both microbes and plants can metabolize it by various enzymes. This chapter comprehensively reviews the metabolic fate of glyphosate in soil microorganisms, endophytes, crops, and weeds.

1.3.3 Chapter 4: Enhanced Metabolic Resistance to Herbicides in *Alopecurus myosuroides* (Black Grass)

Black grass is a major problem weed of winter cereals in Northern Europe, which is difficult to control with herbicides due to the rapid evolution of non-target-site resistance (NTSR) to selective herbicides. Enhanced metabolic resistance (EMR) underpins NTSR in black grass and is correlated with the increased expression of CYPs, GSTs, UDP-glucose-dependent glycosyltransferases (UGTs), and ATP-binding cassette (ABC) transporters that act concertedly to detoxify herbicides. In this chapter, the role of these proteins in herbicide metabolism and EMR in black grass is reviewed, along with an analysis of their respective gene diversity and an assessment of likely members involved in NTSR.

1.3.4 Chapter 5: Herbicide Detoxification in Black Grass

This chapter provides additional insights into the metabolic mechanisms of resistance in black-grass populations in Europe.

1.3.5 Chapter 6: Metabolism-Based Resistance in *Lolium rigidum* and *Raphanus raphanistrum*

L. rigidum Gaud. (annual ryegrass) and *R. raphanistrum* L. (wild radish) are abundant, widespread, and economically damaging weed species across the grain-cropping regions of southern Australia. Multiple herbicide resistance because of sequential selection and/or gene flow is prevalent in populations of these two outcrossing weed species, conferred by both target-site and non-target-site resistance mechanisms. This chapter focuses on recent studies involving these two species that investigated herbicide resistance conferred by enhanced metabolism, which can simultaneously compromise the efficacy of herbicides from different sites of action.

1.3.6 Chapter 7: Management of Metabolism-Based Resistance in *Lolium rigidum* in Australia

This chapter reviews the development and adoption of novel herbicide resistance management strategies to combat resistant *L. rigidum* populations across cropping regions of Australia.

1.3.7 Chapter 8: Metabolic Herbicide Resistance in *Echinochloa phyllopogon*

This chapter reviews almost 20 years of collective research efforts on a multiple herbicide-resistant (MHR) line of *E. phyllopogon* (syn. *Echinochloa oryzicola*), a noxious grass weed in rice fields, which has developed resistance to at least 15 herbicides from 10 chemical classes due to metabolism-based mechanisms.

1.3.8 Chapter 9: Metabolic Resistance to Herbicides in *Echinochloa crus-galli* (and *Echinochloa colona*)

This chapter reviews past and recent research on multiple herbicide-resistant *Echinochloa colona* and *Echinochloa crus-galli* populations in the United States and around the world with a focus on metabolic resistance mechanisms.

1.3.9 Chapter 10: Metabolic Resistance to HPPD-Inhibiting Herbicides in Dioecious *Amaranthus* Species

This chapter discusses the current state of knowledge regarding the occurrences of and metabolic basis for resistance to various 4-hydroxyphenylpyruvate dioxygenase (HPPD)-inhibiting herbicides in dioecious *Amaranthus* spp., including descriptions of molecular genetic, biochemical, and physiological aspects of resistance.

1.3.10 Chapter 11: Overview of Metabolism-Based Resistance to PS-II Inhibitors in Weed Species

Photosystem (PS) II inhibitors are one of the most important and widely used herbicides in modern crop production across the globe. In addition to target-site-based resistance, metabolic resistance to PS II-inhibiting herbicides is common in several weed species, which is the focus of this chapter.

1.3.11 Chapter 12: Fitness and Ecophysiological Cost of Metabolic Herbicide Resistance

This chapter comprehensively discusses the fitness and ecological costs of metabolic herbicide resistance using major weed populations across the world as models.

1.3.12 Chapter 13: Advancing Metabolic Resistance Characterization: Paving the Way for Future Technologies in Weed Management

This chapter systematically reviews past metabolic resistance research, summarizes recent research on metabolic herbicide resistance with emphasis on molecular and genomic tools, and identifies future avenues for training and collaboration by embracing and leveraging the power of interdisciplinary collaboration to effectively address the challenges posed by herbicide resistance in weeds, ensuring sustainable agriculture and enhanced weed control practices for the future.

References

1. Heap I, The International Herbicide-Resistant Weed Database. www.weedscience.org Accessed April 30, 2025.
2. Weed Science Society of America (WSSA), Herbicide resistance and herbicide tolerance defined. *Weed Technol* **12**:789 (1998).
3. Kreuz K, Tommasini R, and Martinoia E, Old enzymes for a new job. *Plant Physiol* **111**:349–353 (1996).
4. Délye C, Unravelling the genetic bases of non-target-site-based resistance (NTSR) to herbicides: a major challenge for weed science in the forthcoming decade. *Pest Manag Sci* **69**:176–187 (2013).
5. Van Eerd LL, Hoagland RE, Zablotowicz RM, and Hall JC, Pesticide metabolism in plants and microorganisms. *Weed Sci* **51**:472–495 (2003).
6. Hatzios KK, and Burgos N, Metabolism-based herbicide resistance: regulation by safeners. *Weed Sci* **52**:454–467 (2004).

2

Synthetic Auxin Herbicide-Tolerant Crops

Dilpreet Riar[1], Carla Yerkes[1], Robert Cicchillo[1], Samantha Griffin[1], Razvan Dumitru[2], and Terry Wright[3]

[1] *Crop Protection Discovery and Development, Corteva Agriscience, Indianapolis, IN, USA*
[2] *Biotechnology, Corteva Agriscience, Johnston, IA, USA*
[3] *Biotechnology, Corteva Agriscience, Indianapolis, IN, USA*

2.1 Introduction to Synthetic Auxin Herbicides

Synthetic auxin herbicides (SAHs), collectively identified as Group 4 Auxin Mimics (legacy group O) by the Weed Science Society of America (WSSA),[1] are a commercially important class of herbicides that mimic the activity of the naturally occurring plant hormone, indole-3-acetic acid (IAA, Fig. 2.1). IAA regulates numerous plant developmental processes, such as cell division, elongation, and differentiation; signal transduction; metabolism; plant growth; and responses to environmental influences.[2] As IAA is a plant growth promoter at low concentrations and plant growth inhibitor at higher concentrations, its cellular concentration is highly regulated by *de novo* IAA synthesis from tryptophan via the indole-3-pyruvic acid (IPA) pathway[3,4] reversible conversion of IAA conjugated with sugars or amino acids, and irreversible catabolism.[2,5] In the late 1930s, three independent research groups identified that 1-naphthalene acetic acid (1-NAA, Fig. 2.1), a synthetic compound, provided a similar level of growth inhibition as noted with IAA.[6]

After the discovery of 1-NAA, multiple laboratories started studying novel synthetic plant growth regulators, leading to the discovery of several important SAHs. Commercial SAHs are subclassified by Herbicide Resistance Action Committee (HRAC) into chemical families (Table 2.1): 6-arylpicolinates, 6-chloropicolinates, benzoates, phenoxy-carboxylates, pyridyloxy-carboxylates, quinoline-carboxylates, and others (nonclassified single actives).[7]

Resistance in Weeds from Herbicide Metabolism, First Edition.
Edited by Vijay K. Nandula and Roland Beffa.

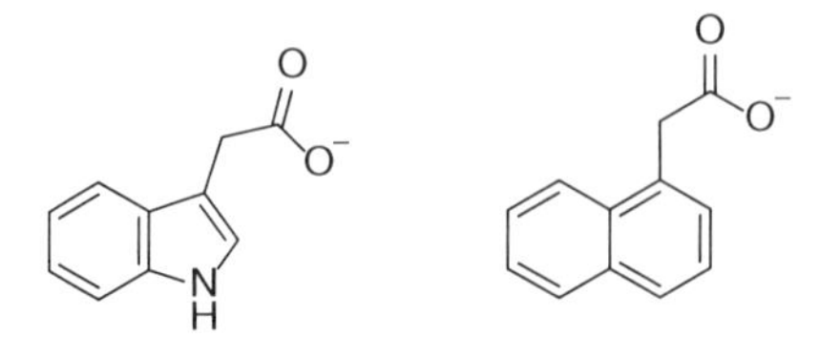

Figure 2.1 Structure of the natural auxin indole-3-acetic acid (IAA) and the synthetic auxin 1-naphthalene acetic acid (1-NAA).

2,4-Dichlorophenoxyacetic acid (2,4-D) and 2,4-DB (4-(2,4-dichlorophenoxy) butyric acid), belonging to the phenoxy-carboxylate family of synthetic auxins, were among the first synthetic organic herbicides discovered and commercialized in the late 1940s.[8,9] Discovery of these first systemic and selective herbicides, which were relatively inexpensive, became the foundation of the present-day weed science.[10] The first SAHs transformed agriculture and were considered among the greatest scientific discoveries of the twentieth century.[6,11] Dicamba, another important SAH from the benzoate chemical family, was also discovered in the early 1940s by Zimmerman and Hitchcock, and was later registered as Banvel for use in the United States by Velsicol Chemical Corporation in the early 1960s.[12]

While studying pyridine-based soil nitrification inhibitors in the late 1950s, scientists from the Dow Chemical Company unexpectedly discovered the 6-chloropicolinate chemical family of SAHs and went on to commercialize picloram in 1963, clopyralid in 1970, and aminopyralid in 2006.[13–15] Structure–activity relationship (SAR) studies on 6-chloropicolinates also led to the discovery of the pyridyloxy-carboxylates triclopyr (1979) and fluroxypyr (1998).[15]

In recent years (2015–2018), Corteva Agriscience (previously Dow Agro-Sciences) launched two new SAHs, halauxifen-methyl and florpyrauxifen-benzyl, that belong to 6-arylpicolinates chemical family. These 6-arylpicolinates differ from previous commercial 6-chloropicolinate in that they possess an aryl- instead of chloro-substituent at position 6 of pyridine ring and are labeled for use rates less than 50 g a.i./ha. Florpyrauxifen is also unique from most other SAHs in its control of several warm-season monocot species (e.g., barnyard grass and sedges).[9]

The other SAH, which exhibits warm-season grass activity, quinclorac, is a quinoline-carboxylate discovered by BASF in the 1980s. It was the first SAH registered for both dicot and monocot weeds with excellent safety to rice; however, its mode of action (MOA) against monocots differs from that in dicots in that it is non-auxinic in mechanism.[9,16]

Aminocyclopyrachlor (registered by Dupont in 2010) is not classified by HRAC into any chemical family and is a selective herbicide that provides pre- and postemergent control of numerous dicot weeds, woody species, and vines. Finally,

Table 2.1 Various synthetic auxin herbicides (SAHs), or auxin mimics are shown with their chemical classes, active moieties, and year of launch.

Chemical class	Commercial form	Active acid moiety	Year of introduction
Phenoxy-carboxylates	2,4-D (various salts and esters)	2,4-D	1945
	2,4-DB	2,4-D	1944
	MCPA ((4-Chloro-2-methylphenoxy)acetic acid) (various salts and esters)	MCPA	1950
	MCPB	MCPA	1960
	Dichlorprop (various salts and esters)	Dichlorprop	1961
	Mecoprop (various salts and esters)	Mecoprop	1957
	Clomeprop	DMPA	1986
Benzoates	Dicamba (various salts and esters)	Dicamba	1963
	Chloramben	Chloramben	1958
6-Chloropicolinates	Picloram (various salts and esters)	Picloram	1963
	Clopyralid (various salts and esters)	Clopyralid	1977
	Aminopyralid (various salts)	Aminopyralid	2005
6-Arylpicolinates	Halauxifen-methyl	Halauxifen	2015
	Florpyrauxifen-benzyl	Florpyrauxifen	2017
Pyridyloxy-carboxylates	Triclopyr (various salts and esters)	Triclopyr	1979
	Fluroxypyr (various salts and esters)	Fluroxypyr	1985
Quinoline-carboxylates	Quinclorac	Quinclorac	1989
	Quinmerac	Quinmerac	1993
Other[a)]	Benazolin-ethyl	Benazolin	1964
	Aminocyclopyrachlor	Aminocyclopyrachlor	2010

2,4-D, 2,4-dichlorophenoxyacetic acid; DMPA, 2,4-dichloro-3-methyl-phenoxypropionic acid; MCPA, 2-methyl-4-chlorophenoxyacetic acid.

a) Single molecules in classes are not classified by Herbicide Resistance Action Committee (HRAC).

Source: Ref: Busi et al.[9], WSSA, (https://wssa.net/wssa/weed/herbicides/), HRAC poster (https://hracglobal.com/tools/2024-hrac-global-herbicide-moa-classification), PPDB: Pesticide Properties DataBase (http://sitem.herts.ac.uk/aeru/ppdb/en/Reports.

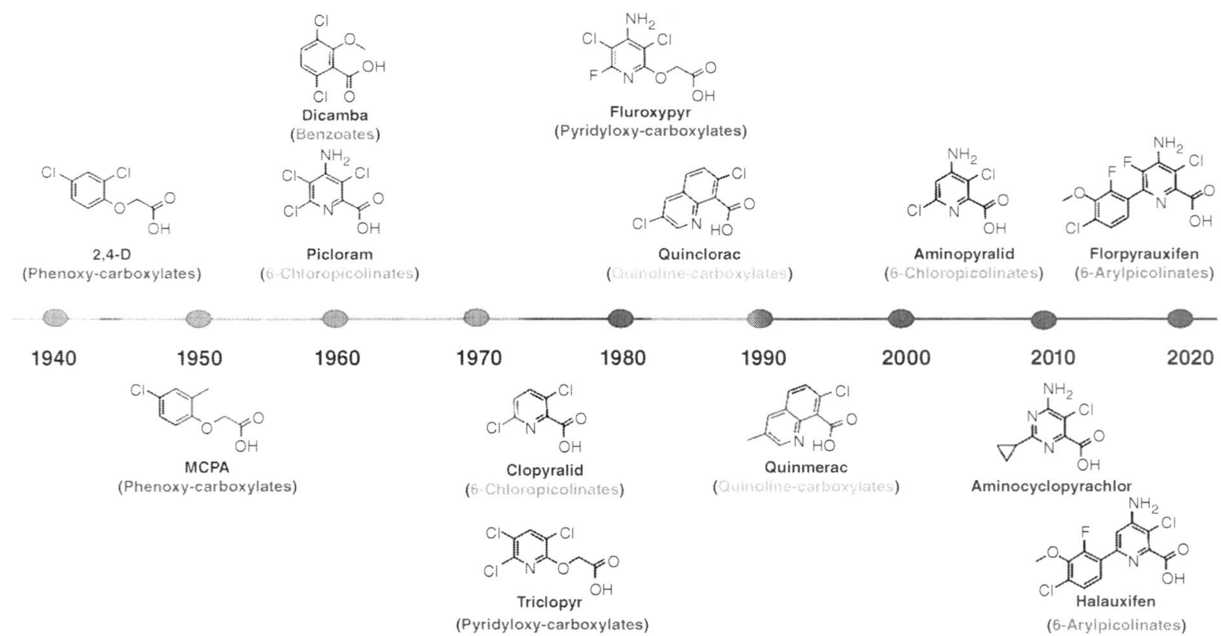

Figure 2.2 Chronology of discovery of representative commercial synthetic auxin herbicides (active moiety is shown with chemical family name in parenthesis). (*See insert for color representation of the figure.*)

benazolin-ethyl, another unclassified SAH by HRAC, discovered by Ciba Geigy in the early 1970s was registered to control annual dicot weeds in cereal, canola, and pastures. The chronology of discovery of important auxin herbicides from each chemical family is shown in Fig. 2.2.

2.2 Synthetic Auxin Herbicide Mode of Action

As with the high dose of natural auxins, SAHs cause deregulation of plant growth by transcriptional and nontranscriptional processes. As reviewed by Grossman,[17] the response of dicot plants to SAHs can be categorized into three phases: stimulation, inhibition, and decay.

The stimulation phase, occurring within five hours after application of the SAH, is represented by abnormal growth such as stem curling, tissue swelling, and leaf epinasty. The stimulation phase is characterized by the de-repression of transcriptional activator proteins, the auxin response factors (ARFs), which activate transcription of auxin-responsive genes such as 1-aminocyclopropane-1-carboxylate (ACC) synthase and 9-*cis*-epoxycarotenoid dioxygenase (NCED). Transcription of these genes results in the activation of membrane ion channels and plasmalemma H^+-ATPases, ethylene biosynthesis, and abscisic acid (ABA) accumulation. At low auxin concentrations, ARFs are repressed by Aux/IAA transcriptional repressor proteins. At higher concentrations, auxin herbicides (and IAA) are perceived by F-box proteins (auxin receptors), including the transport inhibitor response 1 (TIR1) and auxin-binding F-box (AFB; e.g., AFB1-5). The receptors have F-box recognition components marking them as protein substrates of the Skp1-Cullin-F-box (SCF) protein E3 ubiquitin ligase. Auxin binding to $SCF^{TIR1/AFBs}$ E3 ubiquitin ligase complex leads to the degradation of Aux/IAA transcriptional repressor proteins via the ubiquitin-proteasome pathway, releasing ARF genes from transcriptional inhibition (illustrated in Fig. 2.3).

In addition to AFB proteins and TIR1, S-phase kinase-associated protein 2A (SKP2A) is another auxin receptor, which is a regulator of the cell cycle and is involved in degradation of cell-cycle-related transcription factors.[18,19] Some transcription factors and other proteins need to be degraded during the G1-to-S checkpoint in the cell cycle before the commencement of next phase.[20] High auxin levels promote interaction between SKP2A and cell division-related transcription factors E2FC and DPB, but unlike AFBs, SKP2A gets degraded itself allowing cell-cycle progression. Thus, SKP2A can provide a specific mechanism by which plants are regulated by auxins.[21,22]

The inhibition phase occurs within 24 hours after application of SAH and is represented by stunted growth of shoots and roots and intensification of green leaf pigmentation because of stomatal closure, reduced transpiration,

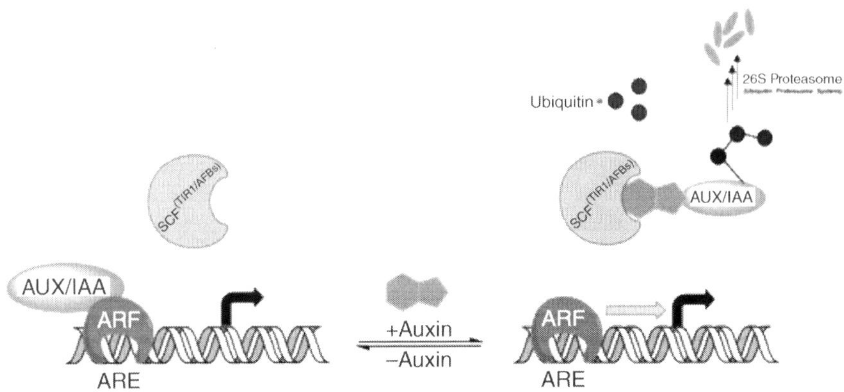

Figure 2.3 Canonical auxin signaling pathway. (*See insert for color representation of the figure.*)

carbon fixation, starch formation, and enhanced reactive oxygen species (ROS) production. The third and final phase of SAH response, the decay phase, occurs at 72 hours after application and beyond and is represented by disruption of chloroplasts, membranes, and the vascular system leading to red discoloration, chlorosis, wilting, and necrosis of plant tissue. Timing of phenotypic symptom appearance depends on the rate of SAH, sensitivity of target species, and environmental conditions.

The auxin response cascade is complex, and researchers continue to identify auxin receptors and various combinations of protein complexes that regulate auxin response. Even more intricate is the specificity to which various SAHs bind to these complexes. For example, *Arabidopsis* plants deficient in functional *AFB5* showed resistance to 6-chloropicolinate SAH, picloram, but not to phenoxy-carboxylate auxin, 2,4-D.[23] The differential response of the *AFB5*-deficient plants to different classes of auxin chemistry suggests potential differences in Aux/IAA perception and gene activation. Split firefly luciferase complementation assay and yeast two-hybrid system, respectively, were used to show that diverse combinations of Aux/IAAs and ARFs define the gene expression profile.[24,25] Binding studies showed that 2,4-D had higher relative binding than picloram to TIR1.[26] Picloram also displaced [^{3}H] IAA from the AFB5-IAA7 complex at a 70-fold higher rate than it displaced [^{3}H] IAA from the TIR1-IAA7 complex. These findings provide additional evidence that SAHs belonging to different chemical groups can have differential perception and response based on F-box protein and Aux/IAA repressor co-selectivity. Differential perception, binding affinities, and transport of auxins can lead to inherent differences in the level of susceptibility/tolerance among different plant species. These differences can be further complicated by selection of specific mutations in auxin receptors or auxin transport genes under field conditions.[23,27]

Most commercial SAHs preferentially provide postemergence control of dicot species only; however, quinclorac and florpyrauxifen-benzyl can also provide selective control of important grass weeds such as *Echinochloa* species in rice. Extensive studies to understand the MOA of quinclorac have revealed that herbicidal activity on broadleaf weed species is similar to other auxin herbicides, while the activity in sensitive grass species is due to a buildup of phytotoxic concentrations of cyanide. It was shown that quinclorac induces 1-aminocyclopropane-1-carboxylic acid (ACC) synthase activity in the roots of sensitive monocot species. ACC is translocated to the shoot and converted to ethylene, which is further converted to cyanide by ACC oxidase.[28] Intracellular cyanide accumulation leads to phytotoxic effects that cause cell and plant death. Key enzymes for detoxification of cyanide in plants are β-cyanoalanine synthase (β-CAS), which catalyzes the reaction to convert cyanide to β-cyanoalanine, and β-cyanoalanine hydrolase, which converts β-cyanoalanine to asparagine. β-CAS activity was only slightly elevated in sensitive species like barnyard grass but was six-fold higher in quinclorac-resistant grass species and rice without any notable changes in ACC synthase activity or ethylene and cyanide production in shoot tissues.[29,30] Enhanced ROS formation and lipid peroxidation, which is independent of ethylene and cyanide accumulation-based phytotoxicity, has also been reported as a mechanism of quinclorac-induced maize root tissue death.[31]

2.3 Synthetic Auxin Herbicide and Herbicide-Tolerant Crop Utility

Historically, the SAHs have been a critical weed control tool utilized in cereals, rice, corn, pastures, rangeland, non-cropland, roadsides, rights-of-way, and turf.[27,32] SAH utility has been attributed to phloem mobility and potency on a wide spectrum of broadleaf weed species.[32,33] In addition to their ability to control tough broadleaf weed species, SAHs have been used for over 75 years with relatively little resistant weed development.[9,27]

As of 2020, 10% of the global herbicide-treated acreage received applications of SAHs,[9] while only 8% of the resistant weed cases included resistance to SAHs[34] (Table 2.2). In contrast, acetolactate synthase (ALS) inhibitors are used on 14% of the global acreage but account for 33% of the resistance cases (Table 2.2). Another important comparison is the utility of glyphosate, which is used on 26% of the global acreage and accounts for 11% of the resistance cases. Interestingly, nearly all the glyphosate resistance cases were reported after the launch of glyphosate-resistant genetically modified crops in 1996. Since 2000, the incident rate of glyphosate resistance in weed species is approximately 13.2 per year,[34] and the resistance development is leaving a gap in producers' battery of economical herbicides.

Table 2.2 Herbicide use and resistance development statistics around synthetic auxin, glyphosate, and ALS-inhibiting herbicides.

Herbicide group	% of global acreage	% of active ingredients	% of resistance cases	# of species with resistance	# of years since herbicide launch	Rate of resistance development since launch (#spp./year)	Rate of resistance development since herbicide tolerant crops (HTC) launch (#cases/year[a])
SAH (Group 4)	10	8	8	41	76	0.54	2.4
5-enolpyruvyl shikimate-3-phosphate synthase (EPSPS) (Group 9)	26	<1	11	55	47	1.17	13.2
ALS (Group 2)	14	17	33	169	31	5.45	NA
All others	50	75	48	244	–	–	NA

SAH, synthetic auxin herbicides.
Data are based on values reported by AgbioInvestor (2020 global acreage values) and Heap 2024.
a) Rate of Resistance Development since HTC Launch includes all reported cases of resistance to Group 4 herbicides since 2016 and all cases of Group 9 since 1996. In this case, HTC is defined as transgenically derived crops.

The discovery and adoption of herbicide-tolerant (HT) trait technology enabled the increase in efficiency of America's agricultural production (Fig. 2.4) and has impacted productivity in other accepting geographies.[35] The launch of the first broadly adopted HT trait technology, glyphosate tolerance, allowed growers to control a broad spectrum of weed species with one nonselective, low-cost molecule, resulting in the increased efficiency for the grower. Likewise, it facilitated the shift of cultural practices from conventional tillage to conservation tillage (no-tillage or reduced-tillage) systems. This shift in weed management practices has contributed to long-term restoration of soil health and improved soil structure through reduced soil erosion and moisture loss, fossil fuel use, carbon emissions, and leaching of nitrogen and pesticides.[35] In short, the advent of genetically modified HT crops made weed control simpler and less costly and facilitated the incorporation of soil conservation practices. Reduced tillage practices would normally increase the predominance of perennial weed species, but the potent glyphosate efficacy on these species resulted in spectral shifts toward annual broadleaf weed species such as *Amaranthus* spp. and *Conyza* spp. on which glyphosate was less effective.[36,37]

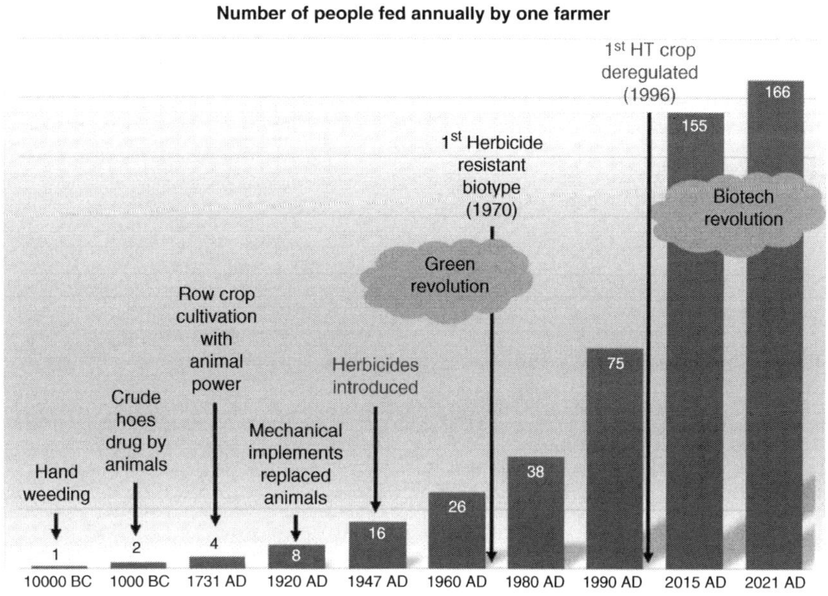

Figure 2.4 Chronology of agricultural production improvement in the United States and major discoveries associated with that growth. Source: Figure adapted from Monaco TJ, Weller SC, Ashton FM. 2002. Weed Science Principles and Practices, 4th Ed, John Wiley & Sons, Inc., NY and Fast Facts About Agriculture & Food, American Farm Bureau Federation, https://www.fb.org/newsroom. (*See insert for color representation of the figure.*)

The adoption of glyphosate-tolerant crops is well-documented,[38] but along with the increased use of glyphosate came intense selection pressure and weed resistance development.[35] The rapid development of glyphosate resistance in economically important weeds created a gap in producers' weed control options. To address these developments, agriculture needed easy-to-use, postemergent herbicide technologies with novel MOAs. However, high regulatory pressure and dearth of new MOAs steered the agrochemical industry toward development of genetically engineered crop technologies with tolerance to existing broad-spectrum herbicide MOA that were originally nonselective to large-acreage crops.[39] Dicamba and 2,4-D, two of the oldest SAHs, had the weed-control spectra to fill the emerging gaps and complement the spectrum of the glyphosate-tolerant crop systems (Table 2.3B).[8,40–42]

The postemergence application utility of SAHs, such as 2,4-D and dicamba, for weed control in dicot crops was limited due to the sensitivity of these species to SAHs at, and below, labeled application rates.[43,44] Even in monocot crops such as maize, postemergent use of 2,4-D and dicamba is limited to application during

Table 2.3 Species that have developed resistance cases to glyphosate. (A) Cases classified by type of weed, dicots, or monocots. (B) Weed genera with the greatest incidents of resistance development (~5% or greater) and ability of 2,4-D and/or dicamba to control native biotypes of species within those genera. A resistance case is defined as an incidence of reported resistance for one weed species to one or more herbicide active ingredients in a single geography.

A. Glyphosate resistance cases by cotyledonary class

Cotyledonary class	# Genera	# Species	# Cases	% Cases
Dicots	19	26	225	65.8
Monocots	15	27	117	34.2
All	34	53	342	100

B. Genera with greatest numbers of glyphosate resistance cases

Genus	% Cases	Control/suppression
Amaranthus	23.4	2,4-D and dicamba
Ambrosia	11.1	2,4-D and dicamba
Conyza	19.6	Dicamba
Eleusine	4.7	
Kochia	5	2,4-D and dicamba
Lolium	14	
Total	77.8	

2,4-D, 2,4-dichlorophenoxyacetic acid.
Source: Ref. Heap 2024.

the crop's early growth stages. The ability of 2,4-D and dicamba to fill critical weed control gaps (Table 2.3B) led the industry to seek genetically modified crop-tolerant solutions for these nonselective, yet effective, molecules.

Nature has employed multiple mechanisms for conferring resistance to herbicides, which include target-site (TS)-based and non-target-site (NTS)-based mechanisms. TS-based mechanisms include altered amino acid sequence of the TS protein, which impacts inhibitor binding, and altered TS protein expression levels. TS-based mechanisms may or may not impair the normal enzymatic function of the TS protein. NTS-based mechanisms include those that reduce the concentration of active herbicide remaining available to interact with the TS protein or any mechanism that allows the plant to cope with TS inhibition. The most reported and studied NTS mechanisms include reduced herbicide absorption, reduced herbicide translocation, increased herbicide sequestration, and/or enhanced degradation or metabolism of the active herbicide to nontoxic molecules.[45]

The low incidence of SAH resistance cases in weed species (Table 2.2) has been attributed to the following factors: (1) SAHs potentially have multiple sites of action (SOA); (2) loss-of-function mechanisms associated with these sites are often recessive and spread more slowly than genetically dominant mechanisms; and (3) reduced fitness under crop competition associated with TS-based SAH resistance.[9] Other theories state that resistance mutations must be either extremely rare or lethal.[46] Fitness penalties associated with SAH resistance has been theorized to be a major reason for the limited occurrence of SAH-resistant weeds.[23,27] Based on potential fitness costs associated with altering the auxin pathway, most industrial strategies to engineer tolerance to SAHs, both experimental and commercial, have been NTS-based, specifically degradation of the active ingredient to a non-phytotoxic metabolite.

2.4 Auxin-Tolerant Crop Discovery

2.4.1 Overarching Approach to the Generation of Auxin-Tolerant Crops

Two widely used SAHs have been the primary focus of auxin-tolerant crop development (e.g., 2,4-D and dicamba). Due to the various reasons discussed in the previous section, the NTS approach was heavily prioritized. Multiple different biodegradation enzymes derived from prokaryotic sources have been adapted for *in planta* use to provide tolerance to herbicides (Fig. 2.5).

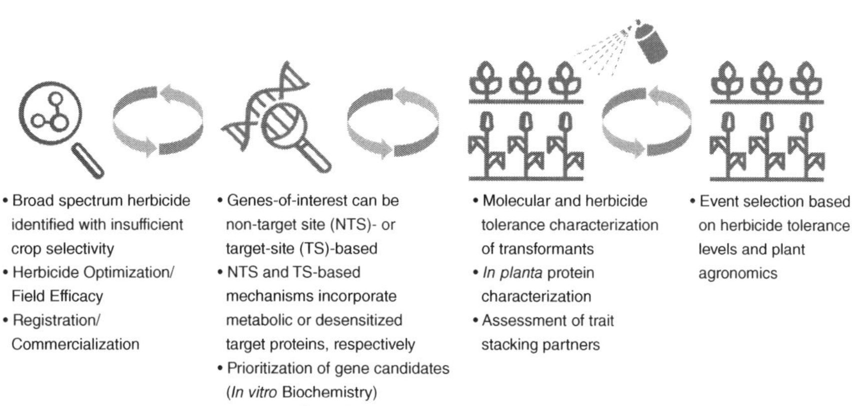

Figure 2.5 General processes for herbicide tolerance trait discovery. (*See insert for color representation of the figure.*)

2.4.2 Microbial Aryloxyalkanoate and Aryloxyphenoxyalkanoate Degradation

The elucidation of pathways involved in the biodegradation of aryloxyalkanoate molecules have been the subject of many studies for more than sixty years.[47] A collection of genetic and biochemical analyses demonstrated that many microbes have evolved the capacity to degrade various natural and xenobiotic molecules.[48,49] Reports have shown that certain microbes possess biodegradation genes that reside on transmissible plasmids (conjugative plasmids).[50–52] Specifically, microbes capable of degrading herbicides containing aryloxy- and aryloxyphenoxyalkanoate moieties within the auxin and acetyl-CoA carboxylase (ACCase) classes of herbicides, respectively, have been characterized (Fig. 2.6).[41,47,53] Whole-organism and purified enzyme studies have allowed for the elucidation of complex catabolic pathways and subsequently prompted investigations into the utility of the enzymes to enable HT within row crops. This section describes relevant catabolic pathways, associated enzymes, and the eventual incorporation of these genes into high-value crops to control troublesome weed species as a part of integrated weed management systems.

As previously mentioned, the degradation of SAHs, primarily aryloxyphenoxyacetate and aryloxyphenoxypropionate (AOPP)-containing compounds, have been described extensively at the whole organismal and enzymatic levels. The most well-characterized pathways for the dissimilation of SAHs, particularly, 2,4-D, 2-(2,4-dichlorophenoxy)propionic acid (dichlorprop), and 2-(4-chloro-2-methylphenoxy)propionic acid (mecoprop), have come from *Cupriavidus pinatubonensis*

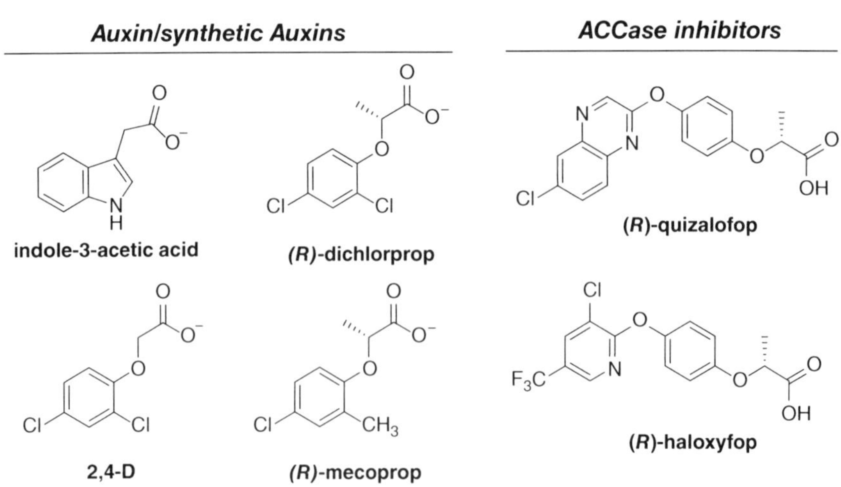

Figure 2.6 Structures of natural auxin IAA, synthetic auxins, and ACCase herbicides.

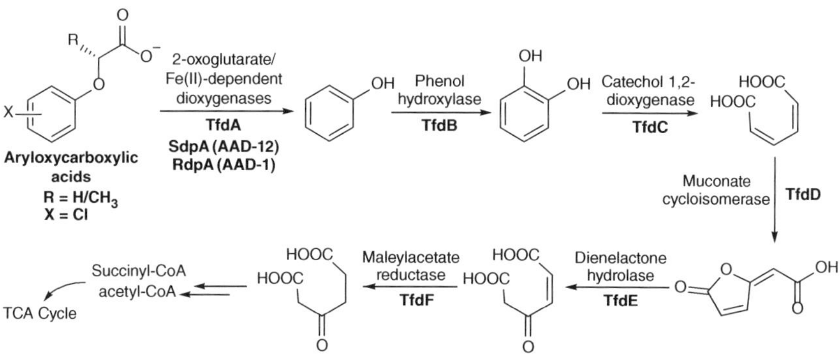

Figure 2.7 General biodegradation pathway of herbicides shared between microorganisms. Representative catabolic enzymes from the *Cupriavidus pinatubonensis* JMP134 pathway are shown below the arrows.

JMP134 (heterotypic synonyms: *Cupriavidus necator* JMP134 and *Alcaligenes eurotrophus* JMP134), *Sphingobium herbicidivorans* MH, and *Delftia acidovorans* MC1.[54,55] While the extrachromosomal context varies between microbes, the requisite genes for herbicide biodegradation are present within these microorganisms. A representative pathway from *C. pinatubonensis* JMP134 is depicted in Fig. 2.7, where a phenoxy-carboxylic acid-containing herbicide is ultimately converted into intermediates that can enter the tricarboxylic acid (TCA) cycle. Similar pathways have been described in other microbes only differing in the initial enzymatic steps.[47,56] Overall, the enzymes involved in the biodegradation of halogenated phenoxy-carboxylic acid-containing herbicides closely resemble those within the β-ketoadipate pathway which is involved in dissimilation of polyphenols like protocatechuic acid.[57]

The characterization of the enzymes involved in herbicide biodegradation pathways has advanced over many years. For the purposes of this review, we will focus on the detailed studies related to elucidation of the first catabolic step and how this has led to the development of herbicide-tolerant crops. Enzymes that catalyze this initial step belong to the 2-oxoglutarate/Fe(II)-dependent dioxygenase (2-ODD) superfamily. Examples are TfdA, aryloxyalkanoate dioxygenase-1 (AAD-1/RdpA), and AAD-12 (SdpA), which accept both achiral and chiral herbicide substrates.[41,55,58] Enzymes in the 2-ODD superfamily contain a mononuclear non-heme iron center within the catalytic site (anchored via a 2-His 1-carboxylate facial triad, Fig. 2.8B). The reaction is initiated via coordination of the co-substrate α-ketoglutarate (α-KG) to the iron center and proteinogenic positioning of the herbicide substrate (enzyme•Fe(II)•αKG•substrate complex). Subsequent binding of molecular oxygen to this complex results in hydrogen atom abstraction from C2 of the herbicide acetate or propionate moieties (Fig. 2.8A). Seminal

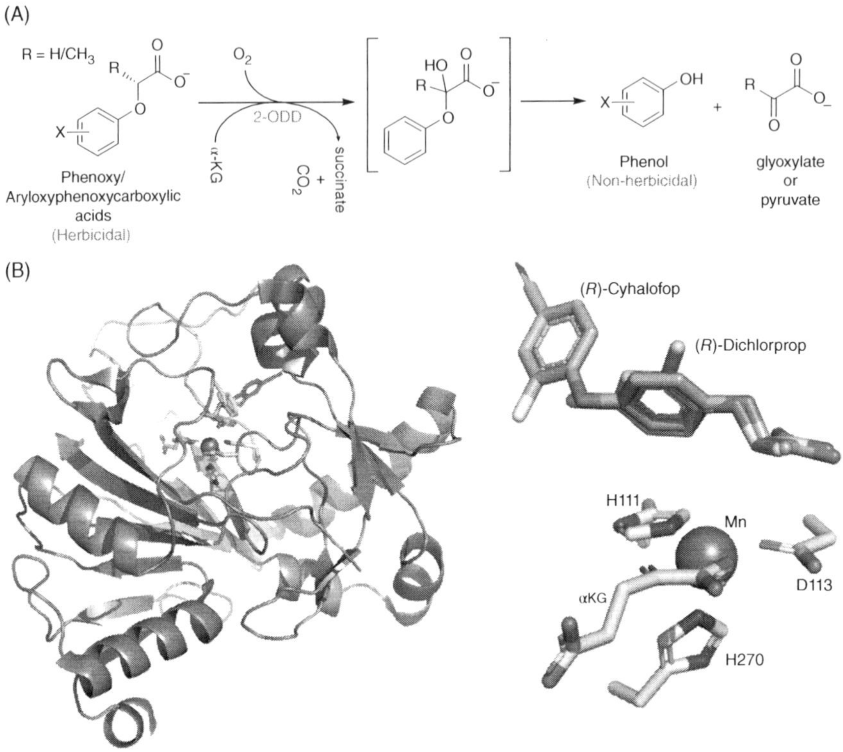

Figure 2.8 (A) General reaction catalyzed by 2-ODD during the initial step of herbicide degradation. (B) AAD-1 co-crystal structures demonstrating binding interactions various herbicide substrates (left, overall structure showing the catalytic site and right, overlay of bound AOPP and auxinic herbicides). (*See insert for color representation of the figure.*)

studies from the laboratories of Bollinger and Krebs have provided evidence that an iron(IV)-oxo (ferryl) intermediate is responsible for hydrogen atom abstraction which results in formation of a substrate radical and subsequent hydroxylation via oxygen rebound from Fe(III)–OH complex. This has been shown to be a canonical mechanism among many other 2-ODD enzymes and was recently demonstrated in studies with AAD-1 from *S. herbicidivorans*.[59] Additionally, crystallographic experiments from the same report support the kinetic mechanism and revealed that the substrate (e.g., *R*-dichlorprop) is positioned for hydrogen atom abstraction above the metal center (Fig. 2.8B, left). Co-crystal structures with AOPP herbicides also demonstrated that AAD-1 can accommodate extended ring structures with similar binding modes (Fig. 2.8B, right) which supported early *in vitro* biochemistry studies.

Recently, structural and *in vitro* biochemical studies using AAD-2, which originates from *Bradyrhizobium diazoefficiens* USDA 110, have also provided insights into AAD enantiospecificity.[59] Like AAD-1 and AAD-12, AAD-2 accepts achiral and chiral substrates from both the auxin and ACCase herbicide classes. The AAD-2 substrate studies demonstrated that only herbicides of (*S*)-enantiomeric configuration were accepted and that this is conferred by the presence of an ~20 amino acid β2–β3 helical insert that is present in enzymes known to be (*S*)-specific (e.g., AAD-12 and TfdA).

2.4.3 Development of Auxin and AOPP-Tolerant Crops Utilizing on AAD-12 and AAD-1

Seminal SAR studies, established by researchers from Corteva Agriscience (previously Dow AgroSciences), demonstrated that AAD enzymes can utilize both auxin and ACCase herbicides as substrates.[41] This finding led to the hypothesis that integration of the enzymes into high-value cropping systems (corn, soybean, and cotton) would be a viable option for herbicide tolerance. Importantly, *in vitro* biochemical studies demonstrated that microbial AADs (SdpA/RdpA) have different substrate enantioselectivity.[41,59–61] The relevance of enhanced enantioselective conversion of herbicides into non-herbicidal metabolites was pivotal as only (*R*)-enantiomers of auxinic and ACCase (specifically AOPP) herbicides are bioactive *in planta*.[62]

Currently, AAD-1 has been deployed by Corteva Agriscience in corn under the trade name Enlist while AAD-12 is being used in soybean and cotton to provide tolerance to 2,4-D (Enlist E3 soybean and Enlist cotton). Enlist corn provides robust tolerance to both 2,4-D and (*R*)-quizalofop.

2.4.4 Dicamba Monooxygenase: A Three-Component Enzyme Providing Microbial Degradation of Dicamba

Dicamba is extensively used as a pre- and postemergent herbicide to control annual and perennial broadleaf weeds. The half-life of dicamba in three soils (cation-exchange capacity, 8.5–12 meq $100\,g^{-1}$ and organic matter, 2.8–3.8%) varied between 17 and 32 days.[63] Microbial biomass and organic matter content have been documented to positively correlate with the dissipation of dicamba in soil.[64] Initial published reports suggested eight microbial species from soil were isolated; however, only three microbial strains (e.g., two *Pseudomonas* species and one *Moraxella* sp.) were further characterized due to rapid metabolism of dicamba in liquid/solid cultures (e.g., strains DI-6, DI 7, and DI-8). The first metabolite of dicamba degradation is herbicidal-inactive compound 3,6-dichlorosalicylic acid

(DCSA); however, complete mineralization to CO_2, H_2O, and Cl^- is observed.[65] It is proposed that dicamba is further metabolized to intermediates that enter Krebs cycle from which the microbe can use it to grow biomass.

Recent findings suggest dicamba can also be degraded under anaerobic conditions and the first step in its inactivation is also formation of DCSA followed by reductive dichlorination to produce 6-chloro salicylic acid (6-CSA). Bacterial (e.g., *Desulfovibrio*, *Exilispiram*, and *Caldisericum*) and archaeal (e.g., *Methanosaeta* and *Methanolinea*) microbes appear to be responsible for biodegradation of dicamba under anaerobic conditions based on abundance and sequencing analysis of sludge sediments after acclimation.[66] Interestingly, metabolite identification profiling suggests different pathways, and microbes are responsible for degradation of dicamba between aerobic and anaerobic conditions. This diversity can be further exploited to understand microbial ecology and identify pathways and subsequently genes responsible for inactivation of xenobiotic compounds that can be used for agricultural biotechnology and as bioremediation tools.

2.4.5 Biochemical, Biophysical, and Structural Characterization of the Dicamba Monooxygenase Enzyme

Dicamba monooxygenase (DMO) enzyme catalyzes an *O*-demethylation reaction, the first step in mineralization of dicamba to carbon dioxide, chloride, and water. The enzyme responsible for inactivation of dicamba to DCSA, water, and formaldehyde was first isolated from microbial sources, namely, *Stenotrophomonas* (formerly *Pseudomonas*) *maltophilia*, strain DI-6 that can grow solely on the herbicide as carbon and energy sources.[67]

The DMO enzyme complex requires three separate components: oxygenase, ferredoxin, and reductase for efficient enzymatic activity (Fig. 2.9). None of the components showed enzyme activity when assayed alone or in the absence of Fe^{2+}, Mg^{2+}, O_2, and NADH. Initially, dicamba *O*-demethylase activity was measured in cell lysates of *Pseudomonas maltophilia*, and a purification scheme was proposed to allow for identification of three separate protein fractions necessary for activity. The components were freely dissociable from each other and

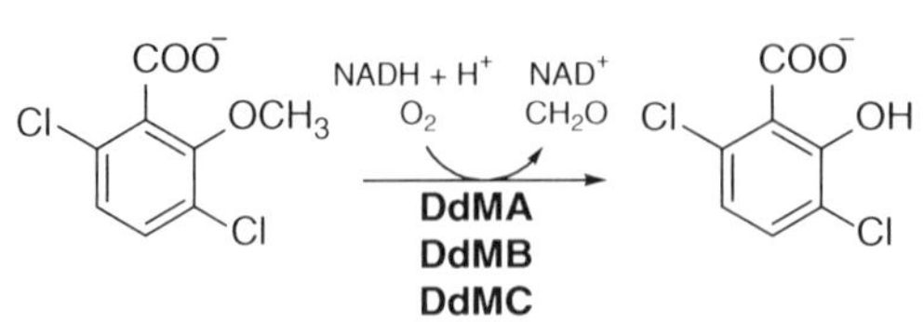

Figure 2.9 Dicamba *O*-demethylase is a three-component complex composed of a reductase (ddmA), ferredoxin (ddmB), and oxygenase (ddmC) responsible for deactivation of synthetic auxin dicamba.

were purified to homogeneity from microbial sources and included a ferredoxin, a reductase, and an oxygenase component. In-depth biochemical and biophysical analyses indicated the ferredoxin component contains an iron sulfur cluster, the reductase contains a flavin cofactor, and the oxygenase contains one iron sulfur cluster and a labile non-heme ferrous iron bound to the active site.[67] Furthermore, N-terminal and internal amino-acid sequencing results from microbially purified proteins were used to identify and clone the four genes that encode the DMO complex: two reductase genes (e.g., *ddmA1* and *ddmA2*), one ferredoxin gene (e.g., *ddmB*), and one oxygenase gene (*ddmC*). Interestingly, Southern blot analysis localized the oxygenase gene on a megaplasmid in the cells of *P. maltophilia*. To further confirm these findings, mixtures of the three recombinantly *Escherichia coli* (*E. coli*)-overexpressed proteins were confirmed to convert dicamba to DCSA with the same catalytic efficiency as the native proteins suggesting that all the components for optimal enzymatic activity have been identified.[68]

The oxygenase component is a homotrimer with a subunit mass of 37.3 kDa and contains a single Rieske [2Fe–2S] cluster and a mononuclear iron in the active site responsible for binding of oxygen. The ferredoxin component is a monomer with an estimated molecular mass of 11.4 kDa and has a single [2Fe–2S] cluster, while the reductase component is a monomer with a molecular mass of around 43.7 or 43.9 kDa and is an FAD-containing cofactor.[68]

Due to important applications for agricultural biotechnology, crystallography studies were deployed to further characterize the oxygenase component of dicamba *O*-demethylase. Two groups, independent of each other, have solved the crystal structure of oxygenase component in the free state, bound to its substrate dicamba, and bound to its product DCSA at 2.10–1.75 Å resolutions.[69,70]

The DMO monomer is a two-domain protein that contains a Rieske domain and a catalytic domain. The Rieske domain harbors the [2Fe–2S] cluster and consists of 12-β strands, whereas the catalytic domain contains the non-heme iron metal and is composed of seven-stranded antiparallel β-sheet flanked by six α-helices and extended loops. Catalysis in Rieske non-heme iron oxygenases requires an intermolecular transfer of electrons from the Rieske cluster of one of monomer to the active-site mononuclear iron in the neighboring monomer, and therefore the observed interaction between DMO monomers is essential for enzymatic activity.[69]

The structures of the DMO-dicamba, DMO-DCSA, and DMO free state play an important role in identifying key residues and understanding the mechanism that catalyzes the *O*-demethylation reaction. These crystal structures showed that (1) dicamba and DCSA bind to the active site in distinct conformations and (2) activated oxygen species attack the exocyclic methyl group rather than carbon in phenyl ring, with the postulated mechanism requiring a hemiacetal intermediate.[71]

2.4.6 Development of Auxin-Tolerant Crops Utilizing DMO

The use of a genetically engineered bacterial enzyme, DMO, capable of inactivating dicamba to DCSA when solely expressed in model and crop plants has been documented.[72] Although the bacterial enzyme complex isolated from *P. maltophilia* strain DI-6 required three proteins to degrade dicamba, *in planta* activity was achieved by only expressing the monooxygenase component. By expressing the protein in either the cytosol or chloroplasts of transgenic plants, the need to overexpress the other two components (e.g., reductase and ferredoxin from bacteria) was eliminated. The ability/versatility of the monooxygenase enzyme to accept electrons for the enzymatic reaction from plant-encoded ferredoxin(s) was successfully exploited to provide resistance and to reach tolerance levels to dicamba spray evaluations in plant model systems like *Arabidopsis thaliana*, tobacco, and broad acre crops like soybeans. In addition, *in vitro* experiments confirmed the ability of the enzyme to accept electrons from ferredoxins from other plants and even other bacterial sources confirming the results initially observed *in planta*. For transgenic events, the tolerance levels to dicamba were also correlated with levels of expression of the gene of interest *in planta*. In fact, higher expression levels were observed when targeting the protein to chloroplast compared with the cytosol.

The main value driver for the dicamba tolerance technology was for use in row crops like soybeans and cotton in which the management of broadleaf weeds is essential to maximize yields, control weed species, and deter the development of resistance to existing HT technologies. This technology has been developed for soybean and cotton varieties by Bayer Crop Science (previously Monsanto Company) for commercial use by farmers in North and South America.[72] In addition, University of Nebraska scientists led by Professor Donald Weeks reported the successful development of transgenic maize events expressing nuclear chloroplast targeting of DMO enzyme that provides high tolerance levels to dicamba when applied either pre- or postemergently.[69]

2.4.7 Additional Herbicide Tolerance Gene Discovery Efforts for Auxins

As described above, TfdA, from *C. pinatubonensis* JMP134, was shown to catalyze the first reaction in the degradation of 2,4-D. The enzyme oxidatively cleaves the acetate moiety of 2,4-D to yield glyoxylate and 2,4-dichlorophenol which are non-herbicidal molecules. The enzyme was integrated into multiple plant species and provided tolerance to 2,4-D.[73–75] Despite early progress, transgenic crops containing *tfdA* as a resistance allele to 2,4-D were not developed further.

While early research efforts focused on the dioxygenases that catalyze the cleavage of aryloxyalkanoates, one might consider alternative enzymatic modifications that could render herbicides inactive *in planta*. The carboxylic acid functional group, that is common to all commercially utilized SAHs, is critical to productive binding of molecular targets like TIR1 and likely AFB5.[76,77] Researchers at the University of Tennessee sought to exploit this by understanding the ability of plant methyltransferases (MTs) from the SABATH (salicylic acid MT, benzoic acid MT, theobromine synthase) family to esterify the acid moiety of synthetic auxins. In particular, a jasmonic acid methyltransferase (JMT) from *Populus trichocarpa* (black cottonwood) was shown to not only catalyze esterification of jasmonic acid but also have low levels of activity on synthetic auxins (Fig. 2.10A).[78] Subsequent work by the same research group showed that a distinct version of the enzyme called PtJBMTm3 had substantially improved activity on 2,4-D and several other SAHs (Fig. 2.10B).[79] To the best of our knowledge, after this work, no further advancements have been made to utilize methyltransferases in crops to confer robust tolerance to SAHs.

As previously mentioned, the carboxylate moiety plays an important role for SAHs in binding to receptors and mimicking IAA effects *in planta*; therefore, enzymatic removal of this moiety could provide an alternative inactivation

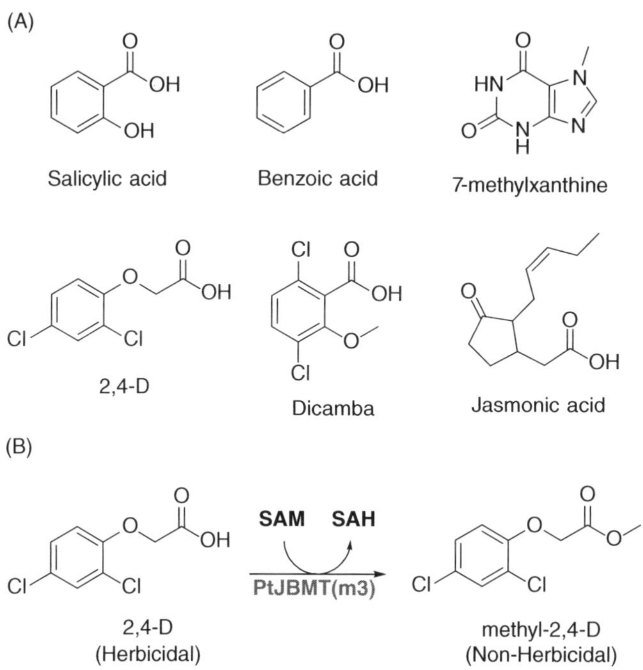

Figure 2.10 (A) Common endogenous substrates and synthetic auxins utilized by the SABATH methyltransferase family of enzymes. (B) PtJBMT (m3) catalyzed esterification of 2,4-D. (*See insert for color representation of the figure.*)

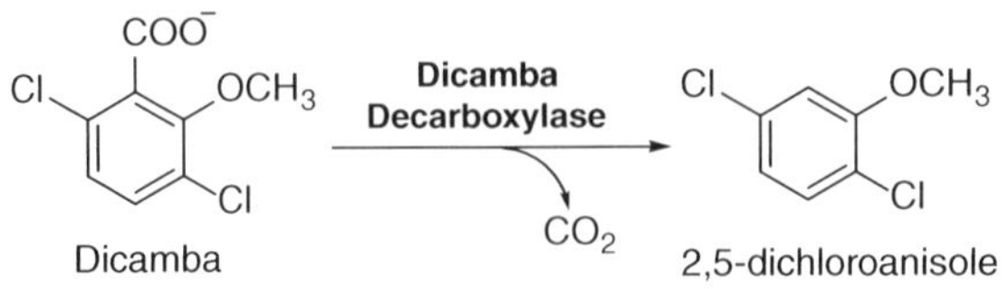

Figure 2.11 Dicamba decarboxylase is responsible for inactivation of dicamba through a different inactivation approach.

mechanism.[77] To this end, research groups from Pioneer Hi-Bred International and Arzeda investigated decarboxylases for initial activity using the herbicide dicamba as a substrate. Using this strategy, several decarboxylases from microbial sources, and subsequently engineered versions of the enzymes, were shown to decarboxylate dicamba *in vitro* resulting in its inactivation to 2,5-dichloroanisole (Fig. 2.11).[77] While incorporation of these genes into whole plants provided some promise, no products providing dicamba tolerance via decarboxylation have been commercialized to date.

2.5 Adoption, Challenges, and Future Considerations

The advent of genetically engineered crops revolutionized weed management by enabling easy postemergent weed control solutions to growers (Fig. 2.4). The percentage of US acres planted in HT crops reached approximately 90% in 2014 (cotton 91%, soy 94%, and corn 89%) (Fig. 2.12).[80] The rapid acceptance of HT crops in the United States can be attributed to the adoption of glyphosate-tolerant trait-enabled crops; however, the rampant glyphosate resistance evolution that followed (Fig. 2.13) brought growers back to using traditional weed control methods and drove the industry to develop offerings with multiple MOAs, such as 2,4-D and dicamba HT crops.[41,72]

Out of the two SAH-tolerant trait technologies available currently (Table 2.4), the dicamba-tolerant (DT) trait was first commercialized with deregulation in US cotton and soybean in 2015. DT soybean was first planted in the United States in 2016 after lifting of import regulations by China[81]; however, registration of dicamba use in crop did not occur until late 2016. In December of 2016, US EPA approved registration of XtendiMax and FeXapan formulations containing the diglycolamine (DGA) salt of dicamba, and Engenia containing the *N*,*N*-bis-(aminopropyl) methylamine (BAPMA) salt of dicamba. Another formulation, Tavium, containing dicamba DGA salt plus *S*-metolachlor was registered by EPA in December of 2019.[39] Programs of residual herbicides applied preemergently followed by postemergence applications of XtendiMax, FeXapan, and Engenia were popular with growers. Phase II of the Agricultural Resource

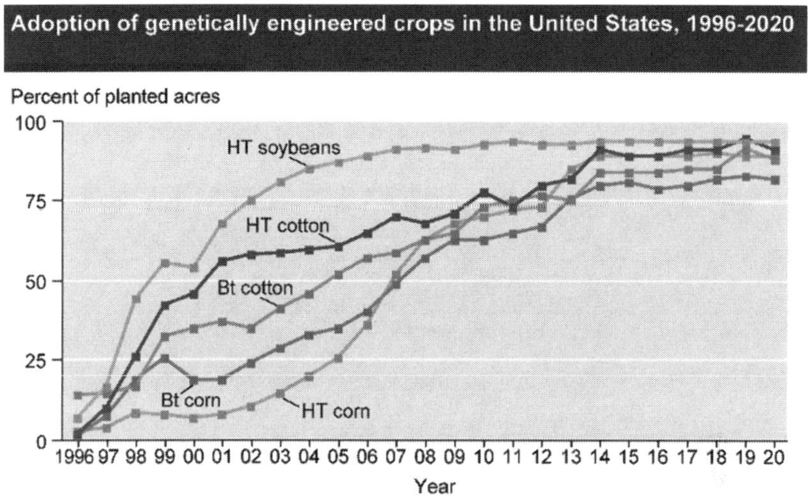

Note: HT indicates herbicide-tolerant varieties; Bt indicates insect-resistant varieties (containing genes from the soil bacterium Bacillus thuringiensis). Data for HT/Bt corn and cotton are not mututally exclusive, as HT and Bt categories include those varieties with overlapping (stacked) HT and Bt traits.
Source: USDA, Economic Research Service using data from the 2002 ERS report, *Adoption of Bioengineered Crops* (AER-810) for the years 1996-99 and National Agricultural Statistics Service, (annual) June Agricultural Survey for the years 2000-20.

Figure 2.12 Adoption of genetically engineered crops in the United States. Source: U.S. Department of Agriculture/CC BY 4.0. (*See insert for color representation of the figure.*)

Management Survey suggested that like glyphosate-tolerant traits, there was a high rate of early adoption of DT trait technology.[81] Approximately 46% of soybean acres in 2018 were planted with DT traits.

Enlist corn and Enlist cotton were developed by Corteva Agriscience (previously Dow AgroSciences) (Table 2.4). Enlist corn was deregulated in the United States in 2014 and offered for commercial sale in 2018. Enlist cotton was deregulated in the United States in 2015 and offered for commercial sale in 2016. Enlist corn and Enlist cotton both confer tolerance to 2,4-D; Enlist corn also provides tolerance to FOP herbicides; and Enlist cotton provides tolerance to glufosinate. When stacked with a glyphosate tolerance trait, Enlist corn and Enlist cotton also provide tolerance to glyphosate. The Enlist E3 soybean trait, jointly developed and owned by Corteva Agriscience and MS Technologies L.L.C, contains a single molecular stack providing tolerance to 2,4-D, glyphosate, and glufosinate.[39,41] Enlist E3 soybean was first offered for commercial sale in the United States in 2019. Enlist E3 soybean is rapidly being adopted in the US soybean seed market and was planted on around 20% soybean acres in 2020, 35% soybean acres in 2021, and over 45% soybean planted acres in 2022.[82,83]

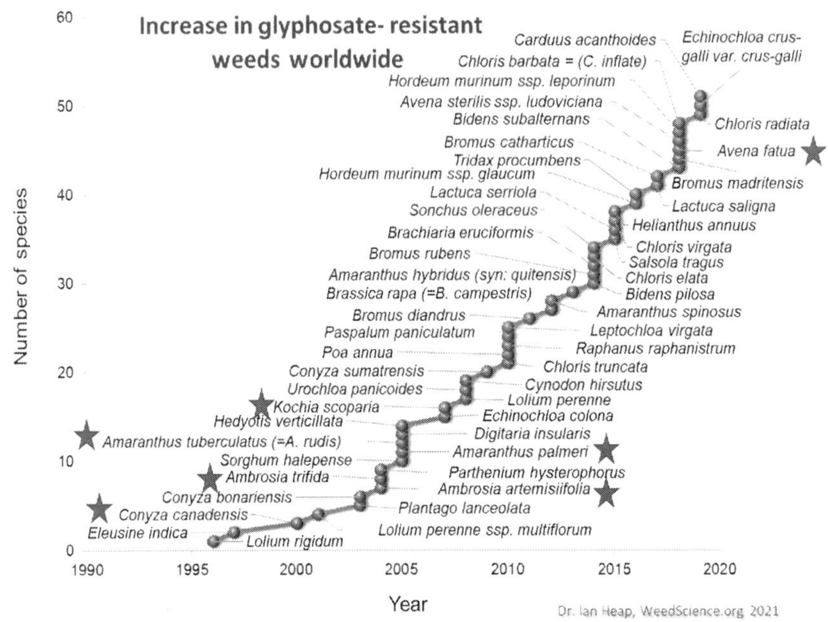

Figure 2.13 Species developing resistance to glyphosate over time. Species indicated with a star are those determined troublesome or most common by poll of growers (2017 WSSA Survey, Canadian Weed Scientists, Farm Journal Pulse Poll). (*See insert for color representation of the figure.*)

2,4-D formulations labeled for use with the Enlist and Enlist E3 technologies are Enlist One (2,4-D choline salt) and Enlist Duo (2,4-D choline salt + dimethylammonium [DMA] salt of glyphosate), both with COLEX-D® Technology. The COLEX-D Technology in combination with 2,4-D choline salt provides a formulation with significant reduction in volatility and drift while providing compatibility between 2,4-D and glyphosate[84]; thus, no other 2,4-D formulations are authorized for use in Enlist crops. Programs of residual herbicides applied PRE followed by POST applications of Enlist One herbicide plus glyphosate provide consistent control of glyphosate-resistant weeds such as Palmer amaranth (*Amaranthus palmeri S. Watson*), tall water hemp (*Amaranthus rudis* J. D. Sauer), giant ragweed (*Ambrosia trifida* L.), and Sumatran fleabane (*Conyza sumatrensis* [Retz.]).[85–89]

The adoption of new technology is not without challenges and need for product stewardship. Two major challenges faced by both SAH-tolerant trait-herbicide technologies include off-target herbicide movement to sensitive crops and management of resistant weed development. Off-target movement of herbicides can be either primary (herbicide drift) or secondary (herbicide volatilization) in nature.[90] Unique herbicide formulations have been introduced to accompany each SAH HT

Table 2.4 Commercially available genetically modified SAH-tolerant trait offerings.

Trait[a)]	Herbicides	HT trade name	Stacked traits[a),b)]	Crops	Commercialization[c)]
ddmC	Dicamba	Xtend	*cp4-epsps* (glyphosate)	Soybean	2016
		XtendFlex	*cp4-epsps* (glyphosate), *pat* (glufosinate)	Soybean	2021
		XtendFlex	*cp4-epsps* (glyphosate), *bar* (glufosinate)	Cotton	2016
AAD1	2,4-D choline, 'fop herbicides	Enlist	*cp4-epsps* (glyphosate), *pat* (glufosinate)	Corn	2018
AAD12	2,4-D choline	Enlist E3	*2mepsps* (glyphosate), *pat* (glufosinate)	Soybean	2019
		Enlist	*cp4-epsps* (glyphosate), *pat* (glufosinate)	Cotton	2016

a) Abbreviations: 2,4-D, 2,4-dichlorophenoxyacetic acid; *AAD*, aryloxyalkanoate dioxygenase; *bar*, bialophos resistance; *epsps*, 5-enolpyruvylshikimate-3-phosphate synthase; *pat*, phosphinothricin acetyltransferase; SAH, synthetic auxin herbicides.
b) Stacked traits can be achieved as a molecular stack single and/or breeding stacks of multiple transgenic events.
c) Year of first commercialization.

trait introduction to mitigate some of these risks. Product label instructions for proper use of these formulations include application rate and volume, application timing (time of day), wind conditions at the time of application (wind speed, wind direction, and downwind buffer zones), weed size, boom height, droplet size, tank cleaning, and required applicator training.[91,92]

The rate of resistance evolution to SAHs is low compared to other herbicide mechanisms of action.[34,41] As discussed previously, the rarity of resistance evolution to SAHs has been attributed to redundancy and complexity of auxin signal perception.[23,27] However, resistance to 2,4-D and dicamba among weed communities does occur.[34] Since discovery of the SAH in the early 1940s, only 42 weed species have been reported to be resistant to SAHs, out of which 25 weed species have been documented with confirmed resistance to 2,4-D and 10 species with confirmed resistance to dicamba.[34] According to HRAC, the most important synthetic auxin-resistant weeds globally are kochia (*Kochia scoparia* [L.] Schrad.), wild radish (*Raphanus raphanistrum* L.), corn poppy (*Papaver rhoeas* L.), and wild mustard (*Sinapis arvensis* L.).[46] Synthetic auxin resistance in three other species, tall water hemp (*Amaranthus tuberculatus* [Moq.] Sauer), palmer amaranth (*A. palmeri S. Watson*), and common lambs-quarters (*Chenopodium album* L.) is not widespread yet but has the potential to become serious problem in the United States if not managed properly.[46,93]

The basis for SAH resistance in weeds is not well understood. For example, in several cases (*S. arvensis*, *K. scoparia*, *Centaurea solstitialis*, *Galium spurium*, and *Galeopsis tetrahit*), the resistance is controlled by one or two genes, but the mechanism is unknown, and the gene(s) have not been identified.[46] Resistance in wild radish,[94] corn poppy,[95] and Waterhemp[96] are likely NTS mechanisms. A Palmer amaranth population from a Kansas conservation tillage field that was resistant to six SOA including 2,4-D was documented to have enhanced metabolism as the primary mechanism of 2,4-D resistance.[93,97] To date, only two cases of resistance due to an altered target site have been reported. Specifically, the alterations in both cases are located in highly conserved degron regions of IAA transcriptional co-receptor/transcriptional repressor proteins: IAA16 in kochia[98] and IAA2 in Indian hedge mustard (*Sisymbrium orientale* Torn.).[99] One of those mutations has a documented fitness penalty,[100] suggesting that TS-based mutations for this MOA may not become prevalent under field conditions.

Despite the low frequency of SAH-resistant biotypes, the development of weed biotypes with resistance to multiple SOAs is particularly threatening to agricultural production. Of the 42 weed species with documented cases of resistance, 18 of those species have developed resistance to more than one SOA. Four species have been identified with biotypes resistant to five or more SOAs. They include Palmer amaranth in Kansas,[97] tall water hemp in Illinois,[101] Sumatran fleabane in Brazil,[102] and wild radish in Western Australia.[103]

Overreliance on a single MOA, even as complex as that of the SAHs, coupled with less diversity in management practices will promote evolution of 2,4-D- and dicamba-resistant weed species. Resistance management stewardship practices need to be adopted by growers to delay or reduce herbicide resistance evolution.[104] These best management practices should employ all available tools at the disposal of the grower, such as, but not limited to, understanding the biology of the weeds present in fields, reducing the soil weed seedbank, planting into and keeping fields weed-free, using weed-free crop seed, routine weed scouting, using multiple herbicide mechanisms of action, applying labeled herbicide rates at recommended weed sizes, using mechanical and biological weed control practices, preventing field-to-field movement of weed seeds/vegetative propagules, preventing influx of weeds into fields by managing field borders, and using cultural practices such as crop rotation, crop trait rotation, stale seedbed, use of cover crops, and soil residual herbicides, and managing weed seed at harvest and after harvest. Genetically engineered stacked traits that allow the use of multiple herbicide mechanisms of action in crops that were originally sensitive to these MOAs can also be of a great utility in resistance management.[105] Xtend® soybean with stacked traits for dicamba and glyphosate and XtendFlex® soybeans and cotton with glufosinate tolerance added to that of dicamba and glyphosate have been powerful tools for managing resistance. Similarly, Enlist E3 soybean with stacked traits for 2,4-D, glyphosate, and glufosinate tolerance is an additional tool for growers to manage resistance evolution. Like Enlist E3 soybeans, Enlist cotton and Enlist corn technologies provide additional tolerance to glyphosate and glufosinate and glyphosate and aryloxyphenoxypropionate (AOPP) herbicides, respectively, when stacked with additional traits.

By stacking SAH HT traits with the existing glyphosate and glufosinate HT traits, growers have been enabled to apply multiple MOAs and readily rotate MOAs while using highly effective herbicides and maintain the convenience and conservation tillage practices enabled by HT crops. Utilizing best management practices with the diversity of HT traits now available will extend the utility of glyphosate and glufosinate HT traits as well as preserve the value of SAH HT traits.

The complexity of the auxin MOA coupled with the specificity in response of diverse SAH molecules to the auxin response pathways yields a MOA somewhat recalcitrant to resistance development. Coupling robust chemistry with SAH HT traits provides agricultural producers an expanded set of resilient tools to meet the need for increased yield and sustainable practices. In addition, adding the flexibility to include SAH chemistry to tank mixes to weed control regimes will help growers manage resistant weed populations as well as reduce the selection pressure of existing chemistries.

Acknowledgments

The authors wish to thank Dr. David Simpson for his valuable review of the manuscript.

References

1 WSSA, Herbicide Site of Action (SOA) Classification List. https://wssa.net/wssa/weed/herbicides/ [Accessed May 7, 2024] (2024).
2 Woodward AW, and Bartel B, Auxin: Regulation, action, and interaction. *Ann Bot-London* **95**:707–735 (2005).
3 Mashiguchi K, Tanaka K, Sakai T, Sugawara S, Kawaide H, Natsume M, *et al.*, The main auxin biosynthesis pathway in Arabidopsis. *P Natl Acad Sci* **108**:18512–18517 (2011).
4 Zhao YD, Auxin Biosynthesis and Its Role in Plant Development. *Annu Rev Plant Biol* **61**:49–64 (2010).
5 Ludwig-Muller J, Auxin conjugates: their role for plant development and in the evolution of land plants. *J Exp Bot* **62**:1757–1773 (2011).

6 Troyer JR, In the beginning: the multiple discovery of the first hormone herbicides. *Weed Sci* **49**:290–297 (2001).

7 HRAC, Mode of Action Classification. https://hracglobal.com/tools/2024-hrac-global-herbicide-moa-classification [Accessed May 7, 2024] (2024).

8 Peterson MA, Shan G, Walsh TA and Wright TR, Utility of aryloxyalkanoate dioxygenase transgenes for development of new herbicide resistant crop technologies. http://www.isb.vt.edu [Accessed May 7, 2024] (2011)

9 Busi R, Goggin DE, Heap IM, Horak MJ, Jugulam M, Masters RA, *et al.*, Weed resistance to synthetic auxin herbicides. *Pest Manag Sci* **74**:2265–2276 (2018).

10 Coupland D, In Powles S, and Holtum JAM (Eds), *Resistance to auxin analog herbicides*, Lewis Publishers, CRC Press, Boca Raton (1994).

11 Fryer JD, Foreward, In Kirby C (Ed.), *Croyden*, British Crop Protection Council, Great Britain (1980).

12 Benbrook C, Dicamba Diary, Part I: Discovery and Development, 1940–2017. https://hh-ra.org/wp-content/uploads/2017/09/Dicamba-Diary-Part-I.pdf [Accessed May 7, 2024] (2017).

13 Johnston H and Tomita MS, Amino-trichloropicolinic acid compounds. US Patent 3285925 (1966).

14 Johnston H, Picolinic acid compounds. US Patent 3317549 (1967).

15 Schmitzer PR, Balko TW, Daeuble JF, Epp JB, SatchivI NM, Siddall TL, Weimer MR, and Yerkes CN, In Maienfisch P, and Stevenson TM (Eds), *Discovery and SAR of halauxifen methyl: a novel auxin herbicide*, American Chemical Society, Washington DC (2015).

16 Grossmann K, Quinclorac belongs to a new class of highly selective auxin herbicides. *Weed Sci* **46**:707–716 (1998).

17 Grossmann K, Auxin herbicides: current status of mechanism and mode of action. *Pest Manage Sci* **66**:113–120 (2010).

18 Del Pozo JC, Diaz-Trivino S, Cisneros N, and Gutierrez C, The balance between cell division and endoreplication depends on E2FC-DPB, transcription factors regulated by the ubiquitin-SCFSKP2A pathway in Arabidopsis. *Plant Cell* **18**:2224–2235 (2006).

19 Jurado S, Abraham Z, Manzano C, Lopez-Torrejon G, Pacios LF, and Del Pozo J, The Arabidopsis Cell Cycle F-Box Protein SKP2A Binds to Auxin. *Plant Cell* **22**:3891–3904 (2010).

20 Jurado S, Diaz-Trivino S, Abraham Z, Manzano C, Gutierrez C, and Del Pozo C, SKP2A, an F-box protein that regulates cell division, is degraded via the ubiquitin pathway. *Plant J* **53**:828–841 (2008).

21 Grones P, and Friml J, Auxin transporters and binding proteins at a glance. *J Cell Sci* **128**:1–7 (2015).

22 Peer WA, From perception to attenuation: auxin signalling and responses. *Curr Opin Plant Biol* **16**:561–568 (2013).

23 Walsh TA, Neal R, Merlo AO, Honma M, Hicks GR, Wolff K, *et al.*, Mutations in an auxin receptor homolog AFB5 and in SGT1b confer resistance to synthetic picolinate auxins and not to 2,4-dichlorophenoxyacetic acid or indole-3-acetic acid in arabidopsis. *Plant Physiol* **142**:542–552 (2006).

24 Li JF, Bush J, Xiong Y, Li L, and McCormack M, Large-Scale Protein-Protein Interaction Analysis in Arabidopsis Mesophyll Protoplasts by Split Firefly Luciferase Complementation. *Plos One* **6**:e27364 (2011).

25 Vernoux T, Brunoud G, Farcot E, Morin V, Van den Daele H, Legrand J, *et al.*, The auxin signalling network translates dynamic input into robust patterning at the shoot apex. *Mol Syst Biol* **7**:508 (2011).

26 Villalobos LIAC, Lee S, De Oliveira C, Ivetac A, Brandt W, Armitage L, *et al.*, A combinatorial TIR1/AFB-Aux/IAA co-receptor system for differential sensing of auxin. *Nat Chem Biol* **8**:477–485 (2012).

27 Mithila J, Hall JC, Johnson WG, Kelley KB, and Riechers DE, Evolution of resistance to auxinic herbicides: Historical perspectives, mechanisms of resistance, and implications for broadleaf weed management in agronomic crops. *Weed Sci* **59**:445–457 (2011).

28 Grossmann K, and Kwiatkowski J, Evidence for a causative role of cyanide, derived from ethylene biosynthesis, in the herbicidal mode of action of quinclorac in barnyard grass. *Pestic Biochem Phys* **51**:150–160 (1995).

29 Grossmann K, and Kwiatkowski J, The mechanism of quinclorac selectivity in grasses. *Pestic Biochem Phys* **66**:83–91 (2000).

30 Abdallah I, Fischer AJ, Elmore CL, Saltveit ME, and Zaki M, Mechanism of resistance to quinclorac in smooth crabgrass (*Digitaria ischaemum*). *Pestic Biochem Physiol* **84**:38–48 (2006).

31 Sunohara Y, and Matsumoto H, Quinclorac-induced cell death is accompanied by generation of reactive oxygen species in maize root tissue. *Phytochem* **69**(12):2312–2319 (2008).

32 Todd OE, Figueiredo MRA, Morran S, Soni N, Preston C, Kubes MF, *et al.*, Synthetic auxin herbicides: finding the lock and key to weed resistance. *Plant Sci* **300**:110631 (2020).

33 Quareshy M, Prusinska J, Li J, and Napier R, A cheminformatics review of auxins as herbicides. *J Exp Bot* **69**:265–275 (2018).

34 Heap IM. The international survey of herbicide resistant weeds. https://www.weedscience.com/ [Accessed May 7, 2024] (2024)

35 Lamichhane JR, Devos Y, Beckie HJ, Owen MDK, Tillie P, Messean A, *et al.*, Integrated weed management systems with herbicide-tolerant crops in the European Union: lessons learnt from home and abroad. *Crit Rev Biotechnol* **37**:459–475 (2017).

36 Johnson WG, Davis VM, Kruger GR, and Weller SC, Influence of glyphosate-resistant cropping systems on weed species shifts and glyphosate-resistant weed populations. *Eur J Agron* **31**:162–172 (2009).

37 Pazuch D, Trezzi MM, Guimaraes ACD, Barancelli MVJ, Pasini R, and Vidal RA, Evolution of Natural Resistance to Glyphosate in Morning Glory Populations. *Planta Daninha* **35**:e017159430 (2017).

38 Green JM, and Owen MDK, Herbicide tolerant crops: Utilities and limitations for herbicide resistant weed management. *J Agric Food Chem* **59**:5819–5829 (2011).

39 Nandula VK, Herbicide resistance traits in maize and soybean: Current status and future outlook. *Plants* **8**:337 (2019).

40 Byker HP, Soltani N, Robinson DE, Tardif FJ, Lawton MB, and Sikkema PH, Control of glyphosate-resistant horseweed (*Conyza canadensis*) with dicamba applied preplant and postemergence in dicamba-resistant soybean. *Weed Technol* **27**:492–496 (2013).

41 Wright TR, Shan G, Walsh TA, Lira JM, Cui C, Song P, *et al.*, Robust crop resistance to broadleaf and grass herbicides provided by aryloxyalkanoate dioxygenase transgenes. *Proc Natl Acad Sci* **107**:20240–20245 (2010).

42 Johnson WG, Bradley K, Young B, Matthews J, Marquardt P, Slack C, *et al.*, Weed control in dicamba-resistant soybeans. *Crop Manage* **9**:1–23 (2010).

43 Egan JF, Barlow KM, and Mortensen DA, A Meta-Analysis on the effects of 2,4-D and dicamba drift on soybean and cotton. *Weed Sci* **62**:193–206 (2014).

44 Johnson VA, Fisher LR, Jordan DL, Edmisten KE, Stewart AM, and York AC, Cotton, peanut, and soybean response to sublethal rates of dicamba, glufosinate, and 2,4-D. *Weed Technol* **26**:195–206 (2012).

45 Gaines TA, Duke SO, Morran S, Rigon CAG, Tranel PJ, Kupper A, *et al.*, Mechanisms of evolved herbicide resistance. *J Biol Chem* **295**:10307–10330 (2020).

46 HRAC, Synthetic Auxin Resistant Weeds. https://hracglobal.com/files/Synthetic-Auxin-Fact-Sheet-June-2016.pdf [Accessed May 7, 2024] (2016).

47 Huang X, He J, Yan X, Hong Q, Chen K, He Q, *et al.*, Microbial catabolism of chemical herbicides: Microbial resources, metabolic pathways and catabolic genes. *Pestic Biochem Physiol* **143**:272–297 (2017).

48 Muller TA, Byrde SM, Werlen C, van der Meer JR, and Kohler HP, Genetic analysis of phenoxyalkanoic acid degradation in Sphingomonas herbicidovorans MH. *Appl Environ Microbiol* **70**:6066–6075 (2004).

49 Schleinitz KM, Kleinsteuber S, Vallaeys T, and Babel W, Localization and characterization of two novel genes encoding stereospecific dioxygenases catalyzing 2(2,4-dichlorophenoxy)propionate cleavage in Delftia acidovorans MC1. *Appl Environ Microbiol* **70**:5357–5365 (2004).

50 Nielsen TK, Rasmussen M, Demaneche S, Cecillon S, Vogel TM, and Hansen LH, Evolution of Sphingomonad Gene Clusters Related to Pesticide Catabolism Revealed by Genome Sequence and Mobilomics of Sphingobium herbicidovorans MH. *Genome Biol Evol* **9**:2477–2490 (2017).

51 Zipper C, Nickel K, Angst W, and Kohler HP, Complete microbial degradation of both enantiomers of the chiral herbicide mecoprop [(RS)-2-(4-chloro-2-methylphenoxy)propionic acid] in an enantioselective manner by Sphingomonas herbicidovorans sp. nov. *Appl Environ Microbiol* **62**:4318–4322 (1996).

52 Harker AR, Olsen RH, and Seidler RJ, Phenoxyacetic acid degradation by the 2,4-dichlorophenoxyacetic acid (TFD) pathway of plasmid pJP4: mapping and characterization of the TFD regulatory gene, tfdR. *J Bacteriol* **171**:314–320 (1989).

53 Muller TA, Fleischmann T, van der Meer JR, and Kohler HP, Purification and characterization of two enantioselective alpha-ketoglutarate-dependent dioxygenases, RdpA and SdpA, from *Sphingomonas herbicidovorans* MH. *Appl Environ Microbiol* **72**:4853–4861 (2006).

54 Fukumori F, and Hausinger RP, Purification and characterization of 2,4-dichlorophenoxyacetate/alpha-ketoglutarate dioxygenase. *J Biol Chem* **268**:24311–24317 (1993).

55 Hausinger RP, and Fukumori F, Characterization of the first enzyme in 2,4-dichlorophenoxyacetic acid metabolism. *Environ Health Perspect* **103**:37–39 (1995).

56 Hausinger RP, Fukumori F, Hogan DA, Assanella TM, Kamagta Y, Takami H, and Saari RE, Biochemistry of 2,4-dichlorophenoxyacetic acid (2,4-D) degradation: evolutionary implications, In Horikoshi K, Fukuda M, and Kudo T (Eds), *Microbial Diversity and Genetics of Biodegradation*, Japan Scientific Societies Press, Tokyo, pp. 35–51 (1997).

57 Parke D, and Ornston LN, Enzymes of the beta-ketoadipate pathway are inducible in *Rhizobium* and *Agrobacterium* spp. and constitutive in *Bradyrhizobium* spp. *J Bacteriol* **165**:288–292 (1986).

58 Saari R, Hogan DA, and Hausinger R, Stereospecific degradation of the phenoxypropionate herbicide dichlorprop. *J Mol Catal B Enzym* **6**:421–428 (1999).

59 Chekan JR, Ongpipattanakul C, Wright TR, Zhang B, Bollinger JM Jr, Rajakovich LJ, *et al.*, Molecular basis for enantioselective herbicide degradation imparted by aryloxyalkanoate dioxygenases in transgenic plants. *Proc Natl Acad Sci* **116**:13299–13304 (2019).

60 Muller RH, and Babel W, Delftia acidovorans MC1 resists high herbicide concentrations – a study of nutristat growth on (RS)-2-(2,4-Dichlorophenoxy) propionate and 2,4-dichlorophenoxyacetate. *BBBIEJ* **68**:622–630 (2004).

61 Muller RH, and Hoffmann D, Uptake kinetics of 2,4-dichlorophenoxyacetate by Delftia acidovorans MC1 and derivative strains: complex characteristics in response to pH and growth substrate. *BBBIEJ* **70**:1642–1654 (2006).

62 Matell M, Stereochemical Studies on Plant Growth Regulators. Optically Active Alpha-(2-methyl-4-chlorophenoxy)-Propionic Acid and Alpha-(2,4-Dichlorophenoxy)-Normal-Butyric Acid and Their Steric Relations. *Arkiv for Kemi* **6**:365–373 (1953).

63 Altom JD, and Stritzke JF, Degradation of dicamba, picloram, and 4 phenoxy herbicides in soils. *Weed Sci* **21**:556–560 (1973).

64 Voos G, and Groffman PM, Relationships between microbial biomass and dissipation of 2,4-D and dicamba in soil. *Biol Fert Soils* **24**:106–110 (1997).

65 Krueger JP, Butz RG, Atallah YH, and Cork DJ, Isolation and Identification of Microorganisms for the Degradation of Dicamba. *J Agr Food Chem* **37**:534–538 (1989).

66 Liu JW, Zhang X, Bao YX, Zhang KY, Qiu JG, He Q, *et al.*, Enhanced degradation of dicamba by an anaerobic sludge acclimated from river sediment. *Sci Total Environ* **777**:145931 (2021).

67 Wang XZ, Li B, Herman PL, and Weeks DP, A three-component enzyme system catalyzes the O demethylation of the herbicide dicamba in *Pseudomonas maltophilia* DI-6. *Appl Environ Microb* **63**(4):1623–1626 (1997).

68 Herman PL, Behrens M, Chakraborty S, Chrastil BM, Barycki J, and Weeks DP, A three-component dicamba O-demethylase from *Pseudomonas maltophilia*, strain DI-6 Gene Isolation, Characterization, and Heterologous Expression. *J Biol Chem* **280**:24759–24767 (2005).

69 Dumitru R, Jiang WZ, Weeks DP, and Wilson MA, Crystal structure of dicamba monooxygenase: A rieske nonheme oxygenase that catalyzes oxidative demethylation. *J Mol Biol* **392**:498–510 (2009).

70 D'Ordine RL, Rydel TJ, Storek MJ, Sturman EJ, Moshiri F, Bartlett RK, *et al.*, Dicamba monooxygenase: Structural insights into a dynamic rieske oxygenase that catalyzes an exocyclic monooxygenation. *J Mol Biol* **392**:481–497 (2009).

71 Jiang WZ, Wilson MA, and Weeks DP, O-demethylations catalyzed by rieske nonheme iron monooxygenases involve the difficult oxidation of a saturated C-H bond. *Acs Chem Biol* **8**:1687–1691 (2013).

72 Behrens MR, Mutlu N, Chakraborty S, Dumitru R, Jiang WZ, LaVallee BJ, *et al.*, Dicamba resistance: Enlarging and preserving biotechnology-based weed management strategies. *Science* **316**:1185–1188 (2007).

73 Bayley C, Trolinder N, Ray C, Morgan M, Quisenberry JE, and Ow DW, Engineering 2,4-D resistance into cotton. *Theor Appl Genet* **83**:645–649 (1992).

74 Last DI, and Llewellyn DJ, A detoxification gene in transgenic Nicotiana tabacum confers 2,4-D tolerance. *Weed Sci* **47**:401–404 (1999).

75 Birru FH, Buffard CJ and Stelpflug RG, Plants and seeds of Hybrid Corn Variety CH600926. US patent US9918447 B1 (2018).

76 Tan X, Calderon-Villalobos LI, Sharon M, Zheng C, Robinson CV, Estelle M, *et al.*, Mechanism of auxin perception by the TIR1 ubiquitin ligase. *Nature* **446**:640–645 (2007).

77 Althoff E, Ban YEA, Castle LA, Grabs D, Lu J, Patten PA, Tao Y and Zanghellini A, Composition having dicamba decarboxylase activity and methods of use. US patent US 2016/0040149 (2016).

78 Zhao N, Yao J, Chaiprasongsuk M, Li G, Guan J, Tschaplinski TJ, *et al.*, Molecular and biochemical characterization of the jasmonic acid methyltransferase gene from black cottonwood (*Populus trichocarpa*). *Phytochem* **94**:74–81 (2013).

79 Chen F, Zhao N and Armel GR, Herbicide resistance gene. US patent US8404933B2 (2010).

80 Dodson L, Recent trends in GE adoption. https://www.ers.usda.gov/data-products/adoption-of-genetically-engineered-crops-in-the-us/recent-trends-in-ge-adoption/ [Accessed May 7, 2024] (2020).

81 Wechsler SJ, Smith D, McFadden J, Dodson L and Williamson S, The use of genetically engineered dicamba-tolerant soybean seeds has increased quickly, benefiting adopters but damaging crops in some fields. https://www.ers.usda.gov/amber-waves/2019/october/the-use-of-genetically-engineered-dicamba-tolerant-soybean-seeds-has-increased-quickly-benefiting-adopters-but-damaging-crops-in-some-fields/ [Accessed May 7, 2024] (2019).

82 Begemann S, Corn and soy seed production on track, 2021 Enlist soybean acreage jump estimated at 15 Percentage Points. https://seedworld.com/corn-and-soy-seed-production-on-track-2021-enlist-soybean-acreage-jump-estimated-at-15-percentage-points/ [Accessed May 7, 2024] (2021).

83 Magro C, Glenn T, King R, Eathington S, Anderson D and Booth K. Corteva Investor relations presentation: Investor Day. https://investors.corteva.com/static-files/0827e852-8751-4c2c-ac3c-0d0f55877386 [Accessed May 7, 2024] (2022).

84 Havens PL, Hillger DE, Hewitt AJ, Kruger GR, Marchi-Werle L, and Czaczyk Z, Field measurements of drift of conventional and drift control formulations of 2,4-D plus glyphosate. *Weed Technol* **32**:550–556 (2018).

85 Ellis JM, Ruen DC, Moechnig MM, Scherder EF, Rosenbaum K, Campbell LA, Haile F and Granke L. Enlist E3™ soybean weed control systems in midwest. *In* Proceedings of Annual Meeting of the North Central Weed Science Society; Minneapolis, MN (2014).

86 Frene RL, Simpson DM, Buchanan MB, Vega ET, Ravotti ME, and Valverde PP, Enlist E3™ soybean sensitivity and Enlist™ herbicide-based program control of sumatran fleabane (*Conyza sumatrensis*). *Weed Technol* **32**:416–423 (2018).

87 Ruen DC, Moechnig MM, Scherder EF, Rosenbaum K, Campbell LA, Haile F, Granke L and Ellis JM, Enlist corn weed control systems in midwest. *In* Proceedings of Annual Meeting of the North Central Weed Science Society, Minneapolis, MN (2014).

88 Manuchehri MR, Dotray PA, and Keeling JW, Enlist™ weed control systems for palmer amaranth (*Amaranthus palmeri*) management in Texas high plains cotton. *Weed Technol* **31**:793–798 (2017).

89 Simpson DM, Rosenbaum K, Ellis JM, Richburg JS, Haile F, Granke L, Thompson GD, Haygood BH and Walton LW. Palmer amaranth weed control with enlist weed control systems. *In* Proceedings of Annual Meeting of the North Central Weed Science Society, Minneapolis, MN (2014).

90 Bish M, Oseland E, and Bradley K, Off-target pesticide movement: a review of our current understanding of drift due to inversions and secondary movement. *Weed Technol* **35**:345–356 (2021).

91 Enlist Weed Control System - Application Guide, https://www.enlist.com/en/herbicides/application-guide.html [Accessed May 7, 2024] (2024).

92 RoundUp Ready Xtend Crop Systems - Overview of Application Requirements, https://www.roundupreadyxtend.com/stewardship/educational-resources/Pages/Application-Requirements.aspx [Accessed May 7, 2024] (2024).

93 Shyam C, Peterson DE, and Jugulam M, Resistance to 2,4-D in Palmer amaranth (*Amaranthus palmeri*) from Kansas is mediated by enhanced metabolism. *Weed Sci* **70**:390–400 (2022).

94 Synthetic Auxin Resistance in Wild Radish, https://hracglobal.com/files/Wild-Radish-Fact-Sheet-June-2016.pdf [Accessed May 7, 2024] (2016).

95 Synthetic Auxin Resistance in Corn Poppy, https://hracglobal.com/files/Corn-Poppy-Fact-Sheet-June-2016.pdf [Accessed May 7, 2024] (2016).

96 Synthetic Auxin Resistance in Waterhemp, https://hracglobal.com/files/Waterhemp-Fact-Sheet-June-2016.pdf [Accessed May 7, 2024] (2016).

97 Shyam C, Borgato EA, Peterson DE, Dille JA, and Jugulam M, Predominance of metabolic resistance in a six-way-resistant palmer amaranth (*Amaranthus palmeri*) population. *Front Plant Sci* **12**:614618 (2021).

98 LeClere S, Wu CX, Westra P, and Sammons RD, Cross-resistance to dicamba, 2,4-D, and fluroxypyr in *Kochia scoparia* is endowed by a mutation in an AUX/IAA gene. *P Natl Acad Sci* **115**:E2911–E2920 (2018).

99 de Figueiredo MRA, Kupper A, Malone JM, Petrovic T, de Figueiredo ABTB, Campagnola G, *et al.*, An in-frame deletion mutation in the degron tail of auxin coreceptor IAA2 confers resistance to the herbicide 2,4-D in *Sisymbrium orientale*. *P Natl Acad Sci* **119**:e2105819119 (2022).

100 Wu CX, LeClere S, Liu K, Paciorek M, Perez-Jones A, Westra P, *et al.*, A dicamba resistance-endowing IAA16 mutation leads to significant vegetative growth defects and impaired competitiveness in kochia (*Bassia scoparia*)(dagger). *Pest Manag Sci* **77**:795–804 (2021).

101 Evans CM, Strom SA, Riechers DE, Davis AS, Tranel PJ, and Hager AG, Characterization of a waterhemp (*Amaranthus tuberculatus*) population from Illinois resistant to herbicides from five site-of-action groups. *Weed Technol* **33**:400–410 (2019).

102 de Pinho CF, Leal JF, dos Santos SA, de Oliveira GF, de Oliveira C, Langaro AC, Machado AF, Christoffoleti PJ, and Zobiole LH, First evidence of multiple resistance of Sumatran Fleabane (*Conyza sumatrensis* [Retz.] E.Walker) to five- mode-of-action herbicides. *Aust J Crop Sci* **13**:1688–1697 (2019).

103 Walsh MJ, Powles SB, Beard BR, and Porter SA, Multiple-herbicide resistance across four modes of action in wild radish (*Raphanus raphanistrum*). *Weed Sci* **52**:8–13 (2004).

104 Norsworthy JK, Ward SM, Shaw DR, Llewellyn RS, Nichols RL, Webster TM, *et al.*, Reducing the risks of herbicide resistance: Best management practices and recommendations. *Weed Sci* **60**:31–62 (2012).

105 Liu C, Neve P, Glasgow L, Wuerffel RJ, Owen MDK, and Kaundun SS, Modeling the sustainability and economics of stacked herbicide-tolerant traits and early weed management strategy for waterhemp (*Amaranthus tuberculatus*) control. *Weed Sci* **68**:195–196 (2020).

3

Metabolic Degradation of Glyphosate in Soil Microbes, Endophytes, Crops, and Weeds

Stephen O. Duke

National Center for Natural Product Research, School of Pharmacy, University of Mississippi, Oxford, Mississippi, USA

3.1 Introduction

Glyphosate (*N*-[phosphonomethyl] glycine) is the most used herbicide worldwide, both in the actual amount used and the land area treated with this postemergence, nonselective herbicide.[1] Its widespread and heavy use is due to several factors, including: (1) it is a high use rate herbicide (0.5–2 kg/ha)[2]; (2) it is a very effective and economical herbicide on almost all weeds[2,3]; and (3) it is heavily used with widely adopted glyphosate-resistant (GR) crops.[3–5] All of glyphosate's herbicidal action is due to inhibition of 5-enolpyruvylshikimate-3-phosphate synthase (EPSPS), a key enzyme of the shikimate pathway, and it is the only commercial herbicide with this molecular target.[2] As of May 2024, there were almost 29,000 publications and patents involving glyphosate in the SciFinder database.[6]

Despite the huge amount of research on glyphosate, relatively little is known about metabolic degradation of glyphosate in plants. As mentioned about 15 years ago,[7] there seem to be more papers on analytical methods to measure glyphosate and its metabolites in plants than on meaningful studies of glyphosate metabolism in plants. For example, there are several papers on measurement of the major degradation product aminomethylphosphonic acid (AMPA) (Fig. 3.1) of glyphosate in rice,[8] but there are no papers showing that rice actually degrades glyphosate to AMPA, even though a gene for its degradation has been reported in rice (see Section 3.4.2).[9] More is known about microbial than plant degradation of glyphosate, and some of the information on microbial degradation is or may be

Resistance in Weeds from Herbicide Metabolism, First Edition.
Edited by Vijay K. Nandula and Roland Beffa.

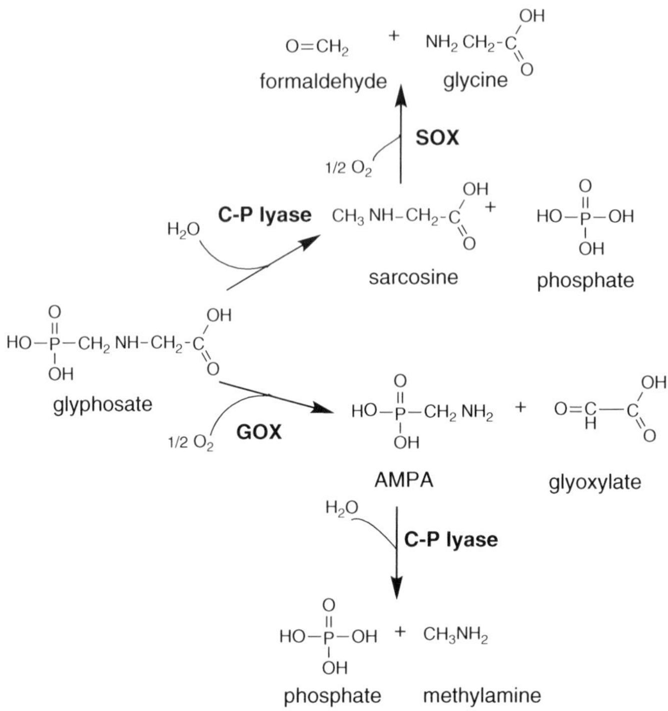

Figure 3.1 The two enzymatic routes of the degradation of glyphosate by C-P lyase, glyphosate oxidoreductase (GOX), and sarcosine oxidase (SOX).

germane to *in planta* degradation. For example, there is circumstantial evidence of endophytes being responsible for at least some glyphosate degradation in plants, and genes from microbes for degradation or chemical modification of glyphosate have been used as transgenes to produce GR crops. Therefore, microbial degradation of glyphosate is briefly reviewed.

There has been one previous review dedicated solely to glyphosate metabolism in plants,[7] as well as more recent reviews that cover metabolism of glyphosate as part of a wider coverage of either herbicide metabolism by plants[10] or glyphosate as a herbicide.[2] Borggaard and Gimsing[11] reviewed microbial degradation of glyphosate in soil, and Zhan et al.,[12] Sviridov et al.,[13] Singh et al.,[14] and Chen et al.[15] reviewed all microbial degradation of the herbicide. Some of what is covered in this chapter will be redundant with these earlier reviews, but the focus is on the most recent and pertinent literature. Several important questions that have not yet been answered are discussed, despite the considerable amount of research focused on this most important herbicide.

3.2 Enzymes that Can Degrade Glyphosate

This section briefly discusses the enzymes that degrade glyphosate for which there is a published gene sequence (Table 3.1). All of these are microbial genes except for the aldo-keto reductase (AKR; EC 1.1) with glyphosate oxidoreductase (GOX; EC 1.5.3.23) activity. GOX is the predominant enzyme that converts glyphosate to AMPA and glyoxylate, a common, natural metabolite (Fig. 3.1). C-P lyase is the other enzyme most involved in glyphosate metabolism, especially in microbes. It degrades glyphosate to sarcosine, a natural product, and phosphate (Fig. 3.1). For a more detailed discussion of the enzymology and biochemistry of most (excluding AKR and glyphosate decarboxylase) of these genes, the study by Pollegioni et al.[16] is recommended.

Most plant species seem to have an enzyme that can act as a GOX, although with very weak activity in most species. Acting as a GOX is clearly not the function of these enzymes in nature. We know appallingly little of this/these enzyme(s). Whether the AKR reported to have GOX activity in rice,[9] *Echinochloa colona*,[17,18] soybean,[19] maize,[19] and *Lolium rigidum*[20] is the same type enzyme in other plant species that clearly metabolize glyphosate to AMPA (Table 3.2) has not been reported. This is the only enzyme that degrades glyphosate in plants that has been isolated and the gene sequenced. AKRs are a super family of enzymes that are involved in redox reactions as part of biosynthetic, intermediary metabolic, and detoxification processes in microbes, animals, and plants.[21,22] Even though microbes possess AKRs, the work of Pan et al.[17] showed that the plant possessed the AKR responsible for the GOX activity that led to glyphosate resistance of *E. colona*. These investigators proposed that $NADP^+$ acts as a cofactor for the

Table 3.1 Enzymes with sequenced genes that are reported to break down or inactivate glyphosate.

Enzyme	Source	Products	References
Aldo-keto reductase	*Echinochloa colona*	AMPA and glyoxylate	16
C-P lyase	*Pseudomonas pseudomallei*	Sarcosine and phosphate	17
Glyphosate oxidase	*Ochrobactrum anthropi*	AMPA and glyoxylate	18
Glycine oxidase	*Bacillus subtilis*	AMPA and glyoxylate	19
	Bradyrhizobium spp.	AMPA and glyoxylate	20
Glyphosate acetyl transferase	*Bacillus licheniformis*	*N*-Acetylglyphosate	21
Glyphosate decarboxylase	Multiple	Phosphonodimethylamine	22

Table 3.2 Examples of weeds and crops reported to metabolize glyphosate.

Species	Degradation reaction	References
Weed		
Agropyron repens	GOX	23
Alternanthera philoxeroides	GOX	24
Amaranthus palmeri	GOX	25
Cassia occidentalis	GOX	26
Convolulus arvensis	GOX and C-P lyase	27
Conyza canadensis	GOX	26
Desmanthus illinoensis	GOX	26
Digitaria insularis	GOX and C-P lyase	28
Echinochloa colona	GOX	16, 29
Eleusine indica	GOX	30
Equisetum arvense	GOX	31
Ipomea pupurea	GOX and C-P lyase	32
Lolium rigidum	GOX	33
Mucuna pruriens	GOX and C-P lyase	34
Pueraria montana	GOX	26
Senna obtusifolia	GOX	26
Sorghum halepense	GOX	35
Crop		
Brassica napus (canola)	GOX	36
Beta vulgaris (sugar beet)	GOX	37
Camellia sinensis (tea)	GOX	38
Glycine max (soybean)	GOX	39
Gossypium hirsutum (cotton)	GOX	40
Nicotiana tabacum (tobacco)	GOX	37
Saccharum officinarum (sugarcane)	GOX	41
Vigna unguiculate (cowpea)	GOX	26
Zea mays (maize)	GOX	42

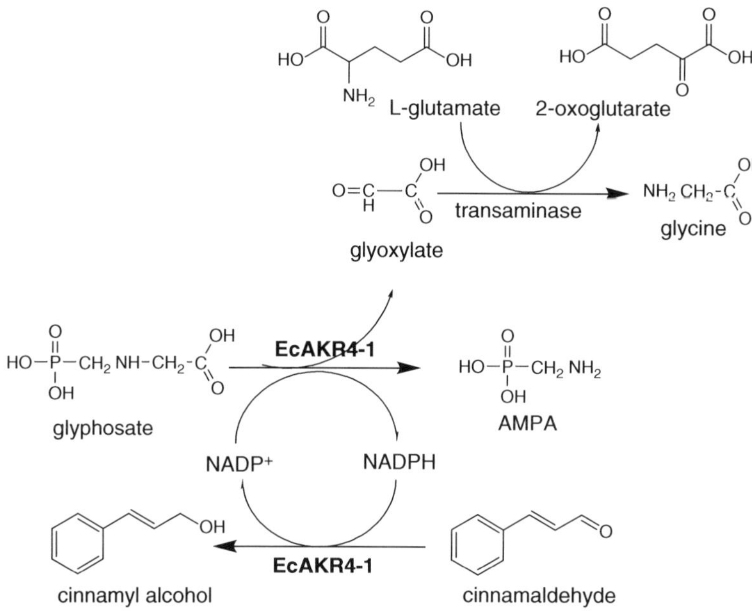

Figure 3.2 Proposed mechanism of *E. colona* AKR enzyme (EcAKR-1) that acts as a GOX, using cinnamaldehyde conversion to cinnamyl alcohol to regenerate NADP+. Source: Pan et al.[17]/Oxford University Press.

enzyme with cinnamaldehyde acting as a redox compound to regenerate NADP+ from NADPH (Fig. 3.2). Metabolomic effects of glyphosate treatment of GR *E. colona* included an increase in cinnamyl alcohol and a decrease in cinnamaldehyde, supporting the view that these compounds are coupled with the oxidation of glyphosate to AMPA and glyoxylate (Fig. 3.2). Thus, the evidence is strong that some plant AKRs can act as a GOX. The AKR of *E. colona* has relatively little homology (23.5% and 27.7%) to the bacterial GOX reported by Barry and Kishore[43] and the glyphosate-oxidizing glycine oxidase found in *Bacillus subtilis* by Pedotti et al.,[29] respectively. Why there has not been more selection for enhanced or more efficient plant AKRs in weed populations is a mystery. Because silencing the *AKR* gene in rice increased this crop's sensitivity to glyphosate and overexpressing it in rice and tobacco imbued these crops with glyphosate resistance,[9] plant AKRs appear to be worthy of further research to reduce glyphosate residues in the food supply.

Several microbially derived glyphosate metabolism transgenes have been proposed for the generation of GR crops. These include transgenes for enzymes that can acetylate, decarboxylate, and oxidize glyphosate. Pollegioni et al. provided a review of the attempts to modify and use some of these bacterial genes to produce GR crops.[16]

The first of these was a GOX from *Ochrobactrum anthropi* (formerly *Achromobacter* sp.).[43,44] This gene was substantially modified to make it more functional in plants before it was used as the transgene found in the first versions of GR canola.[16,45] It is a flavin adenine dinucleotide (FAD)-containing flavoprotein of 430 amino acids. This transgene (*goxv247*) was used in the first commercial GR canola, along with the microbial *cp4 EPSPS* transgene encoding a GR EPSPS. Later commercial varieties contained only the *cp4 EPSPS* transgene.

Considerable effort has gone into optimizing genes for microbial glycine oxidase for use as a *GOX* transgene in crops. A glycine oxidase of *B. subtilis* with very weak GOX activity was improved 210-fold in catalytic efficiency as a GOX by site saturation mutagenesis and rational design.[17] The authors stated that despite this dramatic improvement in GOX activity, their results were far from the limits of optimization. This *B. subtilis*-derived gene for glycine oxidase was further optimized for use in alfalfa, in which it provided moderate glyphosate resistance.[46] The GOX activities of glycine oxidases from other *Bacillus* species have been improved by gene shuffling.[47–49] Han et al. used a gene for glycine oxidase with GOX activity from *Bradyrhizobium japonicum*, with codon optimization for plant expression, to impart glyphosate resistance to *Arabidopsis*.[50] The transformed plants were not significantly affected by a 10 mM glyphosate foliar treatment that killed untransformed plants. Although GOX and glycine oxidase both convert glyphosate to AMPA and glyoxylate, there is little sequence homology between the genes for these enzymes.[17] Unlike GOX, the glycine oxidase reaction results in the formation of hydrogen peroxide to oxidize enzyme-associated FAD (Fig. 3.3). Rapid degradation of glyphosate by a glycine oxidase type reaction might generate enough reactive oxygen species (ROS) to be detrimental, although green plants are highly buffered from ROS.

The gene-shuffled glycine oxidase from *Bacillus cereus* with 326-fold greater GOX activity than the native enzyme was used with a bacterial transgene for a resistant form of EPSPS in rice.[51] The *GOX* transgene resulted in lower glyphosate residues than plants containing only the *EPSPS* transgene, but AMPA levels

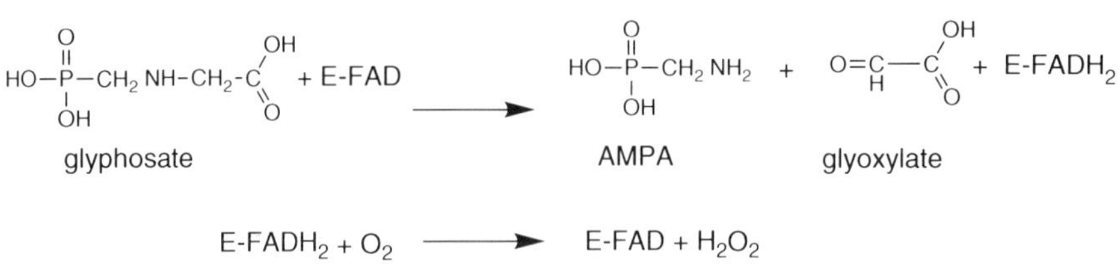

Figure 3.3 The mechanism of microbial glycine oxidase (E).

were not reported. Reduced glyphosate residues were the only advantage of the *GOX* transgene, as the *EPSPS* gene alone provided full resistance to glyphosate at recommended field doses.

A bacterial (*Bacillus licheniformis*)-derived glyphosate acetyltransferase (GAT) was obtained by 11 rounds of gene shuffling to increase its activity several 1000-fold (4 orders of magnitude).[23] Acetylated glyphosate is not phytotoxic, so accumulation of this metabolite in crops would not be a problem for the crop. This transgene produced GR plants,[52] and GR crops with this *GAT* transgene came close to commercialization.[53] A different microbial *GAT* gene was isolated and used without gene shuffling.[54] It was used in conjunction with a GR *EPSPS* gene to produce GR tobacco,[55] soybean,[56] cotton,[57] alfalfa,[58] and maize[59] with low glyphosate residues. The soybeans with this *GAT*/GR *EPSPS* transgene combination did not show the yellow flash syndrome (discussed in Section 3.5.2) found with GR soybeans made resistant with only a transgene for GR EPSPS.

Decarboxylase-type enzymes that inactivate glyphosate have been patented.[24] Such a degradation pathway would theoretically produce a novel glyphosate metabolite, phosphonodimethylamine, a compound apparently not in the literature. The chemical structure of the glyphosate derivative reported in the patent was not provided. Two genes for glyphosate decarboxylase with homology to pyruvate decarboxylase and acetolactate synthase are described in the patent. These genes provided some resistance in *Escherichia coli* to glyphosate. Neither of these transgenes have been used in commercial GR crops, and no peer-reviewed papers are available on these enzymes.

The C-P lyase process requires more than one gene,[60] which is one reason that its activity has not been generated *in vitro* or provided to crops by transgenes. For example, the C-P lyase activity of *E. coli* is encoded by the *phn* operon containing 14 genes.[61] Some of the genes and proteins for this pathway have been described from microbes, but there is no genetic evidence of a C-P lyase in higher plants.

3.3 Microbial Degradation of Glyphosate

3.3.1 Microbes in Soil and Water

Microbial degradation of glyphosate has been more thoroughly studied than degradation in plants. This is partly because of the importance of its relatively short half-life (a few days to a few weeks) of glyphosate in the environment that is known to be due to microbial degradation.[62,63] Glyphosate essentially does not degrade in sterile soil[64,65] or degrades much more slowly than in non-sterile soil.[66] In some soils, degradation of glyphosate is estimated to be 98% by bacterial action.[67] As discussed in Sections 3.3.2 and 3.4.1, there has been speculation that at least some of the degradation of glyphosate in plants is due to microbial contamination and/or microbial endophytes. Also, microbially-mediated degradation of glyphosate in soils is much slower under anaerobic than aerobic conditions,[67] and much of the biodegradation of glyphosate in soil may be due to unculturable microbes.[68]

There are two primary enzymatic routes of glyphosate degradation in microbes, although there appear to be a few minor enzymes found in microbes that can degrade it by other means (see Section 3.2). GOX is the predominant glyphosate degradation enzyme in microbes (Fig. 3.1). Some of the AMPA found in the environment, especially in urban settings, is due to the degradation of certain phosphonate additives to detergents and used by water-treatment plants.[69–71] The water-treatment compound diethylenetriaminepenta(methylenephosphonic) acid (DTPMP, trade name Dequest®) can be degraded to AMPA by microbes via a C-P lyase-independent pathway.[72]

A GOX-encoding gene (*goxv247*) from the soil microbe *O. anthropi* was used as a glyphosate resistance transgene in the first commercialized GR canola cultivars (see Section 3.5.2).[45] The other major enzymatic means of glyphosate degradation in microbes is through a C-P lyase (Fig. 3.1).[e.g.,73,74] Some C-P lyases can also break down AMPA to inorganic phosphate and methylamine (Fig. 3.1). The enzymatic C-P lyase steps in Fig. 3.1 are abbreviated versions of a complicated process that involves several steps.[60] C-P lyase-mediated degradation of glyphosate by soil microbes is considered less common than by GOX. However, sarcosine may have a shorter half-life as a natural product than AMPA, masking the importance of C-P lyase as a microbial glyphosate metabolism route in soil and water. The C-P lyase of some or all microbes could be considered a phosphotransferase, as the gene for the C-P lyase that degrades glyphosate in *Pseudomonas pseudomallei* has considerable homology with the hygromycin phosphotransferase gene of *E. coli*.[75] Furthermore, hygromycin phosphotransferase genes increase glyphosate resistance of *E. coli*.[75]

Considering the myriad of soil microbes, there are no extensive surveys of the soil microbes that degrade glyphosate, but three reviews list bacteria and fungi that degrade glyphosate.[12–15] Table 3.3 provides a listing of some of the microbes and their degradation pathways from these reviews and other sources. Some studies have shown degradation of glyphosate by specific microbes but have not determined the degradation pathway.[91] Most of the microbes are bacteria that degrade glyphosate via GOX activity. Some fungi, such as *Penicillium citrinum*,[92] *Alternaria* sp., and *Trichoderma* spp.,[88] are listed. Most of the fungi also appear to degrade glyphosate with GOX; however, one study reports fungal conversion of glyphosate to sarcosine.[93] Also, the finding that *Fusarium* strains can utilize glyphosate as a sole source of phosphorus indicates that they have a C-P lyase.[94] There is some evidence that the GOX activity of at least some fungi is associated with oxidative ligninolytic enzymes such as peroxidases and laccase.[95] However, bacteria are considered to be the leading microbes responsible for glyphosate degradation

Table 3.3 Examples of bacterial and fungal microbes that metabolize glyphosate.

Species	Degradation reaction	References
Bacteria		
Achromobacter sp. LW9	GOX	76
Achromobacter sp. MPK 7A	C-P lyase	77
Agrobacterium radiobacter	GOX	76
Alcaligenes sp.	C-P lyase	78
Arthrobacter atrocyaneus	GOX	79
Bacillus cereus	GOX and C-P lyase	80
Flavobacterium sp.	GOX	81
Geobacillus caldoxylosilyticus	GOX	82
Ochrobactrum sp.	GOX	83
Pseudomonas pseudomallei	C-P lyase	17
Pseudomonas sp.	95% GOX and 5% C-P lyase	74
Rhizobium meliloti	C-P lyase	84
Streptomyces sp.	C-P lyase	85
Fungi		
Aspergillus niger	GOX	86
Aspergillus oryzae	GOX	87
Mucor sp.	C-P lyase	88
Penicillium chrysogenum	GOX	89
Penicillium notatum	GOX	90
Scopulariopsis sp.	C-P lyase	86
Trichoderma harzianum	C-P lyase	86

in soil.[68] Evidence of C-P lyase-mediated degradation of glyphosate in soil is sparse, as sarcosine is not found in soil-containing microbes degrading glyphosate.[11,96,97] Wang et al. attributed this to the rapid oxidation of sarcosine to glycine.[98] They found initial microbial degradation of glyphosate in sediment to be via a C-P lyase, but later degradation to be dominated by GOX. The issue over which degradation pathway predominates is confounded by the finding that some microbes have both GOX and C-P lyase activities,[77,81] and the relative activities of the two enzymes in these microbes in soil in an agricultural environment is unknown.

Some microbes break down both glyphosate and AMPA with a C-P lyase, using both compounds as a sole source of phosphorus.[99] However, microbes that can break the C-P bond of glyphosate cannot always do the same with AMPA.[100] La Cecilia and Maggi estimated that only 9% of AMPA in soil is microbially degraded, whereas 98% of glyphosate is degraded.[67] Certain minerals like birnessite ($[Mn^{4+}, Mn^{3+}]_2O_2$) can mediate chemodegradation of both glyphosate and AMPA,[67] but whether chemodegradation of glyphosate occurs *in planta* has not been proven. However, the very low level of degradation of glyphosate in some plant species indicates that this route of degradation is insignificant. Hove-Jensen et al. reviewed the biochemistry and genetics of C-P lyases from microbes that break down both glyphosate and AMPA.[60] Although C-P lyase converts AMPA to inorganic phosphate and methylamine (Fig. 3.1), the finding that AMPA is more environmentally persistent than glyphosate in soils, as evidenced by many findings that AMPA levels are higher than glyphosate levels within a short time after glyphosate use in the field[101,102] and after longer periods following application,[102,103] supports the view that this route of glyphosate degradation is not as prevalent as degradation by GOX. Sun et al. reported that AMPA is three to six times more resistant than glyphosate to degradation in soil-water systems.[97] However, the half-life of sarcosine in soils is unknown, and it has not been reported in glyphosate-treated soil, although the work of Wang et al. indicates that it is very rapid.[98] Some microbes have very active sarcosine oxidases that convert sarcosine to glycine and formaldehyde (Fig. 3.1).[104]

A complicating factor is that the presence of phosphate in soil can affect degradation by either the GOX or C-P lyase pathway, depending on various other factors.[60,105] The presence of inorganic phosphate in cells of *Arthrobacter* sp. with a C-P lyase inhibits breakdown of glyphosate by this enzyme.[106] With one glyphosate-degrading microbe, metabolism of glyphosate via C-P lyase was inhibited by the presence of AMPA.[100] This particular microbe uses AMPA better than glyphosate as a substrate. GOX activity itself is independent of inorganic phosphorous levels.[74]

In summary, the dominant route of degradation of glyphosate in all soils has not been clearly determined, but most of the data suggest that formation of AMPA via GOX activity is much more common than direct cleavage of the C-P bond.[11] An exception is a paper by Aslam et al. who found C-P lyase to be the predominant route of glyphosate degradation in a German soil with a history of glyphosate and organic fertilizer applications.[107] It is clear that AMPA has a much longer half-life than glyphosate in soil.

There are several papers and patents on the use of glyphosate-degrading microbes to remove glyphosate from soil[108] and water.[109] Considering the relatively short half-life of glyphosate in soil and water[110,111] and the lack of significant (magnitude or duration) adverse biological activity of glyphosate in soil and water at the concentrations found with normal use,[2,112] there is usually no need for microbial bioremediation except in cases of significant spills.

Repeated use of pesticides can lead to the selection for soil microbiota that degrade the chemicals, leading to shorter half-lives in that soil, a phenomenon known as accelerated degradation.[113] Accelerated degradation should be expected in crops where glyphosate has been used for decades, especially because it is a highly used herbicide that can be used by many microbes as a source of phosphorous and by some as a carbon and/or nitrogen source. Microbes able to use glyphosate as a sole phosphorous source were more commonly isolated from soils with previous exposure to glyphosate than those without exposure.[100] Gimsing et al.[105] found higher glyphosate degradation in soils from organic farms than from conventional farms that had some history of glyphosate use, although the different soil types, lack of statistical analysis, and missing glyphosate use history made interpretation of the results inconclusive with regard to accelerated degradation. Degradation of glyphosate was compared in similar Argentine soils with and without long-term (11 continuous years) glyphosate use.[114] There were differences in soil microbe responses to glyphosate (e.g., glyphosate caused more soil respiration in the soil with no glyphosate use history), but, unfortunately, glyphosate degradation rates were not measured. A laboratory study showed that repeated application of glyphosate over a short time can shift soil microbial community composition, but there was no evidence of accelerated degradation.[115] Glyphosate can influence the degradation rate of other herbicides in soil. For example, its use increased atrazine degradation in a Brazilian Oxisol soil.[116,117] The effect was attributed to glyphosate supplying the microbes with nutrients as they were adapting to atrazine.[117] Glyphosate applied to Weswood silty clay loam soil increased its production of CO_2 and decreased its production as influenced by fluometuron and aldicarb.[118]

There is only one adequate field study to determine whether there is accelerated degradation of glyphosate in soils with a history of annual glyphosate use.[102] Glyphosate degradation in soils with 5 and 10 years of glyphosate use was compared with that in a soil with no record of herbicide use. The half-lives of glyphosate were 8.6 versus 16.1 days in the soils with 10 years versus 0 years of glyphosate use, respectively. Before glyphosate treatment with 3 mg/kg of soil, there were traces of glyphosate in the soils in which glyphosate had been previously used (7.3 and 8.1 mg/kg of soil used for 5 and 10 years of glyphosate-treated soils, respectively) and none in the naïve soil, but AMPA levels were high in the previously exposed soils (up to 20 mmol/kg [3.38 mg/kg] of soil). Glyphosate application only raised AMPA levels incrementally (less than 20%) in previously exposed soils, whereas, in the naïve soil, levels rose from 0 to almost 10 mmol/kg (1.7 mg/kg) of soil. There appeared to be no accelerated degradation of AMPA in soils with a history of glyphosate use. A later study by Guijarro et al.[33] further compared the glyphosate biodegradation potential of soils that had no previous glyphosate exposure with a similar soil with a history of 10 years of glyphosate use. The study quantitated copy number of the *Bradyrhizobium* spp. gene *thiO*, a glycine oxidase with GOX activity. The relative abundance (gene copies/μmoles of DNA) was greater in the soil with a history of glyphosate use, but the relative abundance of the *thiO* gene went down in the soil with a history and up in the naïve soil after glyphosate application. The abundance of *Rhizobium* spp. was the same in the two soils before and after glyphosate application; however, the *Rhizobium* species differed in the two soils. The potential role of other microbial taxa was not determined. Considering the importance of glyphosate and the many papers published on its real and/or speculated environmental effects, the fact that there has been so little research to establish whether there is accelerated degradation of glyphosate in soils of GR crops treated annually with glyphosate over many years is surprising.

However, in a study to simulate the effects of potential accumulation of glyphosate and AMPA in soil on growth of wheat, field peas, and canola when glyphosate is used over multiple years, Blackshaw and Harker found no effects on crops at recommended rates.[119] Depending on the crop and location, they found that application rates many times higher (17.6–77 kg a.e./ha) than recommended rates would be required to add enough glyphosate to soil to cause any crop injury. Considering the concerns of whether glyphosate significantly accumulates in soil and, if so, how this can be prevented,[120] more studies to determine whether accelerated degradation occurs with long-term use of glyphosate should be made.

There are rigorous papers on the generally transient effects of glyphosate on soil microflora under field conditions under which glyphosate is used yearly in GR crops,[112] but the finding of microbial (both bacterial and fungal) enzymes that are highly active in degrading glyphosate from soil and water sources with high glyphosate contamination[76,81,87] indicates that microbes with enhanced ability to degrade glyphosate should be selected for in-field situations in which glyphosate is heavily used for long time periods, such as GR crops.

3.3.2 Endophytes and Rhizoplane Microbes

Plants contain endosymbiotic bacteria and fungi that do not cause disease. Many of these microbes are beneficial to the plant by improving nutrient accumulation and enhancing responses to various stresses. They can also be involved in degradation of xenobiotics such as synthetic herbicides.[121] Whether endophytes are involved in degradation of glyphosate *in planta* has not been established. Many endophytes are sensitive to glyphosate. Endophytes from soybean include *B. japonicum*, *Burkholdia gladioli*, and *Pseudomonas oryzihabitans*.[122,123] Sensitivity to glyphosate varies between strains of *B. japonicum*[122] and between different endoyphyte species.[123] After a lag phase of 24 hours, *Burkholderia gladioli* grew normally in 20 mM glyphosate.[123] Zabaloy and Gómez found the growth of most strains of *Sinorhizobium* sp. and *Rhizobium* sp., isolated from nodules of several legume species, to be little or unaffected by 1.5 mM glyphosate.[124] Likewise, they found most strains of *Mesorhizobium* sp. and *Bradyrhizobium* sp. to be unaffected or slightly inhibited. They found only a *Bradyrhizobium* isolate to grow well in a medium with glyphosate as the sole carbon and energy source. *Bradyrhizobium* spp. have a glycine oxidase that acts as a GOX.[33] *B. japonicum* also has genes for a C-P lyase,[60] but these authors state that there are no data for use of phosphonates (e.g., glyphosate or AMPA) as a source of phosphorus. Earlier work had found glyphosate degradation via a C-P lyase by Rhizobiaceae to be common[84]; however, sarcosine has not been found to be a glyphosate degradation product in legumes, perhaps because it is rapidly metabolized.

Glyphosate is translocated to nodules of GR soybean, where it can accumulate in plants treated with 0.84 kg a.e./ha to levels up to ca. 150 ng/g dry weight.[125] This concentration is similar to that reported in seeds (ca. 200 ng/g dry weight) of glyphosate-treated (0.84 kg/ha) GR soybean.[41] Unfortunately, the immunochemical assay used by Reddy and Zablotowicz could not detect AMPA.[125] But, considering the amount of glyphosate found in the nodules and the presence of a GOX in *B. japonicum*,[50] one would expect some degradation of glyphosate to AMPA. AMPA applied to GR soybeans at rates meant to simulate the amount accumulated after field rates of glyphosate was phytotoxic to leaf tissue, but nodule function was relatively unaffected.[126] This could mean that although *B. japonicum* is sensitive to glyphosate, it may be tolerant to AMPA. There are apparently no reports on the direct effects of AMPA on *B. japonicum*.

The very beneficial endophyte *B. japonicum* has a glycine oxidase with GOX activity.[50] Since soybean is a plant with relatively high metabolic degradation of glyphosate via GOX,[41] much or all of this metabolism could be due to metabolism by *B. japonicum*. However, an argument against this is that although this endophyte exports nitrogen-containing compounds, there is no evidence that it would export AMPA *in planta*. *B. japonicum* also contains the genes for C-P lyase, but there are no data for growth of this endophyte on a phosphonate compound as a source of phosphorous.[60]

Colonization by endophytes can vary considerably within the same plant species, depending on many factors. For example, Haron et al. found the differences (species and bacterial loads) in bacterial communities colonizing different samples of the medicinal plant *Echinacea purpurea* accounted for the 8000-fold difference in biological activity (monocyte stimulation).[127] Similarly, if endophytes are involved in metabolism of glyphosate in plants, some or all of the differences reported in the same species (e.g., see discussion of maize below) could be due to differences in the types and amounts of endophytes of the samples studied.

A relatively simple experiment is needed to find some indication of whether an endophyte or endophytes is/are involved in some or all of the metabolism of glyphosate to AMPA in a plant such as soybean. GR forms of such plants should be grown with and without inoculum in sterile substrate. The plants should then be treated with a high, but not phytotoxic, dose of glyphosate (e.g., 3 kg a.e./ha) by carefully painting the leaves with glyphosate solution. All glyphosate and AMPA should be determined in residues left on the leaves and in residues in the shoots, roots, and soil at different times after application. The amount of glyphosate in the soil in such an experiment could be significant, in that some of the foliar-applied glyphosate is exuded from roots (reviewed by Ghanizadeh and Harrington[128]). The kinetics of a balance sheet of glyphosate and its principle metabolite(s) in plants enriched in a glyphosate-degrading endophyte (e.g., *B. japonicum*) should provide a definitive answer to the question of the role of endophytes in the metabolism of glyphosate.

Infection of a crop with an endophyte capable of rapid and efficient metabolism of glyphosate could be a method of producing a non-GMO, GR crop. Instead of using the genes from microbes to impart resistance, the microbes expressing these genes could be used, as suggested by Tétard-Jones and Edwards.[121] Microbes with genes altered to generate enzymes with greater activity in degrading glyphosate might be needed to make such a strategy successful. If such microbes degraded both glyphosate and AMPA, the harvested crop would be free of these residues. With a herbicide such as glyphosate, the endophytes would have access to the herbicide in foliar tissues to be effective in protecting the plant.

Because glyphosate is readily exuded from roots of foliar-treated plants,[128] microbes of the rhizoplane (on the outside of root epidermal cells, but in contact with soil) could also contribute to breakdown of glyphosate from the plant. Kryuchkova et al. found that such a microbe (*Enterobacter cloacae* C7), associated with both sunflower and sorghum, uses glyphosate as a phosphorous source via a C-P lyase.[129] No one has examined the influence of this or similar rhizoplane microbes on degradation of root-exuded glyphosate. Whether glyphosate degradation products in soil, formed as the result of exudation of glyphosate from roots into soil, are taken up by roots of the plants from the soil in the rhizoplane has not been determined. Roots of GR maize have been shown to take up AMPA from soil, although calcium montmorillonite clay reduced the amount taken up.[130] Glyphosate applied only to soil (2.37 kg/ha) and AMPA, apparently from microbial degradation in soil, is taken up by peach tree roots, resulting in about five-fold and eight-fold more AMPA than glyphosate in the tissues of leaves and fruit, respectively.[66] These ratios were found one day after application to soil, suggesting that rapidly generated AMPA from the soil was taken up, but the results are only circumstantial, as there was no proof that some or all of the AMPA was formed in the plant.

3.4 Glyphosate Degradation in Weeds

3.4.1 Glyphosate-Susceptible Weeds

Before GR weeds were first reported in the mid-1990s,[131] all weeds were considered susceptible to glyphosate, although some were clearly more susceptible than others (see Section 3.4.2). For a slow-acting, nonselective herbicide like glyphosate to be active on essentially all weeds, significant metabolic degradation was not expected. In most early studies of glyphosate in weeds, investigators found little or no metabolism of glyphosate.[132,133] Some even speculated that the small amounts of metabolites found in glyphosate-treated weeds might be due to microbial contamination at some point in the studies.[35] This view was bolstered by the finding that there was slow degradation of glyphosate to AMPA in hydroponic media in which plants were grown and treated with glyphosate via the roots.[34] However, the findings of AMPA,[25,26,28,34] and occasionally both AMPA and sarcosine[30,36] (Table 3.2), in plants treated with glyphosate strongly suggested that some plants can metabolize this herbicide; even though at the time of the publication of the Monsanto scientists-authored monograph on glyphosate,[134] there were more papers finding no metabolism of glyphosate than those reporting metabolism.[135] In one of these early papers, 85% of the glyphosate that had been in the shoots of *Equisetum arvense* (horsetail) had been metabolized to AMPA (53%) and unknown metabolites (32%) within one week of application of glyphosate.[34] This high level of degradation may explain the tolerance of this primitive plant species to glyphosate. The lack of detection of glyphosate metabolism in early studies of some other plant species might be due to the early methods of AMPA analysis not being as sensitive as those used now.[136] However, even some recent studies have found no metabolism of glyphosate in some plant species. For example, glyphosate, but not AMPA, was found in tissues of the perennial shrub *Rosa acicularis* two years after application of glyphosate to a Canadian forest site.[137] The apparent lack of GOX activity may account for the longevity of glyphosate in these plant tissues. Even with the capacity to metabolize glyphosate, how a plant is grown and treated with glyphosate can also influence whether and/or how much of the herbicide is metabolized. This is evidenced from the work with maize that is discussed in Section 3.5.2.

As in soil, AMPA is reported much more often than sarcosine as a degradation product in glyphosate-treated plants. Also, as in soil, this finding may be due to a very short half-life of sarcosine in plants. Plants have sarcosine oxidase,[138] but it is a little-explored enzyme in plants. Its role in degrading sarcosine derived from glyphosate in plants has not been determined.

3.4.2 Glyphosate-Tolerant and -Resistant Weeds

Weeds have been evolving resistance to glyphosate rapidly over the past 30 years, with examples of at least 60 different species being resistant.[139] The resistance mechanisms are many, including both one and two codon changes in the gene for EPSPS, gene amplification (duplication) of *EPSPS* genes, reduced uptake and translocation of glyphosate, and enhanced degradation of glyphosate.[140] A recent significant finding related to evolved glyphosate resistance is that a GR *E. colona* biotype has an upregulated ABC transporter, probably located on the plasma membrane, that pumps glyphosate from the plasma membrane to the plant apoplast.[141] This and other mechanisms of resistance or tolerance that reduce the amount of glyphosate in the cytoplasm would also limit the exposure of glyphosate to enzymatic degradation. The role of metabolism of glyphosate in evolved glyphosate resistance has been covered in a recent overall review of mechanisms of evolved herbicide resistance in weeds[142] and in more detailed reviews of evolved glyphosate resistance.[140,143]

Some weed species are naturally less susceptible (tolerant) to glyphosate without selection by glyphosate for this trait. For example, morning glory (*Ipomoea* spp.) species require higher rates of glyphosate to kill than that required for most other weed species.[144,145] Considering that there is no reported metabolism in some weed species and others can readily metabolize it,[28] more rapid metabolic loss of taken up glyphosate might be expected in at least some glyphosate-tolerant weeds.

To properly compare glyphosate metabolism between species, the metabolism could be compared in plants exposed to rates that are equally phytotoxic to the different species, as toxicity could limit metabolic degradation. In the only study that compares metabolism between species at an I_{50} rate (the dose that reduces growth by 50%), there was no correlation between I_{50} values and the relative amount of glyphosate converted to AMPA by different species.[28] For example, at seven days after treatment, no AMPA formation was found in Italian ryegrass (*Lolium perenne*) with an I_{50} of 220 g a.e./ha, whereas about a 20% of the glyphosate in the legume sicklepod (*Senna obtusifolia*), with an I_{50} of 250 g a.e./ha, was converted to AMPA after two days (Fig. 3.4). The amount of glyphosate found in the tissues of the two species was similar. Thus, the I_{50} value for glyphosate in these species appears to be unrelated to metabolism of glyphosate. In the same experiment, only a trace of AMPA was found in the legume hemp sesbania (*Sesbania herbacea*), which was the most tolerant species (I_{50} = 456 g a.e./ha). Other weed and crop species were included in this study, and there was no indication that a relatively high degradation rate provided tolerance (Fig. 3.4). The species included in Fig. 3.4 are all legumes, except for horseweed (*Conyza canadensis*). Other nonlegumes (maize, Italian ryegrass [*Lolium perrene*], and velvet leaf

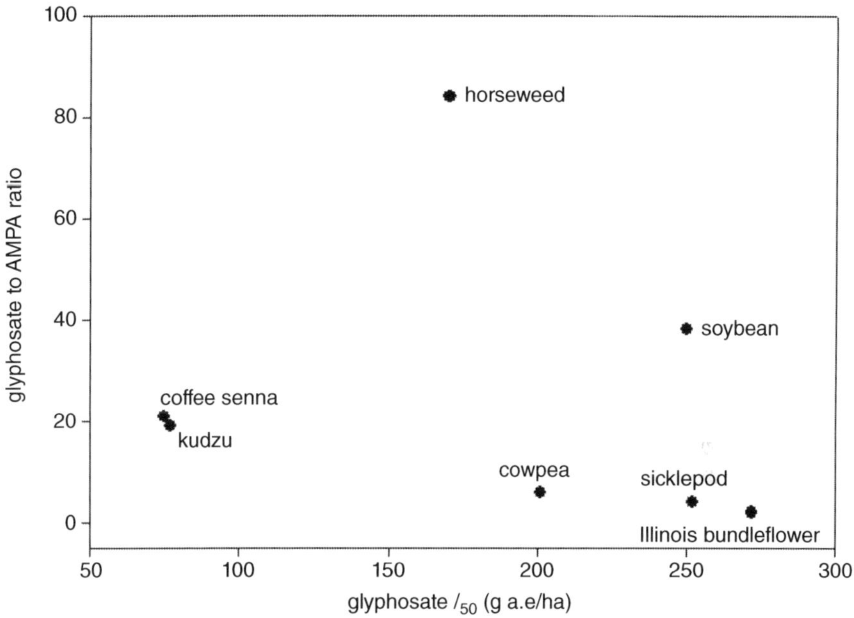

Figure 3.4 The ratio of glyphosate to AMPA (Y-axis) in seven species seven days after treatment with the glyphosate rate that provides 50% growth inhibition (X-axis). The soybean is non-GR soybean. Source: Adapted from Reddy et al.[28]

[*Abutilon threophrasti*]) in this study could not be plotted because no AMPA was detected. This study strongly indicates that glyphosate metabolism to AMPA is more common in legumes than other species, but the amount of metabolism was not related to the level of tolerance in these species. In general, there have been fewer reports of GOX-type metabolism of glyphosate in grass (Poaceae) weed and crop species, but there was similar degradation of glyphosate to AMPA in both GR and susceptible biotypes of johnsongrass (*Sorghum halepense*),[37] and maize does metabolize glyphosate to AMPA at a very slow rate (see Section 5.2).

Comparison of metabolism of glyphosate in the most tolerant and least tolerant of 14 *Ipomoea lacunosa* accessions (2.3-fold difference) found no difference in degradation of glyphosate to AMPA.[145] No metabolism to sarcosine was found. Tolerance levels were associated with differences in translocation of glyphosate.

Since metabolic degradation of herbicides is a common mechanism of evolved resistance to herbicides in general,[142] one might assume that enhanced metabolism of glyphosate degradation would be a common evolved mechanism or resistance to glyphosate, especially for a herbicide that took more than 20 years of intensive use and selection pressure for resistance to evolve in weeds and for which enzymes for degradation apparently already exist in some species.[7] Because glyphosate is a very slow-acting herbicide, evolution of enhanced degradation as a resistance mechanism would seem likely. But, out of many cases of evolved resistance to glyphosate,[140] there have been relatively few claims of enhanced metabolism of glyphosate in a few GR weed species,[146] and such claims have been met with skepticism[143] and have not been corroborated. Whether endophytes could contribute to glyphosate resistance in weeds has not been shown, but the higher levels of endophyte infection of GR *Lolium multiflorum* than in non-GR biotypes[147] is worth for further examination for differences in glyphosate metabolism.

Until recently, neither the enzyme nor the gene for the GOX that is considered responsible for glyphosate degradation in most plants had been identified. AMPA is very weakly phytotoxic (see Section 3.5.2),[148] so sufficiently rapid degradation of glyphosate to AMPA without further degradation should provide resistance. However, most studies have found no differences in glyphosate degradation between GR and glyphosate-susceptible weed biotypes of the same species.[140] Pan et al.[17] recently reported that the mechanism of glyphosate resistance in an *E. colona* biotype involves elevated levels of AKR, due to upregulation of two *AKR* genes, resulting in more rapid metabolism of glyphosate to AMPA and glyoxylate (Fig. 3.2). There was some degradation of glyphosate in the susceptible biotype, but three days after treatment there was more than four times as much glyphosate left in tissues of the susceptible plants than that of the resistant plants (Fig. 3.5). This was

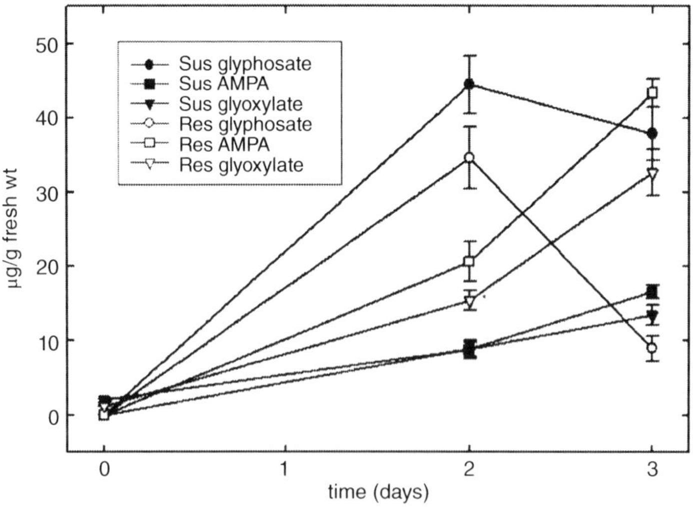

Figure 3.5 Change in glyphosate, AMPA, and glyoxylate contents of leaf tissues of susceptible (Sus) and GR (Res) 3- to 4-leaf *E. colona* plants after spraying with 67.5 g/ha of glyphosate. Error bars are 1 SE. Source: Adapted from Pan et al.[17]

a landmark paper, as it identified an enzyme and a gene for it that is responsible for GOX-type activity in plants. Earlier, overexpression of an *AKR* gene from rice provided glyphosate resistance to tobacco, and silencing this gene in rice caused hypersensitivity to glyphosate.[9] Unfortunately, glyphosate degradation products were not measured in this study.

Another paper reported that the same GR biotype of *E. colona* with elevated AKR GOX activity also contains a resistant form of EPSPS.[149] It concluded that this mutation and the enhanced AKA activity work in concert to provide a more robust glyphosate resistance than either mechanism alone. Whether AKRs are the GOX enzymes that cause accumulation of AMPA in other weeds or crops remains to be determined. This first finding of evolved enhanced metabolic degradation of glyphosate associated with evidence of the degradation enzyme involved established this mechanism among the many other glyphosate resistance mechanisms that apparently took less time to evolve.[150]

This well-documented case of evolved glyphosate resistance in *E. colona* is probably a case of what is called "creeping resistance" by some scientists.[151] This phenomenon is accumulation of mutations that each produce weak resistance, but together provide more robust resistance. In this case, the plant apparently evolved a weak form of resistance with a one codon change in EPSPS (proline 106 to serine), which allows enough plants in a population to survive in glyphosate-treated fields for selection of a second mechanism, enhanced metabolic degradation. The combined mechanisms provide stronger resistance than either mutation alone. Such creeping evolution of resistance is also responsible for the evolution of the two-codon mutations of EPSPS in GR *Eleusine indica*, which could not evolve simultaneously.[152] Each of these cases of creeping resistance took many years to occur or to be detected – 45 years in the case of the AKR involvement in enhanced metabolism of glyphosate as a glyphosate resistance mechanism.

No one has identified a C-P lyase protein or a gene or genes for one in plants. However, the few reports of breakdown of glyphosate to sarcosine in plants[31,36] suggest that such an enzyme might exist in some plant species or their endophytes. One might expect that plants with a C-P lyase with even weak activity with glyphosate as a substrate would, under the extreme selection pressure of intense glyphosate use, would evolve higher levels of C-P lyase gene expression or a C-P lyase mutant with greater activity with glyphosate as a substrate. There is no evidence that this has happened. There could be a fitness penalty for high levels of a C-P lyase in plants. For example, Baek et al.[140] speculated that the hydrogen peroxide and formaldehyde (Fig. 3.1) associated with sarcosine degradation are phytotoxic[138,153]; however, plants have good defenses against hydrogen peroxide and formaldehyde. Even selection for endophytes with C-P lyase[60] would be expected. Since the presence of phosphate can inhibit the expression of genes encoding C-P lyase,[60,105] the amount of endophyte C-P lyase *in planta* could vary considerably, depending on the phosphate status of the plant.

There are no good studies of glyphosate metabolism in GR weeds with a high-resistance factor. As discussed below, GR crops are ideal for the study of natural glyphosate metabolism because, even at relatively high application rates, there should be no adverse effect on processes that would metabolize the herbicide. Highly GR *E. indica* with a double mutation in the *EPSPS* gene[152] or *Amaranthus palmeri* with many extra copies of the *EPSPS* gene[154] would also be ideal for such studies, as their resistance factors are closer to those of GR crops than resistant weeds with a single *EPSPS* mutation or reduced uptake and/or translocation of glyphosate.

3.5 Glyphosate Degradation in Crops

Glyphosate degradation in weeds and crops are separated for two reasons. First, there is intense interest in the contamination of edible crops, both GR and conventional, with glyphosate and its metabolites.[111] Second, GR crops, due to their high (50–100-fold) resistance levels,[155,156] are the ideal plants for studies of metabolism of glyphosate without interference by the phytotoxicity of glyphosate.

3.5.1 Glyphosate-Susceptible Crops

In studies with sterile tobacco and sugar beet leaves, glyphosate was found to be metabolically degraded to AMPA, with no evidence of a C-P lyase involvement.[39] Microbe-free cell suspension cultures of wheat, soybean, maize, and soybean metabolized glyphosate to AMPA.[157] Soybean was most efficient, with about 50% degradation to AMPA by 48 hours of treatment, with maize and wheat having less than 5% AMPA. In the same study, rootless shoots of non-sterile spruce (*Picea abies*), tobacco, wheat, soybean, and maize also metabolized glyphosate to AMPA, with soybean being the most active. Microbe-free sycamore (*Acer pseudoplatanus*) cell cultures also readily degraded glyphosate to AMPA, reaching concentrations of 0.3–0.7 mM after incubation in 1 mM glyphosate for 50 hours.[158] All of the AMPA in this study appeared to be localized in the cytoplasm, although others have found AMPA in plant vacuoles.[159,160] These studies with microbe-free cell cultures proved that some plant species can readily metabolize glyphosate without the aid of endophytes. Nevertheless, some of the endophytes mentioned in Section 3.3.2 might enhance the metabolism of glyphosate by plants.

There is no proof in the Vemanna et al.'s study mentioned above that glyphosate is degraded to AMPA or sarcosine in rice.[9] Furthermore, the fact that all previous reports of glyphosate degradation in the Poaceae, the plant family to which rice belongs, indicate that glyphosate degradation is very low or nonexistent in members of this family. This seems inconsistent with the finding that silencing the putative gene for degradation of glyphosate in rice results in greater sensitivity of rice to the herbicide.[9] The genes identified for glyphosate-degrading enzyme in rice were homologues for the *igrA* gene of a *Pseudomonas* sp. strain that apparently encodes a glyphosate-degrading C-P lyase.[161] Yet, Pan et al. found a similar gene in *E. indica* to encode AKR that acts with GOX activity.[17]

Before transgenic, GR crops, glyphosate was not used directly on growing agronomic crops because of their susceptibility. Glyphosate is commonly used before planting non-GR crops and in mature, growing orchards and vineyards, where directed applications to lower-growing weeds can be made while minimizing contact with the foliage of these crops. It is occasionally used in row crops with devices that minimize contact with the crop.[162–164] Some contamination of the "item of commerce" part of the crop could be expected with all of these methods.

Glyphosate is used directly on some non-GR crops as a harvest aid, particularly soybeans and small grains.[2,165] This has caused concerns over residues of glyphosate and its metabolites in the harvested food products, especially when glyphosate is applied before the crop seeds are mature. Cessna et al. found both glyphosate and AMPA in harvested seed of field pea, barley, and flax when glyphosate was used as a harvest aid.[166] The levels were lower when it was applied at later stages of development. When properly used as a harvest aid in wheat, most of the glyphosate ends up in the straw, with very little in the seed, and relatively little AMPA is found.[167] AMPA was found in some of the non-GR crop food samples of the papers reviewed by Xu et al.,[168] indicating that translocated glyphosate was degraded in the plant or at some point in the food supply chain.

Sugarcane is perhaps the only non-GR crop on which glyphosate is applied directly as a growth regulator. Low doses (0.16–0.47 kg a.e./ha) of glyphosate are routinely used about eight weeks before harvesting sugarcane to enhance sugar yields.[169] AMPA has been found in sugarcane two days after treatment with 0.72 kg a.e./ha of glyphosate.[170] The same authors found significant AMPA levels (ca. 0.2 μg/g dry weight) in sugarcane treated with only 0.18 kg a.e./ha of glyphosate at seven days after treatment.[171]

In summary, glyphosate appears to be degraded in non-GR crops via a GOX activity, and evidence for this is found for both members of the Poaceae and other crop plant families.

3.5.2 Glyphosate-Resistant Crops

GR crops with a GR form of EPSPS are ideal for studying plant metabolism of glyphosate because they are not harmed significantly by high doses of glyphosate that would be toxic to other plants, causing inhibited or altered metabolic processes, including reduced enzymatic degradation of glyphosate. Therefore, if a GR crop has even a weak enzyme that degrades glyphosate, degradation should be easily determined, considering the high doses of glyphosate that these plants can resist. The first GR crop to be examined for glyphosate metabolism was soybean.[41] Metabolism of glyphosate to AMPA is significant in GR soybean, with levels of ~2–3 and 7–25 μg of glyphosate and AMPA g^{-1} of dry harvested seed, respectively, from plants treated with 1.26 kg a.e./ha of glyphosate at full bloom or eight weeks after planting. This is not surprising, as glyphosate is phloem mobile, translocating to metabolic sinks such as meristems and developing seeds.[133] Others have also found similar levels of glyphosate and AMPA in seed of GR soybean.[42,172–174] In another study, in which GR soybean was treated with 0.87 kg/ha glyphosate at both five and seven weeks after planting, high levels of glyphosate (8–15 μg/g) but very little AMPA (~0.1 μg/g) were detected in leaf tissues.[175] Harvested seed had much lower levels of glyphosate (0.5–0.9 μg/g), but relatively more AMPA (0.2–0.3 μg/g). Whether the AMPA in the seed was translocated there or was formed by metabolism of glyphosate in the seed is unknown.

The only significant metabolic effect of high glyphosate dose might be due to the accumulation of AMPA. AMPA is mildly phytotoxic compared to glyphosate,[126,148,176] and genetic and physiological responses to glyphosate and AMPA are different.[177] Furthermore, GR crops are not resistant to AMPA,[126,178] indicating that it has one or more molecular targets other than EPSPS. Indeed, GR soybeans can accumulate high levels of AMPA, and this accumulated AMPA has been associated with "yellow flash," a chlorosis phenomenon sometimes seen in the field 4–20 days after GR soybeans are sprayed with full, recommended glyphosate rates under certain environmental conditions.[179,180] This effect is not seen in all GR soybean varieties.[181] The differences between soybean varieties could be due to different AMPA levels accumulating in the different varieties, but differences in glyphosate degradation between varieties have not been determined under the same conditions. Amounts of AMPA or glyphosate applied to GR soybeans that result in the same levels of AMPA within the plant tissues result in similar phytotoxicity symptoms,[178] providing strong evidence that yellow flash is due to AMPA phytotoxicity and not to a secondary effect of glyphosate. The fact that the resistance factor for GR soybeans is about 50-fold[156] indicates that glyphosate itself does not cause the effect. Also, the fact that yellow flash is not seen in GR maize, which produces little or no AMPA (see below) when treated with high doses of glyphosate, indicates that it is not a direct effect of glyphosate. The 100-fold resistance factor of GR maize[155] versus the 50-fold resistance factor for GR soybean and canola may be due to insufficient AMPA accumulation in maize to cause phytotoxicity (yellow flash) at high glyphosate application rates.

The yellow flash effect is temporary and has not been found to affect the yield of the crop.[179,180,182] Similar adverse effects (e.g., yellowing) of AMPA applied to GR soybean at a dose calculated to simulate the amount accumulated by plants treated with glyphosate were temporary.[126] Reddy et al. concluded that yellow flash in glyphosate-treated soybean is due to the phytotoxicity of accumulated AMPA.[178] Later work by Smedbol et al. found that the amount of AMPA in roots of glyphosate-treated, GR soybeans was better correlated with phytotoxicity indicators in leaves such as reduced chlorophyll and stomatal conductance and increased hydrogen peroxide levels than AMPA levels of leaves.[183] Guo et al. found no yellow flash syndrome in transgenic soybeans with a GR EPSPS and a transgene for GAT that should prevent high AMPA levels from accumulating.[56] Unfortunately, AMPA levels were not measured in this study. Under the same treatment (3.6 kg glyphosate/ha), plants with only a GR EPSPS developed yellow flash. As noted below, glyphosate-treated canola with the *GOX* transgene accumulates high levels of AMPA, and chlorosis occurs at higher doses of glyphosate,[184] although this effect has not been tied to AMPA accumulation. The glyphosate resistance factors of GR canola and GR soybean are essentially the same (ca. 50-fold).[156]

GR maize metabolizes glyphosate to AMPA, but at a much lower rate than GR soybean does.[42,185,186] In a field study using recommended glyphosate rates (0.87 kg a.e./ha over the top, followed by a later 0.87 kg a.e./ha application to the base of the plants in Mississippi and a single application of 1.68 kg a.e./ha at the V4-5 growth stage in Illinois), low levels of AMPA were found in glyphosate-treated GR maize leaves in one field site (Mississippi, USA) two years in a row, but not in another site (Illinois, USA) with different GR cultivars.[185] However, the harvested seeds at the Mississippi site had no AMPA, whereas a very low AMPA level (ca. 30 ng/g dry weight) was found in seeds in one of two years in Illinois. Similarly, Rodrigues et al. could not detect AMPA from seeds of 9 of 17 GR maize field studies, and in 8 from which it could be detected, only 2 had enough for quantification.[42] In a similar study with GR cotton, they had comparable results. In two extensive studies, no glyphosate or AMPA was found in seed of GR maize treated once with 980 g a.e./ha, and low levels of only glyphosate (20–120 ng/g) were sometimes found in some in seeds of plants given sequential glyphosate treatments (520 followed by 980 g a.e./ha).[187,188] Bernal et al. also reported levels of AMPA in glyphosate-treated (1.6 kg/ha) GR maize leaves that were 65-fold less than the glyphosate levels at a week after treatment.[186] Both glyphosate and AMPA concentrations decreased with time after spraying, but the ratio of glyphosate to AMPA decreased with time, indicating that AMPA degrades and/or translocates more slowly than glyphosate *in vivo*. AMPA and glyphosate compete for movement into the vacuole and the cell and perhaps the plastid,[159] so these processes will also influence the ratio, depending on where the GOX activity occurs. Hearon et al. reported AMPA to be readily taken up by GR maize from soil treated with either AMPA or glyphosate, so all AMPA found in GR maize is not necessarily from degradation of glyphosate in the plant.[130] They also claimed conversion of glyphosate to AMPA *in planta*, but this was not rigorously proven. The finding that microbe-free cell cultures of maize can metabolize glyphosate to AMPA[157] proves that maize has an enzyme that can act as a GOX at relatively low *in vivo* activity levels compared to some other species. In general, however, members of the Poaceae (Gramineae) like maize have relatively little capability for degrading glyphosate.[7]

The first GR canola commercially grown was the only GR crop that contained a transgene for a microbial GOX, in addition to a gene for a microbial GR EPSPS.[45] Very little glyphosate is converted to AMPA in non-GR canola, whereas virtually all of the glyphosate applied to GR canola is converted to AMPA within seven days, when glyphosate is supplied in small amounts (Fig. 3.6).[38] Only AMPA and no glyphosate was found in leaves of treated plants that were not directly treated. Whether the AMPA was translocated to untreated leaves or formed by glyphosate oxidation in those leaves was not determined. The high levels of AMPA found in the leaves of glyphosate-treated plants could be the reason for chlorosis reported by Schilling et al. in GR canola treated with three applications of 450 g a.e./ha.[184] The relative contributions to resistance of the *GOX* and GR *EPSPS* genes are unknown in GR canola, but current commercial varieties of GR canola contain only a GR *EPSPS* transgene, as do all other commercial GR crops. If the high

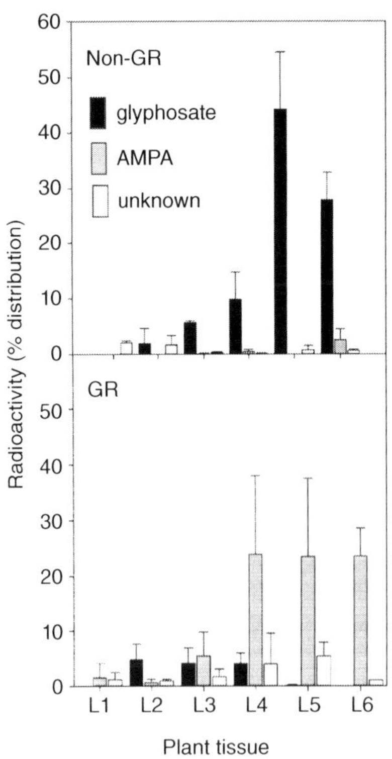

Figure 3.6 Metabolism of [^{14}C]-glyphosate to AMPA and unknown metabolites seven days after treatment with glyphosate in non-GR and GR canola. The data are expressed as percent distribution of total radioactivity in each leaf from the bottom (L1 = leaf one) to the top (L6 = leaf six) of the plant. Only leaves 2, 3, and 4 were treated with glyphosate. Error bars are ±1 SE. Source: From Corrêa et al. [38]/American Chemical Society.

levels of AMPA[38] and the chlorosis reported at higher rates[184] were related, as with soybeans,[178] a *GOX* gene is not desirable if a GR *EPSPS* gene is sufficient for resistance. Whether the canola *GOX* gene alone would provide adequate resistance for a commercial GR crop is unknown, but other research groups have explored the use of other transgenes for glyphosate-degrading enzymes to impart glyphosate resistance (see Section 3.2).

Issues around the toxicology of glyphosate, whether real or exaggerated,[189] have led to public concern about glyphosate and AMPA in the food supply. Thus, crops that metabolize glyphosate and AMPA to less toxicologically suspect compounds are desirable. Several of the genes for microbial enzymes that would accomplish this are discussed in Section 3.2.

3.6 Summary

Glyphosate is readily degraded by soil microbes, both bacterial and fungal, via oxidation to AMPA and glyoxylate with a GOX activity and/or by breaking the C-P bond with a C-P lyase to form sarcosine and phosphate, contributing to its relatively short environmental half-life. Other enzymes (e.g., glyphosate decarboxylase and GAT) that degrade glyphosate to other compounds are very weak degradation enzymes as they are found in microbes, but they have had their activity increased by genetic manipulation to generate thus far uncommercialized transgenes for GR crop production. In addition to soil microbes, both rhizoplane and endophyte microbes degrade glyphosate. Both of the main degradation pathways have also been reported in plants, but to a much lesser extent than in microbes. Members of the Leguminaceae (legumes) are more frequently reported to degrade glyphosate and to a greater extent than species of other plant families. Some endophytes of legumes have enzymes for glyphosate degradation. Poaceae (grasses) are often reported to not degrade glyphosate or to degrade it very slowly compared to other genera. Inherent degradation of glyphosate in plants is best studied in GR biotypes or GR crops with a high resistance factor, not related to enhanced degradation. The GR legume soybean with a GR EPSPS readily degrades glyphosate by oxidation to AMPA, but the grass crop GR maize has a much lower capacity to degrade glyphosate to AMPA. The only plants with an enzyme with a sequenced gene for a GOX-type gene are rice and a GR *E. colona* biotype with an AKR that acts as a GOX. Whether AKRs are the GOX enzymes responsible for degradation of glyphosate in other plant species is unknown. The role of endophytic degradation of glyphosate in plants is also unresolved. Considering the fact that many plant species are able to degrade glyphosate, the relatively small role of enhanced glyphosate degradation in GR weeds as a resistance mechanism is an enigma.

Acknowledgments

Writing of this review was funded in part by USDA Cooperative Agreement 58-6060-6-015 grant to the University of Mississippi. I thank Kirsten Hawking for her careful proofreading of the review.

References

1. Benbrook CM, Trends in glyphosate herbicide use in the United States and globally. *Environ Sci Europe* **28**:1–15 (2016).
2. Duke SO, Glyphosate: uses other than in glyphosate-resistant crops, mode of action, degradation in plants, and effects on non-target plants and agricultural microbes. *Rev Environ Contam Toxicol* **155**:1–65 (2021).
3. Duke SO, The history and current status of glyphosate. *Pest Manag Sci* **74**:1027–1034 (2018).
4. Duke SO, Biotechnology: herbicide-resistant crops, In Van Alfen N (Ed.), editor-in-chief*Encyclopedia of Agriculture and Food Systems*, Vol. **2**, Elsevier, San Diego, pp. 94–116 (2014).
5. Green JM, and Siehl DL, History and outlook for glyphosate-resistant crops. *Rev Environ Contam Toxicol* **155**:67–91 (2021).
6. SciFinder, https://www.n.cas.org/products/scifinder/content, [Accessed April 1, 2025].
7. Duke SO, Glyphosate degradation in glyphosate-resistant and -susceptible crops and weeds. *J Agric Food Chem* **59**:5835–5841 (2011).
8. Tseng S-H, Lo Y-W, Chang P-C, Chou S-S, and Chang H-M, Simultaneous quantification of glyphosate, glufosinate, and their major metabolites in rice and soybean sprouts by gas chromatography with pulsed flame photometric detector. *J Agric Food Chem* **52**:4057–4063 (2004).
9. Vemanna RS, Vennapusa AR, Easwaran M, Chandrashekar BK, Rao H, Ghanti K, Sudhakar C, Mysore KS, and Makarla U, Aldo-keto reductase enzymes detoxify glyphosate and improve herbicide resistance in plants. *Plant Biotechnol J* **15**:794–804 (2017).
10. Nandula VK, Riechers DE, Ferhatoglu Y, Marrett M, Duke SO, Dayan FE, *et al.*, Herbicide metabolism: crop selectivity, bioactivation, resistance mechanisms, and regulation. *Weed Sci* **67**:149–175 (2019).
11. Borggaard OK, and Gimsing AL, Fate of glyphosate in soil and the possibility of leaching to ground and surface waters: a review. *Pest Manag Sci* **64**:441–456 (2008).
12. Zhan H, Feng Y, Fan X, and Chen S, Recent advances in glyphosate biodegradation. *Appl Microbiol Biotechnol* **102**:5033–5043 (2018).

13 Sviridov AV, Shushkova TV, Ermakova IT, Ivanova EV, Epiktekov DO, and Leontievsky AA, Microbial degradation of glyphosate herbicides. *Appl Biochem Microbiol* **51**:188–195 (2015).

14 Singh S, Kumar V, Datta S, Wani AB, Dhanjal D, Romero R, and Singh J, Glyphosate uptake, translocation, resistance emergence in crops, analytical monitoring, toxicity and degradation: a review. *Environ Chem Lett* **18**:663–702 (2020).

15 Chen Y, Chen W-J, Huang Y, Li J, Zhong J, Zhang W, Zou Y, Mishra S, Bhatt P, and Chen S, Insights into the microbial degradation and resistance mechanisms of glyphosate. *Environ Res* **215**(Part 1):114153 (2022).

16 Pollegioni L, Schonbrunn E, and Siehl D, Molecular basis of glyphosate resistance – different approaches through protein engineering. *FEBS J* **278**:2758–2766 (2011).

17 Pan L, Yu Q, Han H, Mao L, Nyporko A, Fan L, Bai L, and Powles SB, Aldo-keto reductase metabolizes glyphosate and confers glyphosate resistance in *Echinochloa colona*. *Plant Physiol* **181**:1519–1534 (2019).

18 Li H, Yang Y, Chen C-C, Huang J-W, Min J, Dai L, and Guo R-T, Structural analysis and engineering of aldo-keto reductase from glyphosate-resistant *Echinochloa colona*. *J Hazard Mater* **436**:129181 (2022).

19 Chen R, Wang S, Sun Y, Li H, Wan S, Lin F, and Xu H, Comparison of glyphosate-degradation ability of aldo-keto reductase (AKR4) proteins in maize, soybean and rice. *Int J Mol Sci* **24**:3421 (2023).

20 Zhou F-Y, Han H, Han Y-J, Nyporko A, Yu Q, Beckie HJ, and Powles SB, Aldo-keto reductase may contribute to glyphosate resistance in *Lolium rigidum*. *Pest Manag Sci* **79**:1528–1537 (2023).

21 Barski OA, Tipparaju SM, and Bhatnagar A, The aldo-keto reductase superfamily and its role in drug metabolism. *Drug Metab Rev* **40**:553–624 (2008).

22 Simpson PJ, Tantitadapitak C, Reed AM, Mather OC, Bunce CM, White SA, and Ride JP, Characterization of two novel aldo-keto reductases from Arabidopsis: expression patterns, broad substrate specificity, and an open active-site structure suggest a role in toxicant metabolism following stress. *J Mol Biol* **392**:465–480 (2009).

23 Castle LA, Siehl DL, Gorton R, Patten PA, Chen YH, Bertain S, *et al.*, Discovery and directed evolution of a glyphosate tolerance gene. *Science* **304**:1151–1154 (2004).

24 Hammer PE, Hinson TK, Duck NB and Koziel M, Methods to confer herbicide resistance. US Patent Application 20070107078 A1, 1–53 (2007).

25 Coupland D, The effect of temperature on the activity and metabolism of glyphosate applied to rhizome fragments of *Elymus repens* (= *Agropyron repens*). *Pestic Sci* **15**:226–234 (1984).

26 Eberbach PL, and Bowmer KH, Conversion of C^{14}-glyphosate to carbon dioxide by alligator weed. *J Aquat Plant Manage* **33**:27–29 (1995).

27 Domínguez-Valenzuela JA, Gherekhloo J, Fernández-Moreno PT, Cruz-Hipolito HE, Alcántara-de la Cruz R, Sánchez-González E, and De Prado R, First confirmation and characterization of target and non-target site resistance to glyphosate in palmer amaranth (*Amaranthus palmeri*). *Plant Physiol Biochem* **115**:212–218 (2017).

28 Reddy KN, Rimando AM, Duke SO, and Nandula VK, Aminomethylphosphonic acid accumulation in plant species treated with glyphosate. *J Agric Food Chem* **56**:2125–2130 (2008).

29 Pedotti M, Rosini E, Molla G, Moschetti T, Savino C, Vallone B, and Pollegioni L, Glyphosate resistance by engineering the flavoenzyme glycine oxidase. *J Biol Chem* **284**:36415–36423 (2009).

30 Sprankle P, Sandberg CL, Meggitt WF, and Penner D, Separation of glyphosate and possible metabolites by thin-layer chromatography. *Weed Sci* **26**:673–674 (1978).

31 De Carvalho LB, Alves PLCA, González-Torrala F, Cruz-Hipolito HE, Rojano-Delgado AM, de Prado R, Gil-Humanes J, Barro F, and de Castro MDL, Pool of resistance mechanisms to glyphosate in *Digitaria insularis*. *J Agric Food Chem* **60**:615–622 (2012).

32 Gherekhloo J, Fernández PT, Alcántara-de la Cruz R, Sánchez-González E, Cruz-Hipolito HE, Domínguez-Valenzuela JA, and De Prado R, Pro-106-ser mutation and EPSPS overexpression acting together simultaneously in glyphosate-resistant goosegrass (*Eleusine indica*). *Sci Rep* **7**:6702. DOI: 10.1038/s41598-017-06772-1 (2017).

33 Guijarro KH, De Gerónimo E, and Erijman L, Glyphosate biodegradation potential in soil based on glycine oxidase (*thiO*) from *Bradyrhizobium*. *Curr Microbiol* **78**:1991–2000 (2021).

34 Marshall G, Kirkwood RC, and Martin DJ, Studies on the mode of action of asulam, aminotriazole and glyphosate in *Equisetum arvense* (field horsetail). II: the metabolism of [^{14}C] asulam, [^{14}C] aminotriazole and [^{14}C] glyphosate. *Pestic Sci* **18**:65–77 (1987).

35 Sandberg CL, Meggitt WF, and Penner D, Absorption, translocation and metabolism of ^{14}C-glyphosate in several weed species. *Weed Res* **20**:195–200 (1980).

36 Rojano-Delgado AM, Cruz-Hipolito H, De Prado R, de Castro MDL, and Franco AR, Limited uptake, translocation and enhance metabolic degradation contribute to glyphosate tolerance in *Mucuna pruriens* var. *utilis* plants. *Phytochemistry* **73**:34–41 (2012).

37 Vazquez-Garcia JG, Palma-Bautista C, Rojano-Delgado AM, De Prado R, and Menendez J, The first case of glyphosate resistance in johnsongrass (*Sorghum* *halepense* [L.] Pers.) in Europe. *Plants* **9**:313. DOI: 10.3390/plants9030313 (2020).

38 Corrêa EL, Dayan FE, Owen DL, Rimando AM, and Duke SO, Glyphosate-resistant and conventional canola (*Brassica napus* L.) responses to glyphosate and aminomethylphosphonic acid (AMPA) treatment. *J Agric Food Chem* **64**:3508–3513 (2016).

39 Stock M, Hieke M, Weidhase RA, and Schütte HR, Degradation of glyphosate in excised leaves of tobacco and sugar beet. *Plant Physiol* **139**:171–174 (1991).

40 Tong M, Gao W, Jiao W, Zhou J, Li Y, He L, and Hou R, Uptake, translocation, metabolism, and distribution of glyphosate in nontarget tea plant (*Camellia sinensis* L.). *J Agric Food Chem* **65**:7638–7646 (2017).

41 Duke SO, Rimando AM, Pace PF, Reddy KN, and Smeda RJ, Isoflavone, glyphosate, and aminomethylphosphonic acid levels in seeds of glyphosate-treated, glyphosate-resistant soybean. *J Agric Food Chem* **51**:340–344 (2003).

42 Rodrigues NR, De Sousa APF, Morais PPP, Braga DPV, Crivellari AC, Favoretto LRG, and Berger GU, Residues of glyphosate and aminophosphonic acid (AMPA) in genetically modified glyphosate tolerant soybean, corn and cotton crops. *Cienc Rural* **51**:doi.org/10.1590/0103-8478cr20190244 (2021).

43 Barry GF and Kishore GM, Glyphosate tolerant plants. United States Patent Application No. US5463175A (1995).

44 Hallas LE, Hahn EM, and Korndofer C, Characterization of microbial traits associated with glyphosate biodegradation in industrial activated sludge. *J Ind Microbiol* **3**:377–385 (1988).

45 Green JM, Evolution of glyphosate-resistant crop technology. *Weed Sci* **57**:109–117 (2009).

46 Nicolia A, Ferradini N, Molla G, Biagetti E, Pollegioni L, Veronesi F, and Rosellini D, Expression of an evolved engineered variant of a bacterial glycine oxidase leads to glyphosate resistance in alfalfa. *J Biotechnol* **184**:201–208 (2014).

47 Yao P, Lin Y, Wu G, Lu Y, Zhan T, Kumar A, Zhang L, and Liu Z, Improvement of glycine oxidase by DNA shuffling, and site saturation mutagenesis of F247 residue. *Int J Biol Macromol* **79**:965–970 (2015).

48 Zhan T, Zhang K, Chen Y, Lin Y, Wu G, Zhang L, Yao P, Shao Z, and Liu Z, Improving glyphosate oxidation activity of glycine oxidase from *Bacillus cereus* by directed evolution. *PLoS One* **8**:e79195 (2013).

49 Zhang K, Guo Y, Yao P, Lin Y, Kumar A, Liu Z, Wu G, and Zhang L, Characterization and directed evolution of BliGO, a novel glycine oxidase from *bacillus licheniformis*. *Enzyme Microbial Technol* **85**:12–19 (2016).

50 Han H, Zhu B, Fu X, You S, Wang B, Li Z, Zhao W, Peng R, and Yao Q, Overexpression of D-amino acid oxidase from *Bradyrhizobium japonicum*, enhances resistance to glyphosate in *Arabidopsis thaliana*. *Plant Cell Rep* **34**:2043–2051 (2015).

51 Wen L, Zhong J, Cui Y, Duan Z, Zhou F, Li CY, Ma W, Yin C, Chen H, and Lin Y, Coexpression of *I. Variabilis-EPSPS** and *WBceGO-B3S1* genes contributes to high glyphosate tolerance and low glyphosate residues in transgenic rice. *J Agric Food Chem* **69**:7388–7398 (2021).

52 Siehl DL, Castle LA, Gorton R, and Keenan RJ, The molecular basis of glyphosate resistance by an optimized microbial acetyltransferase. *J Biol Chem* **15**:11446–11455 (2007).

53 Green JM, Hazel CB, Forney DR, and Pugh LM, New multiple-herbicide crop resistance and formulation technology to augment the utility of glyphosate. *Pest Manag Sci* **64**:332–339 (2008).

54 Dun BQ, Lu W, Ping SZ, Zhang W, Chen M, Xu YQ, Jin D, Zhao ZL, and Lin M, Isolation of a novel glyphosate tolerance N-acetyltransferase gene and expression in *E. Coli*. *Chin High Technol Lett* **16**:943–947 (2006).

55 Dun B, Wang X, Lu W, Chen M, Zhang W, Ping S, Wang Z, Zhang B, and Lin M, Development of highly glyphosate-tolerant tobacco coexpression of glyphosate acetyltransferase *gat* and EPSPS G2-*aro*A genes. *The Crop J* **2**:164–169 (2014).

56 Guo B, Guo Y, Hong H, Jin L, Zhang L, Chang RZ, Lu W, Lin M, and Qiu L-J, Co-expression of *G2-EPSPS* and glyphosate acetyltransferase *GAT* genes conferring high tolerance to glyphosate in soybean. *Front Plant Sci* **6**:847 (2015).

57 Liang C, Sun B, Meng Z, Meng Z, Wang Y, Sun G, *et al.*, Co-expression of *GR79 EPSPS* an *GAT* yields herbicide-resistant cotton with low glyphosate residues. *Plant Biotechnol J* **15**:1622–1629 (2017).

58 Meng Y, Zhang W, Wang Z, Yuan F, Gu S, Lin H, and Niu L, Co-expression of *GR79-EPSPS* and *GAT* generates high glyphosate-resistant alfalfa with low glyphosate residues. *aBIOTECH* **4**:352–358.

59 Li S, Li P, Li X, Wen N, Wang Y, Lu W, Lin M, and Lang Z, In maize, co-expression of *GAT* and *GR79-EPSPS* provides high glyphosate resistance, along with low glyphosate residues. *aBIOTECH* **4**:277–290 (2023).

60 Hove-Jensen B, Zechel DL, and Jochimsen B, Utilization of glyphosate as phosphate source: biochemistry and genetics of bacterial carbon-phosphorous lyase. *Microbiol Molec Biol Rev* **78**:176–197 (2014).

61 Stosiek N, Talma M, and Klimek-Ochab M, Carbon-phosphorous lyase – the state of the art. *Appl Biochem Biotechnol* **190**:1525–1552 (2020).

62 European Food Safety Authority, Conclusion on the peer review of the pesticide risk assessment of the active substance glyphosate. *EFSA* **J13**:4302 (2015).

63 Duke SO, Glyphosate: environmental fate and impact. *Weed Sci* **68**:201–207 (2020).

64 Sprankle P, Meggitt WF, and Penner D, Adsorption, mobility and microbial degradation of glyphosate in the soil. *Weed Sci* **23**:229–234 (1975).

65 Torstensson NTL, and Aamisepp A, Detoxification of glyphosate in soil. *Weed Res* **17**:209–212 (1977).

66 Qiao C, Wang C, Pang R, Tian F, Han L, Guo L, *et al.*, Environmental behavior and influencing factors of glyphosate in peach orchard system. *Ecotox Environ Safety* **206**. DOI: 10.1016/j.ecoenv.2020.111209 (2020).

67 La Cecilia D, and Maggi F, Analysis of glyphosate degradation in a soil microcosm. *Environ Pollut* **233**:201–207 (2018).

68 Forlani G, Mangiagalli A, Nielson E, and Suardi CM, Degradation of the phosphonate herbicide glyphosate in soil: evidence for possible involvement of unculturable microorganisms. *Soil Biol Biochem* **31**:991–997 (1999).

69 Kolpin DW, Thurman EM, Lee EA, and Meyer MT, Furlong ET and Glassmeyer, urban contributions of glyphosate and its degradate AMPA to streams in the United States. *Sci Total Environ* **354**:191–197 (2006).

70 Botta F, Lavison G, Couturier G, Alliot F, Moreau-Guigon E, Fauchon N, Guery B, Chevreuil M, and Blanchoud H, Transfer of glyphosate and its degradate AMPA to surface waters through urban sewerage systems. *Chemosphere* **77**:133–139 (2009).

71 Kuhn R, Jensch R, Bryant IM, Fischer T, Liebsch S, and Martienssen M, Photodegradation of ethylenediamine(methylenephosphonic acid) – the effect of the system configuration. *J Photochem Photobiol A Chem* **388**:112192 (2020).

72 Drzyzga D, Forlani G, Vermander J, Kafarski P, and Lipok J, Biodegradation of the aminopolyphosphonate DTPMP by the cyanobacterium *anabaena vriabilis* proceeds via a C-P lyase-independent pathway. *Environ Microbiol* **19**:1065–1076 (2017).

73 Kishore GM, and Jacob GS, Degradation of glyphosate by *Pseudomonas* sp. PG2982 via a sarcosine intermediate. *J Biol Chem* **262**:12164–12168 (1987).

74 Jacob GS, Garbow JR, Hallas LE, Kimack NM, Kishore GM, and Schaefer J, Metabolism of glyphosate in *pseudomonas* sp. strain LBr. *Appl Environ Microbiol* **54**:2953–2958 (1988).

75 Peñaloza-Vazquez A, Mena GL, Herrera-Estrella L, and Bailey AM, Cloning and sequencing of the genes in involved in glyphosate utilization by *pseudomonas pseudomallei*. *Appl Envron Microbiol* **61**:538–543 (1995).

76 McAuliffe KS, Hallas LE, and Kulpa CF, Glyphosate degradation by *agrobacterium radiobacter* isolated from activated sludge. *J Indust Mircrobiol* **6**:219–221 (1990).

77 Ermakova IT, Shushkova TV, Sviridov AV, Zelenkova NF, Vinokurova NG, Baskunov BP, and Leontievsky AA, Organophosphonates utilization by soil strains of *Ochrobactrum anthropi* and *Achromobacter* sp. *Arch Microbiol* **199**:665–675 (2017).

78 Lerbs W, Stock M, and Parthier B, Physiological aspects of glyphosate degradation in *Alcaligenes* spec. Strain GL. *Arch Microbiol* **153**:146–150 (1990).

79 Pipke R, and Amrhein N, Degradation of the phosphonate herbicide glyphosate by *Arthrobacter atrocyaneus* ATCC 13752. *Appl Environ Microbiol* **54**:1293–1296 (1988).

80 Fan J, Yang G, Zhao H, Shi G, Geng Y, Hou T, and Tao K, Isolation, identification and characterization of a glyphosate-degrading bacterium, *Bacillus cereus* CB4, from soil. *J Gen Appl Microbiol* **58**:263–271 (2012).

81 Balthazor TM, and Hallas LE, Glyphosate-degrading microorganisms from industrial activated sludge. *Appl Environ Microbiol* **51**:432–434 (1986).

82 Obojska A, Ternan NG, Lejczak B, Kafarski P, and McMullan G, Organophosphonate utilizatjon by the thermophile *Geobacillus caldoxyylosilyticus* T20. *Appl Environ Microbiol* **68**:2081–2084 (2002).

83 Hadi F, Mousavi A, Noghabi KA, Tabar HG, and Salmanian AH, New bacterial strain of the genus *Ochrobactrum* with glyphosate-degrading activity. *J Environ Sci Health* **48b**:208–213 (2013).

84 Liu C-M, McLean PA, Sookdeo CC, and Cannon FC, Degradation of the herbicide glyphosate by members of the family Rhizobiaceae. *Appl Environ Microbiol* **57**:1799–1804 (1991).

85 Obojska A, Lejczak B, and Kubrak M, Degradation of phosphonates by *Streptomycete* isolates. *Appl Microbiol Biotechnol* **51**:872–876 (1999).

86 Krzyśko-Lupicka T, Strof W, Kubs K, Skorupa M, Wieczorek P, Lejczak B, and Kafarski P, The ability of soil-borne fungi to degrade organophosphate carbon-to-phosphorus bonds. *Appl Microbiol Biotechnol* **48**:549–552 (1997).

87 Fu G-M, Chen Y, Li R-Y, Yuan X-Q, Liu C-M, Li B, and Wan Y, Pathway and rate-limiting step of glyphosate degradation by *Aspergillus oryzae* A-FO2. *Prep Biochem Biotech*. DOI: 10.1080/10826068.2017.1342260 (2017).

88 Krzyśko-Łupicka T, and Orlik A, The use of glyphosate as the sole source of phosphorus or carbon for the selection of soil-borne fungal strains capable to degrade this herbicide. *Chemosphere* **34**:2601–2605 (1997).

89 Klimek M, Lejczak B, Kafarski P, and Forlani G, Metabolism of the phosphonate herbicide glyphosate by a non-nitrate-utilizing strain of *Penicillium chrysogenum*. *Pest Manag Sci* **57**:815–812 (2001).

90 Bujacz B, Wieczorek P, Krzysko-Lupicka T, Golab Z, Lejczak B, and Kavfarski P, Organophosphonate utilization by the wildtype strain of *Penicillium notatum*. *Appl Environ Microbiol* **61**:2905–2910 (1995).

91 Yu XM, Yu T, Yin GH, Dong QL, An M, Wang HR, and Ai CX, Glyphosate biodegradation and potential soil bioremediation by *Bacillus subtilis* strain Bs-15. *Gen Mol Res* **3**:14717–14730 (2015).

92 Zboińska E, Maliszewska I, Lejczak B, and Kafarski P, Degradation of organophosphonates by *Penicillium citrinum. Lett App Micobiol* **15**:269–272 (1992).

93 Adelowo FE, Olu-Arotiowa OA, and Amuda OS, Biodegradation of glyphosate by fungal species. *Adv Biosci Bioeng* **2**:104–118 (2014).

94 Castro JV, Peralba MCR, and Ayub MAZ, Biodegradation of the herbicide glyphosate by filamentous fungi in platform shaker and batch bioreactor. *J Envron Sci Health* **42**:883–886 (2007).

95 Pizzul L, del Pilar CM, and Stenström J, Degradation of glyphosate and other pesticides by ligninolytic enzymes. *Biodegradation* **20**:751–759 (2009).

96 Al-Rajab AJ, Amellal S, and Shiavon M, Sorption and leaching of ^{14}glyphosate in agricultural soils. *Agron Sustain Dev* **28**:419–428 (2008).

97 Sun M, Li H, and Jaisi DP, Degradation of glyphosate an bioavailability of phosphorus derived from glyphosate in a soil-water system. *Water Res* **163**. DOI: 10.1016/j.watres.2019.07.007 (2019).

98 Wang S, Seiwert B, Kastner M, Miltner A, Schaffer A, Reemtsma T, Yang Q, and Nowak KM, (Bio)degradation of glyphosate in water-sediment microcosms – a stable isotope co-labeling approach. *Water Res* **99**:91–100 (2016).

99 Selvapanidiyan A, and Bhatnagar RK, Isolation of a glyphosate-metabolizing *pseudomonas*: detection, partial purification and localization of carbon-phosphorous lyase. *Appl Microbiol Biotechnol* **40**:876–882 (1994).

100 Dick RE, and Quinn JP, Glyphosate-degrading isolates from environmental samples: occurrence and pathways of degradation. *Environ Biotechnol* **43**:545–550 (1995).

101 Simonsen L, Fomsgaard IS, Svensmark B, and Spliid NH, Fate and availability of glyphosate and AMPA in agricultural soil. *J Environ Sci Health* **43B**:365–375 (2008).

102 Guijarro KH, Aparicio V, De Gerónimo E, Castellote M, Figuerola EL, Costa JL, and Erijman L, Soil microbial communities and glyphosate decay in soils with different application history. *Sci Total Envion* **634**:974–982 (2018).

103 Silva V, Montanarella L, Jones A, Fernández-Ugalde O, Mol HGJ, Ritsema CJ, and Giessen V, Distribution of glyphosate and aminomethylphosphonic acid (AMPA) in agricultural topsoils of the European Union. *Sci Total Environ* **621**:1352–1359 (2018).

104 González-Valenzuela LE, and Dussán J, Molecular assessment of glyphosate-degradation pathway via sarcosine intermediate in *Lysinibacillus sphaericus. Environ Sci Pollut Res* **25**:22790–22796 (2018).

105 Gimsing AL, Borggaard OK, Jacobsen OS, Aamand J, and Sørensen J, Chemical and microbial characteristics controlling glyphosate mineralization in Danish surface soils. *Appl Soil Ecol* **27**:233–242 (2004).

106 Pipke R, Schulz A, and Amrhein N, Uptake of glyphosate by an *Arthrobacter* sp. *Appl Environ Microbiol* **53**:974–978 (1987).

107 Aslam S, Jing Y, and Nowak KM, Fate of glyphosate and its degradation products AMPA, glycine and sarcosine in an agricultural soil: implications for environmental risk assessment. *J Hazard Mater* **447**:13–847 (2023).

108 Ermakova IT, Kiseleva NI, Shushkova T, Zharikov M, Zharikov GA, and Leontievsky AA, Bioremediation of glyphosate-contaminated soils. *Appl Microbiol Biotechnol* **88**:585–594 (2010).

109 Hallas LE, Adams WJ, and Heitkamp MA, Glyphosate degradation by immobilized bacteria: field studies with industrial wastewater effluent. *Appl Environ Microbiol* **58**:1215–1219 (1992).

110 Blake R, and Pallett K, The environmental fate and ecotoxicology of glyphosate. *Outlooks Pest Manag* **29**:266–269 (2018).

111 Rodríguez-Gil JL, Prosser RS, Duke SO, and Solomon KR, Ecotoxicology of glyphosate, its formulants, and environmental degradation products. *Rev Environ Contam Toxicol* **155**:129–205 (2021).

112 Kepler RM, Epp-Schmidt DJ, Yarwood SA, Cavigelli MA, Reddy KN, Duke SO, Bradley CA, Williams MM, Buyer JS, and Maul JE, Soil microbial communities in diverse agroecosystems exposed to the herbicide glyphosate. *Appl Environ Microbiol* **86**:e01744–e01719. DOI: 10.1128/AEM.01744-19 (2020).

113 Yale RL, Sapp M, Sinclair CJ, and Moir JWB, Microbial changes linked to the accelerated degradation of the herbicide atrazine in a range of temperate soils. *Environ Sci Res Int* **24**:7359–7374 (2017).

114 Zabaloy MC, Gómez E, Garland JL, and Gómez MA, Assessment of microbial community function and structure in soil microcosms exposed to glyphosate. *Appl Soil Ecol* **61**:333–339 (2012).

115 Lancaster SH, Hollister EB, Senseman SA, and Gentry TJ, Effects of repeated glyphosate applications on soil microbial community composition and mineralization of glyphosate. *Pest Manag Sci* **66**:59–64 (2010).

116 Bonfleur EJ, Lavorenti A, and Tornisielo VL, Mineralization and degradation of glyphosate and atrazine applied in combination in a Brazilian Oxisol. *J Environ Sci Health* **46B**:69–75 (2010).

117 Bonfleur EJ, Tornisielo VL, Regitano JB, and Lovorenti A, The effects of glyphosate and atrazine mixture on soil microbial population and subsequent impacts on their fate in a tropical soil. *Water Air Soil Pollut* **226**:21. DOI: doi.org/10.1007/s11270-014-2190-8 (2015).

118 Lancaster SH, Haney RL, Senseman SA, Kenerley CM, and Hons FM, Microbial degradation of fluometuron is influenced by roundup WeatherMAX. *J Agric Food Chem* **56**:8588–8593 (2008).

119 Blackshaw RE, and Harker KN, Wheat, field pea, and canola response to glyphosate and AMPA soil residues. *Weed Technol* **30**:985–991 (2016).

120 Shushkova T, Ermakova I, and Leontievsky A, Glyphosate bioavailability in soil. *Biodegradation* **21**:403–410 (2010).

121 Tétard-Jones C, and Edwards R, Potential roles for microbial endophytes in herbicide tolerance in plants. *Pest Manag Sci* **72**:203–209 (2016).

122 Moorman TB, Becerril JM, Lydon J, and Duke SO, Production of hydroxybenzoic acids by *Bradyrhizobium japonicum* strains after treatment with glyphosate. *J Agric Food Chem* **40**:289–293 (1992).

123 Kuklinsky-Sobral J, Araujo WL, Mendes R, Pizzirani-Kleiner AA, and Azevedo JL, Isolation and characterization of endophytic bacteria from soybean (*Glycine max*) grown in soil treated with glyphosate herbicide. *Plant Soil* **273**:91–99 (2005).

124 Zabaloy MC, and Gómez MA, Diversity of rhizobia isolated from an agricultural soil in Argentina based on carbon utilization and effects of herbicides on growth. *Biol Fertil Soils* **42**:83–88 (2005).

125 Reddy KN, and Zablotowicz RM, Glyphosate-resistant soybean response to various salts of glyphosate and glyphosate accumulation in soybean nodules. *Weed Sci* **51**:496–502 (2003).

126 Ding W, Reddy KN, Zablotowicz RM, Bellaloui N, and Bruns HA, Physiological responses of glyphosate-resistant and glyphosate-sensitive soybean to aminomethylphosphonic acid, a metabolite of glyphosate. *Chemosphere* **83**:593–598 (2011).

127 Haron MH, Tyler HL, Pugh ND, Moraes RM, Maddox VL, Jackson CR, and Pasco DS, Activities and prevalence of proteobacteria members colonizing *Echinacea purpurea* fully account for macrophage activation exhibited by extracts of this botanical. *Planta Med* **82**:1258–1265 (2016).

128 Ghanizadeh H, and Harrington KC, Perspective: root exudation of herbicides as a novel mode of herbicide resistance in weeds. *Pest Manag Sci* **76**:2543–2547 (2020).

129 Kryuchkova YV, Burygin GL, Gogoleva NE, Gogolev YV, Chernyshova MP, Makarov OE, Fedorov EE, and Turkovskaya OV, Isolation and characterization of a glyphosate-degrading rhizosphere strain, *Enterobacter cloacae* K7. *Microbiol Res* **169**:99–105 (2014).

130 Hearon SE, Wang M, TJ MD, and Phillips TD, Decreased bioavailability of aminomethylphosphonic acid (AMPA) in genetically modified corn with activated carbon or calcium montmorillonite clay inclusion in soil. *J Environ Sci* **100**:131–143 (2021).

131 Powles SB, Lorraine-Colwill DF, Dellow JJ, and Preston D, Evolved resistance to glyphosate in rigid ryegrass (*Lolium rigidum*) in Australia. *Weed Sci* **46**:604–607 (1998).

132 Coupland D, Metabolism of glyphosate in plants, In Grossbard E, and Atkinson D (Eds), *The Herbicide Glyphosate*, Butterworth, London, pp. 25–34 (1985).

133 Duke SO, Glyphosate, In Kearney PC, and Kaufman DD (Eds), *Herbicides-Chemistry, Degradation and Mode of Action*, Vol. **III**, Marcel Dekker, Inc., New York, pp. 1–70 (1988).

134 Franz JE, Mao MK, and Sikorski JA, *Glyphosate: A Unique Global Herbicide*, American Chemical Society Mongraph 189, Washington, DC, p. 653 (1997).

135 Devine MD, Bandeen JD, and McKersie BD, Temperature effects on glyphosate absorption, translocation, and distribution in quackgrass (*Agropyron repens*). *Weed Sci* **31**:461–464 (1983).

136 Koskinen WC, Marek LJ, and Hall KE, Analysis of glyphosate and aminomethylphosphonic acid in water, plant materials and soil. *Pest Manag Sci* **72**:423–432 (2016).

137 Golt A, and Wood LJ, Glyphosate-based herbicides alter the reproductive morphology of *Rosa acicularis* (prickly rose). *Front Plant Sci* **12**. DOI: doi.org/10.3389/fpls.2021.698202 (2021).

138 Goyer A, Johnson TL, Olsen LJ, Collakova E, Shachar-Hill Y, Rhodes D, and Hanson AD, Characterization and metabolic function of a peroxisomal sarcosine and pipecolate oxidase from *Arabidopsis*. *J Biol Chem* **279**:16947–16953 (2004).

139 Heap I, The International Herbicide-Resistant Weed Database. Online. www.weedscience.org. [Accessed May 1, 2025]

140 Baek HY, Bobdilla LK, Giacomini DA, Montgomery JS, Murphy BP, and Tranel PJ, Evolution of glyphosate-resistant weeds. *Rev Environ Contam Toxicol* **155**:93–128 (2021).

141 Pan L, Yu Q, Wang J, Han H, Mao L, Nyporko A, Maguza A, Fan L, Bai L, and Powles SB, An ABCC-type transporter endowing glyphosate resistance in plants. *Proc Natl Acad Sci USA* **118**. DOI: 10.1073/pnas.2100136118 (2021).

142 Gaines TA, Duke SO, Morran S, Rigon C, Tranel PJ, Küpper A, and Dayan FE, Mechanisms of evolved herbicide resistance. *J Biol Chem* **295**:10307–10330 (2020).

143 Sammons RD, and Gaines TA, Glyphosate resistance: state of knowledge. *Pest Manag Sci* **70**:1367–1377 (2014).

144 Norsworthy JK, Burgos NR, and Oliver LR, Differences in weed tolerance to glyphosate involve different mechanisms. *Weed Technol* **15**:725–731 (2001).

145 Ribiero DN, Nandula VK, Dayan FE, Rimando AM, Duke SO, Reddy KN, and Shaw DR, Possible glyphosate tolerance mechanism in pitted morningglory (*Ipomoea lacunosa*). *J Agric Food Chem* **63**:1689–1697 (2015).

146 González-Torralva F, Rojano-Delgado AM, Luque de Castro MD, Müller N, and De Prado R, Two non-target mechanisms are involved in glyphosate-resistant horseweed (*Conyza canadensis* L. Cronq.) biotypes. *J Plant Physiol* **169**:1673–1679 (2012).

147 Handayani VDS, Tanno Y, Yamashita M, Tobina H, Ichihara M, Ishida Y, and Sawada H, Influence of weed management measures on glyphosate resistance and endophyte infection in naturalized Italian ryegrass (*Lolium multiflorum*). *Weed Biol Manag* **17**:84–90 (2017).

148 Hoagland RE, Effects of glyphosate on metabolism of phenolic compounds: VI. Effects of glyphosine and glyphosate metabolism on phenylalanine ammonia-lyase activity, growth, and protein, chlorophyll and anthocyanin levels in soybean (*Glycine max*) seedlings. *Weed Sci* **28**:393–400 (1980).

149 McElroy JS, and Hall ND, *Echinochloa colona* with reported resistance to glyphosate conferred by aldo-keto reductase also contains a Pro-106-Thr EPSPS target-site mutation. *Plant Physiol* **183**:447–450 (2020).

150 Duke SO, Enhanced metabolic degradation of glyphosate: the last evolved resistance mechanism? *Plant Physiol* **181**:1401–1403 (2019).

151 Gressel J, Evolving understanding of the evolution of herbicide resistance. *Pest Manag Sci* **65**:1164–1174 (2009).

152 Yu Q, Jalaludin A, Han H, Chen M, Sammons RD, and Powles SB, Evolution of a double amino acid substitution in the 5-enolpyruvylshikimate-3-phosphate synthase in *Eleusine indica* conferring high-level glyphosate resistance. *Plant Physiol* **167**:1440–1447 (2015).

153 Mutters RG, Madore M, and Bytnerowicz A, Formaldehyde exposure affects growth and metabolism of common bean. *Air Waste* **43**:113–116 (1993).

154 Gaines TA, Zhang W, Wang D, Bukun B, Chisholm ST, Shaner DL, *et al.*, Gene amplification confers glyphosate resistance in *Amaranthus palmeri*. *Proc Natl Acad Sci USA* **107**:1029–1034 (2010).

155 Hetherington PR, Reynolds TL, Marshall G, and Kirkwood RC, The absorption, translocation and distribution of the herbicide glyphosate in maize expressing the CP-4 transgene. *J Exp Bot* **50**:1567–1576 (1999).

156 Nandula VK, Reddy KN, Rimando AM, Duke SO, and Poston DH, Glyphosate-resistant and -susceptible soybean (*Glycine max*) and canola (*Brassica napus*) dose response and metabolism relationships with glyphosate. *J Agric Food Chem* **55**:3540–3545 (2007).

157 Komoßa GI, and Sandermann H, Plant metabolism of herbicides with C-P bonds. *Pestic Biochem Physiol* **43**:85–94 (1992).

158 Gout E, Bligny R, Genix P, Tissut M, and Douce R, Effect of glyphosate on plant cell metabolism. ^{31}P and ^{13}C NMR studies. *Biochimie* **74**:875–882 (1992).

159 Ge X, d'Avignon A, Ackerman JJH, Ostrander E, and Sammons RD, Application of ^{31}P-NMR spectroscopy to glyphosate studies in plants: insights into cellular uptake and vacuole sequestration correlated to herbicide resistance,

In *Handbook on Herbicides: Biological Activity, Classification and Health and Environmental Implications*, Nova Science Publishers, Inc., pp. 55–83 (2013).

160 Ge X, d'Avignon DA, Ackerman JJH, and Sammons RD, In vivo ^{31}P-nuclear magnetic resonance studies of glyphosate uptake, vacuolar sequestration, and tonoplast pump activity in glyphosate-resistant horseweed. *Plant Physiol* **166**:1255–1268 (2014).

161 Fitzgibbon JE, and Braymer HD, Cloning of a gene from *pseudomonas* sp. strain PG2982 conferring increased glyphosate resistance. *Appl Environ Microbiol* **56**:3382–3388 (1990).

162 Derting CW, Wiper application, In McWhorter CG, and Gebhardt MR (Eds), *Methods of Applying Herbicides*, Weed Science Society of America Monograph 4, Champaign, IL, pp. 207–229 (1987).

163 Harrington KC, and Ghanizadeh H, Herbicide application using wiper applicators – a review. *Crop Prot* **102**:56–62 (2017).

164 Westerman RB, and Murray DS, Silverleaf nightshade (*Solanum elaeagnifolium*) control in cotton (*Gossipium hirsutum*) with glyphosate. *Weed Technol* **8**:720–727 (1994).

165 Griffin JL, Boudreaux JM, and Miller DK, Herbicides as harvest aids. *Weed Sci* **58**:355–358 (2010).

166 Cessna AJ, Darwent AL, Townley-Kent L, Harker KN, and Kirkland KJ, Residues of glyphosate and its metabolite AMPA in field pea, barley and flax seed following preharvest applications. *Can J Plant Sci* **82**:485–489 (2002).

167 Cessna AJ, Darwent AL, Kirkland KJ, Townley-Smith L, Harker KN, and Lefkovitch LP, Residues of glyphosate and its metabolite AMPA in wheat seed and foliage following preharvest applications. *Can J Plant Sci* **74**:653–661 (1994).

168 Xu J, Smith S, Smith G, Wang W, and Li Y, Glyphosate contamination in grains and foods: An overview. *Food Control* **106**:106710 (2019).

169 Dalley CD, and Richard EP, Herbicides as ripeners for sugarcane. *Weed Sci* **58**:329–333 (2010).

170 Gomes GLGC, Carbonari CA, Velini ED, Trindade MLB, and Silva JRM, Extraction and simultaneous determination of glyphosate, AMPA and compounds of the shikimic acid pathway in plants. *Planta Daninha* **33**:295–304 (2015).

171 Carbonari CA, Gomes GLGC, Velini ED, Machado RF, Simões PS, and Macedo GC, Glyphosate effects on sugarcane metabolism and growth. *Amer. J Plant Sci* **5**(24) (2014).article ID 51972

172 Arregui MC, Lenardón A, Sanchez D, Maitre MI, Scotta R, and Enrique S, Monitoring glyphosate residues in transgenic glyphosate-resistant soybean. *Pest Manag Sci* **60**:163–166 (2003).

173 Bohm GMB, Rombaldi CV, Genovese MI, Castilhos D, Alves BJR and, Rumjanek NG, Glyphosate effects on yield, nitrogen fixation, and seed quality in glyphosate-resistant soybean. *Crop Sci* **54**:1737–1743 (2014)

174 Bøhn T, Cuhra M, Traavik T, Sanden M, Fagan J, and Primicerio R, Compositional differences in soybeans on the market: glyphosate accumulates in roundup ready GM soybeans. *Food Chem* **153**:207–215 (2014).

175 Duke SO, Rimando AM, Reddy KN, Cizdziel JV, Bellaloui N, Shaw DR, Williams MM, and Maul JE, Lack of transgene and glyphosate effects on yield, and mineral and amino acid content of glyphosate-resistant soybean. *Pest Manag Sci* **74**:1166–1173 (2018).

176 Gomes MP, Smedbol E, Chalifour A, Hénault-Ethier L, Labrecque M, Lepage L, Lucotte M, and Juneau P, Alteration of plant physiology by glyphosate and its by-product aminomethylphosphonic acid: an overview. *J Exp Bot* **65**:4691–4703 (2014).

177 Serra A-A, Nuttens A, Larvor V, Renault D, Couée I, Sulmon C, and Gouesbet G, Low environmentally relevant levels of bioactive xenobiotics and associated degradation products cause cryptic perturbations of metabolism and molecular stress in *Arabidopsis thaliana*. *J Exp Bot* **64**:2753–2766 (2013).

178 Reddy KN, Duke SO, and Rimando AM, Aminomethylphosphonic acid, a metabolite of glyphosate, causes injury in glyphosate-treated, glyphosate-resistant soybean. *J Agric Food Chem* **52**:5139–5143 (2004).

179 Sammons RD, and Tran M, Examining yellow flash in roundup ready soybean. *Proc North Central Weed Sci Sci* **63**:120 (2008).

180 Duke SO, Lydon J, Koskinen WC, Moorman TB, Chaney RL, and Hammerschmidt R, Glyphosate effects on plant mineral nutrition, crop rhizophere microbiota, and plant disease in glyphosate-resistant crops. *J Agric Food Chem* **60**:10375–10397 (2012).

181 Cerny L, Wu K and Floyd T, Molecular markers associated with yellow flash in glyphosate tolerant soybeans. US Patent 9617605 B2. April 11, 2017 (2017).

182 Ebelhar SA, Varsa EC and Hart CD, Soil pH and manganese effect on yield of Roundup Ready soybeans, In *Illiitois Fertilizer Conference Proceedings*, pp. 54–65 (2006).

183 Smedbol W, Lucotte M, Maccario S, Gomes MP, Paquet S, Moingt M, Mercier LLC, Sobarzo MRP, and Blouin M-A, Glyphosate and aminomethylphosphonic acid content in glyphosate-resistant soybean leaves, stems, and roots associated with phytotoxicity following a single glyphosate-based herbicide application. *J Agric Food Chem* **67**:6133–6142 (2019).

184 Schilling BS, Harker KN, and King JR, Glyphosate can reduce glyphosate-resistant canola growth after individual or sequential applications. *Weed Technol* **20**:825–830 (2006).

185 Reddy KN, Cizdziel JV, Williams MM, Maul JE, Rimando A, and Duke SO, Glyphosate resistance technology has minimal effect on maize mineral nutrition and yield. *J Agric Food Chem* **66**:10139–10146 (2018).
186 Bernal J, Martin MT, Soto ME, Nozal MJ, Marotti I, Dinelli G, and Bernal JL, Development and application of a liquid chromatography-mass spectrometry method to evaluate the glyphosate and aminomethylphosphonic acid dissipation in maize plants after foliar treatment. *J Agric Food Chem* **60**:4017–4025 (2012).
187 Bidóia VS, JCD SN, de Goes Maciel CD, Tropaldi L, Carbonari CA, Duke SO, and de Carvalho LB, Lack of significant effects of glyphosate on glyphosate-resistant maize in different field locations. *Agronomy* **13**:1071 (2023).
188 Costa FR, Rech R, Duke SO, and Carvalho LB, Lack of effects of glyphosate and glufosinate on growth, mineral content, and yield on glyphosate- and glufosinate-resistant maize. *GM Crops Food* **9**:180–198 (2018).
189 Williams GM, Aardema M, Acquavella J, Berry Sir C, Brusick D, Burns MM, *et al.*, A review of the carcinogenic potential of glyphosate by four expert panels and comparison to the IARC assessment. *Crit Rev Toxicol* **46**:S3–S20 (2016).

4

Enhanced Metabolic Resistance to Herbicides in *Alopecurus myosuroides* (Black Grass)

Alina Goldberg Cavallieri, Sara Franco Ortega, Stewart Brown, Nawaporn Onkokesung, and Robert Edwards

School of Natural and Environmental Sciences, Newcastle University, Newcastle Upon Tyne, UK

4.1 Introduction

Black grass (*Alopecurus myosuroides*) is an obligate outcrossing diploid wild grass initially reported as a problem weed in the 1950s.[1] As a member of the family Poaceae, which includes more than 10,000 species,[2] it is an annual wild grass that germinates in the autumn and sheds its seeds between late June and late August with a peak in July, making it a perfectly adapted competitive weed of winter wheat.[3,4] Black grass probably originated in the Mediterranean "fertile crescent" from where it was dispersed across Europe and Asia by migrant Neolithic (4000–2000 BCE) farmers.[5] While its range continues to expand, it is currently most problematic as a weed in the wheat-growing areas of Northern Europe.

Black grass is a tall (30–90 cm in height) annual grass with purple or greenish leaf sheaths (Fig. 4.1A), with a single plant producing up to 150 tillers,[6] bearing 20–300 heads per plant that can produce over 1000 seeds/month.[2,3] Emerging in the autumn, black grass competes directly for both nutrients and light with winter cereals causing crop yield losses of up to 66% in the most severely infested fields.[7] The high reproductive capacity, genetic diversity, and phenotypic plasticity of black grass make it a perfect weed to evolve protective defenses toward many selective pressures, most notably in evolving complex herbicide resistance traits. Since the 1970s, winter cereal production in northern Europe has been associated with the continuous use of herbicides and a reduction in crop rotations, conditions that have led to the widespread evolution of resistance to postemergence selective herbicides in wild grasses.[8,9] In the case of herbicide resistance in black grass, in the United Kingdom the associated loss of weed control has led to annual losses in winter wheat production valued at approximately 0.5 billion Great British Pound (GBP) per annum.[10]

Resistance in Weeds from Herbicide Metabolism, First Edition.
Edited by Vijay K. Nandula and Roland Beffa.

(A)

(B)

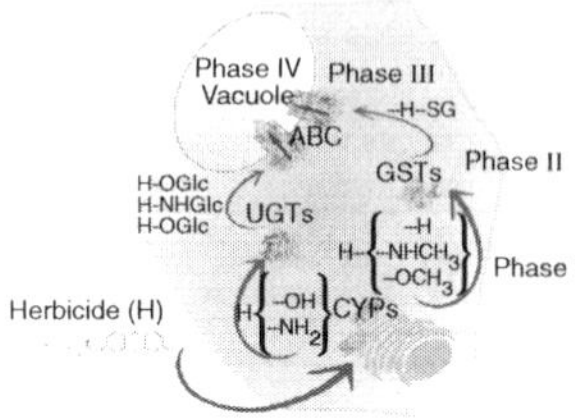

Figure 4.1 (A) Black grass and (B) herbicide detoxification in four phases by the xenome. ABCs=ATP-binding cassette transporters, CYPs=cytochrome P450s, GSTs=glutathione-*S*-transferases, UGTs=UDP-glucose-dependent glycosyltransferases. (*See insert for color representation of the figure.*)

The control of black grass in northern Europe has been historically heavily dependent upon postemergence selective herbicides, notably, phenylureas acting on photosystem II (PSII), the aryloxyphenoxypropionates (FOPs), phenylpyrazolines (DENs), and cyclohexanediones (DIMs), all of which inhibit acetyl CoA carboxylase (ACCase) and sulfonylureas, that target acetolactate synthase (ALS).[11] As resistance to these compounds has steadily developed, the use of selective preemergence herbicides has expanded to include pendimethalin (tubulin inhibitor), S-metolachlor and flufenacet (fatty acid elongase inhibitors), and prosulfocarb (lipid synthesis inhibitor). Nonselective herbicides, including glyphosate, are also extensively used as chemical mulching agents prior to cereal drilling. An overview of the resistance status to herbicides used in black grass control is summarized in Table 4.1.

Resistance to herbicides can arise through two global mechanisms termed target-site resistance (TSR) or non-target-site resistance (NTSR). TSR in black grass is most commonly seen with ALS and ACCase inhibiting herbicides but has rarely developed with herbicides targeting microtubule assembly, or PSII.[12] Single nucleotide polymorphisms are the most common cause of TSR, leading to a reduced ability of the herbicide to bind to the target enzyme. In the case of ACCase and ALS, these mutations are found at a number of sites in the respective coding sequences, with their frequency varying in different regions.[13,14] TSR in other wild grasses, such as ryegrass (*Lolium rigidum*), is also known to include other mechanisms, notably an increased copy number of EPSP (5-enolpyruvylshikimate-3-phosphate) synthase, the glyphosate target gene.[15,16] However, to date, TSR toward glyphosate has not been reported in black grass. While TSR gives very high levels of resistance to herbicides acting on the respective protein target, it can be counteracted by rotating the application of compounds with differing modes of action.

Table 4.1 Herbicide resistance in black grass *(Alopecurus myosuroides)*.

Herbicide	Mode of action	Herbicide chemistry	Confirmed cases of NTSR
Pregermination			
Pendimethalin	Microtubule assembly inhibition	Dinitroaniline	Yes
Flufenacet	Inhibition of very long-chain fatty acid	Oxyacetamide	Yes
Prosulfocarb	Lipid synthesis inhibitor		Yes
Postgermination			
Mesosulfuron-methyl+ iodosulfuron-methyl sodium	Inhibition of acetolactate synthase	Sulfonylurea	Yes
Pyroxsulam	Inhibition of acetolactate synthase	Triazolopyrimidine	Yes
Isoproturon	Inhibition of photosynthesis at photosystem II	Urea	Yes
Chlorotoluron	Inhibition of photosynthesis at photosystem II	Urea	Yes
Fenoxaprop-*P*-ethyl	Inhibition of acetyl CoA carboxylase	Aryloxyphenoxy-propionate	Yes
Clodinafop-propargyl	Inhibition of acetyl CoA carboxylase	Aryloxyphenoxy-propionate	Yes
Pinoxaden	Inhibition of acetyl CoA carboxylase	Phenylpyrazoline	Yes
Cycloxydim	Inhibition of acetyl CoA carboxylase	Cyclohexanedione	No

Source: Adapted from Heap.[12]

In contrast to TSR, NTSR-based resistance in black grass is not linked to the mode of action of herbicides, and extends to multiple chemistries, and cannot be counteracted by herbicide rotation alone. Unlike TSR that behaves as a Mendelian inheritable trait, NTSR behaves as a complex quantitative trait, regulated by multiple genes and presenting phenotypes varying from slight tolerance to high levels of resistance to multiple herbicides.[17] Suboptimally, its diagnosis in black grass is often indirectly inferred due to the absence of any known mutations in herbicide target giving rise to TSR.[14] As a further complication, black grass can exhibit both NTSR and TSR resistance mechanisms, giving rise to biotypes varying in their tolerance to multiple herbicides.[18,19] In the United Kingdom, 35 black grass multiple resistance biotypes have been identified to date, predominantly associated with winter wheat (http://www.weedscience.org/Pages/Species.aspx). The best characterized black grass resistant biotypes are named after their place of origin, with those referred to in this review being collected from Peldon (Essex), Lincs E1 (Lincolnshire), Oxford, Notts (Nottinghamshire), Clay (Claycroft field at Rothamsted), Kent, and Suffolk.[4,20] As resistance to multiple herbicides is currently so common, herbicide-sensitive (HS) populations have become important as a point of reference. As such, the HS population Rothamsted, originating from the Broadbalk experimental site at Rothamsted, Hertfordshire (UK), which has never been exposed to herbicides is a very valuable resource.[11] In contrast, the highly resistant population Peldon, which presents both TSR and NTSR, arises from a site that has been continuously treated with herbicides since the 1970s. The Peldon population has been widely used as a model to study multiple herbicide resistance in black grass.[21] It is characterized by its high level of tolerance to different herbicide classes, including FOPS (diclofop-methyl, fenoxaprop-*P*-ethyl, and fluazifop-*P*-butyl), DIMS (tralkoxydim), phenylureas (chlorotoluron), and sulfonylureas (mesosulfuron).[11]

4.2 NTSR and Enhanced Metabolic Resistance to Herbicides

By its nature, while TSR arises from impaired binding of the herbicide to its targeted protein, NTSR can arise through multiple mechanisms that can be broadly split into (1) reducing the bioavailability of the toxiphore at its site of action and (2) reducing the consequences of toxic injury through the elicitation of general cellular defense mechanisms. In practice, these mechanisms are unlikely to operate in isolation from one another. While the majority of attention in NTSR research has been focused on limiting herbicide bioavailability, in black grass there is now a growing body of evidence that general cytoprotection from herbicide injury involving the accumulation of antioxidants and the suppression and scavenging of reactive oxygen and electrophile species is also important.[22–24]

The focus of this review is on the importance of NTSR and the role of enhanced metabolic resistance (EMR) that influences the bioavailbility of herbicides. Other factors, such as reduced herbicide penetration and translocation, or increased sequestration in non-targeted tissues, which have been shown to be important in NTSR in other weeds,[25,26] have not been identified in studies conducted on black grass to date. As an example, studies with radiolabeled fenoxaprop-*P*-ethyl in black grass showed identical rates of herbicide uptake in HS and nine populations exhibiting both NTSR and TSR that had been collected from both the United Kingdom and northern Europe.[20] Similarly, no differences in herbicide uptake and translocation were reported in resistant and HS black grass populations treated with diclofop-methyl[27] or chlorotoluron.[28] Instead, greater attention has been paid on the role of enhanced herbicide detoxification and NTSR.[22,29] The types of enzymes and transporters responsible for herbicide detoxification appear to be identical in black grass as those described in crops and weeds. The detoxification of herbicides and other xenobiotics occurs in a series of biotransformations involving the introduction, or exposure of reactive chemical groups (phase 1), bioconjugation (phase 2), and vacuolar transport (phase 3). As will be discussed below, black grass contains a large portfolio of genes encoding enzymes of phase 1 metabolism, notably cytochrome P450s (CYPs), oxidoreductases, and hydrolases; phase 2 conjugating enzymes such as glutathione-*S*-transferases (GSTs) and UDP-glucose-dependent glycosyltransferases (UGTs); and phase 3 ATP-binding cassette (ABC) and multidrug and toxic compound extrusion (MATE) transporter proteins. Collectively, together with the associated regulatory network of genes and proteins, we have termed this detoxification system as the 'xenome'[30] with a schematic of black grass shown in Fig. 4.1B.

While black grass possesses a full complement of xenome proteins, their respective importance in conferring EMR and associated NTSR has only been reported with a limited number of herbicides. A widely used method to indirectly demonstrate EMR is through the use of xenome inhibitors to act as synergists, with compounds like malathion, piperonyl butoxide (PBO), and aminobenzotriazole (ABT) used to inhibit CYP-based detoxification, and tridiphane to reduce GST-mediated metabolism. Using these inhibitors in different combinations, it was possible to identify distinct EMR phenotypes toward flupyrsulfuron and the FOPs clodinafop-propargyl, haloxyfop, and fenoxaprop-*P*-ethyl in three herbicide-resistant black-grass populations.[31] While the use of herbicide synergists can indeed counteract NTSR mechanisms, their specificity as metabolic inhibitors is not always well-defined. Instead, the "gold standard" method for determining EMR is to monitor herbicide uptake and metabolism using either radiochemical or quantitative mass spectrometry-based methods.[32,33]

Several studies have reported the metabolism of the phenylurea herbicide chlorotoluron in NTSR and HS black grass. Initial studies determined the rates of disappearance of chlorotoluron in the resistant Peldon and Lincs E1 populations as compared with HS plants.[11] Having eliminated TSR as a potential resistance mechanism, the relative rates of % dose that had disappeared over 24 hours was in the order: Peldon (94%) > Lincs E1 (48%) > HS (14%). As the loss of herbicide was markedly reduced by co-application with ABT, the potential role of CYPs in detoxification was inferred. Subsequent studies focused on the biotransformation reactions involved.[34,35] Phase 1 metabolism of chlorotoluron was found to be involved in either *N*-demethylation (both mono- and di-demethylation) or hydroxylation of the ring methyl group (Fig. 4.2), followed by rapid glycosylation.[35] This primary metabolism of chlorotoluron is typical of the reactions catalyzed by CYPs and was found to be critical in determining the effective detoxification of this particular herbicide in black grass. Thus, while *N*-demethylated metabolites retained their ability to inhibit PSII, ring methyl hydroxylation abolished toxic activity.[35] Intriguingly, ABT was more effective in inhibiting ring methyl hydroxylation as compared with *N*-demethylation, suggesting that different CYPs were involved in these two reactions. Further studies showed that a black grass population from Spain showing EMR toward chlorotoluron also showed enhanced metabolism and tolerance toward the FOP herbicide diclofop-methyl.[27] The rate of detoxification of diclofop-methyl in the population exhibiting EMR was double than that seen in HS plants, with the principle metabolites being the glucose ester of diclofop acid and a glycosylated conjugate of an aryl hydroxylated intermediate. Again, CYPs were strongly implicated in the aryl hydroxylation, with herbicide synergism demonstrated in the presence of ABT. Studies with the UK black grass populations Peldon and Lincs E1, which were known to show EMR toward chlortoluron, were also found to demonstrate enhanced metabolism and tolerance toward diclofop-methyl, with partial resistance to fluazifop-*P*-ethyl and tralkoxydim being determined.[11] Significantly, all three herbicides are metabolized by CYPs. In contrast, while Peldon plants were moderately resistant to fenoxaprop-*P*-ethyl due to EMR, enhanced metabolism could not explain the high levels of resistance to this herbicide determined in Lincs E1 black grass, and it was concluded that an additional NTSR mechanism was at work in this population. Additional studies with a wider range of populations showing NTSR toward fenoxaprop-*P*-ethyl showed that resistance was associated with an increased rate of herbicide disappearance in the majority of cases.[20] However, one highly resistant population (Clay) showed no such EMR trait, confirming that NTSR is not always associated with enhanced herbicide metabolism. Unlike the other FOP herbicides, in most cereals and wild grasses, fenoxaprop-*P*-ethyl is metabolized by the action of GSTs to yield a glutathionylated cleavage product.[11,36] NTSR black grass populations carried out this primary detoxification more rapidly than HS plants (Fig. 4.2), processing the glutathione conjugate to the respective cysteine derivative.[37] However, it is also likely that alternative routes of detoxification of fenoxaprop-*P*-ethyl contribute to resistance. Thus, the formation of the CYP-derived metabolite chlorobenzoxazol-2(3H)-one was subsequently linked to NTSR in other black grass populations that also showed EMR to the

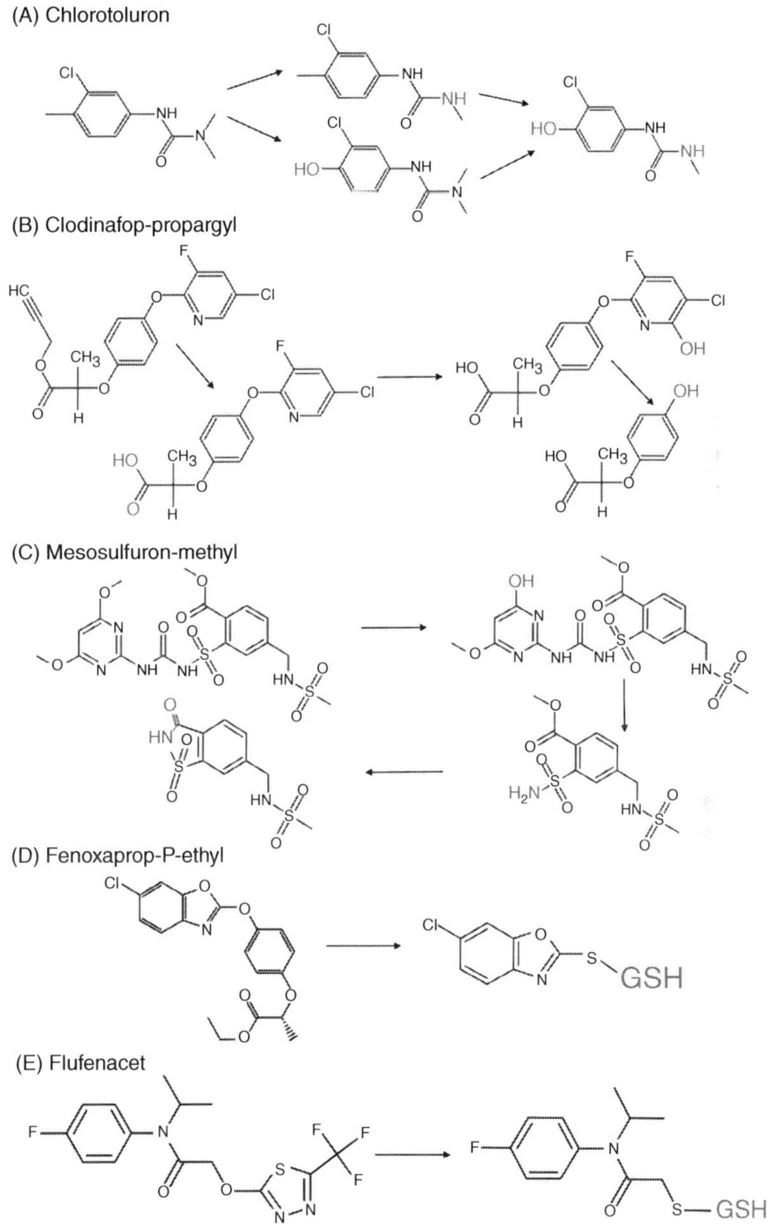

Figure 4.2 Black grass herbicides and their primary metabolites. Modifications are highlighted in red. (A) Chlorotoluron, *N*-demethylation, ring methyl hydroxylation, and *N*-demethylation plus ring methyl hydroxylation; (B) clodinafop-propargyl, hydrolysis, ring hydroxylation, and ether cleavage; (C) mesosulfuron-methyl, demethylation, cleavage, and cyclization; (D) fenoxaprop-*P*-ethyl and glutathione conjugation; and (E) flufenacet and glutathione conjugation. (*See insert for color representation of the figure.*)

sulfonylurea herbicides mesosulfuron-methyl + iodosulfuron.[38] Similarly, the toxicity of fenoxaprop-*P*-ethyl and mesosulfuron-methyl in EMR populations of the related species *Alopecurus aequalis* was synergized with the CYP inhibitors, namely, PBO and malathion.[39]

In addition to postemergence herbicides, recent studies have also implicated EMR as a resistance mechanism with flufenacet, a compound used for preemergence control of black grass.[40] Increased tolerance toward flufenacet was observed in several black grass populations, and this was associated with the enhanced detoxification of the herbicide by glutathionylation (Fig. 4.2). The involvement of GSTs in flufenacet metabolism was inferred by the herbicide synergism observed in the NTSR population with the GST inhibitors, namely, tridiphane and ethacrynic acid.[41] It is therefore clear that both CYPs and GSTs are implicated in EMR in black grass toward a range of herbicides. In the following sections, the respective potential candidate enzymes and genes are identified and the role of other largely unstudied xenome components is explored.

4.3 Potential for Esterases and Herbicide Bioactivation to be Linked to NTSR

Several herbicides used in selective weed control that act on ACCase and ALS, notably FOPs and sulfonylureas, respectively, are applied as the pro-herbicidal esters that are bioactivated to the phytotoxic acids following foliar uptake through the action of esterases as the first step in their metabolism.[27] Therefore, there has been interest in the potential for herbicide bioactivation to contribute to NTSR through resistant plants having a reduced capacity to hydrolyze the pro-herbicides.[42,43] When a range of carboxylic acid xenobiotic esters were tested in wheat and black grass, higher activities toward model substrates such as naphthyl acetate were determined in the crop as compared with the weed, whereas the opposite was the case with diclofop-methyl.[44] These studies with crude plant extracts revealed no differences in esterase activities toward either herbicides or model substrates when NTSR (Peldon) and HS black grass were assayed. Subsequent purification of the esterase responsible for the activation of a range of FOP esters in black grass revealed the presence of a highly specific 40 kDa GDSL hydrolase, termed *Am*GDSH1, an enzyme that was predicted to be expressed in the apoplast.[45] As such, no empirical evidence for the involvement of esterases in NTSR in black grass was determined in these studies, but the dependence of this herbicide bioactivation on a single enzyme means a loss of function mutation in *Am*GDSH1 would be a powerful new evolutionary route to NTSR in the future.

4.4 Cytochrome P450s Linked to NTSR in Black Grass

As is the case with all higher plants where CYPs (EC 1.14) can represent up to 1% of the total protein-coding genes in the genome,[46] transcriptome studies have shown that black grass contains a diverse range of these enzymes.[21,33] While all CYPs share a common catalytic center,[47] conservation in sequence in the other domains is restricted, resulting in the diversification of the respective genes into 127 families in plants.[46,48] Derived from common gene ancestors into distinct clans, the complex nomenclature of the CYP superfamily is based on their sequence relatedness, with members of the same families showing 40% similarity and subfamilies 55%.[49]

While there is a growing body of transcriptome data relating to CYPs in black grass, our knowledge on their functional activities and tissue localization is restricted. CYPs are known to be active toward a range of natural and xenobiotic substrates, catalyzing a diverse range of redox reactions, including dealkylation, decarboxylation, dehydration, desaturation, dimerization, C—C cleavage, isomerization, reduction, and ring hydroxylation.[48] In the case of herbicides, as shown in Fig. 4.2, metabolism studies on black grass point to CYPs catalyzing ring hydroxylation, demethylation, and oxidative cleavage reactions. As with other higher plants, CYPs in black grass are likely to be anchored to the endoplasmic reticulum through a short hydrophobic segment in their N terminus. However, signal peptides directing these enzymes to the plastids, or mitochondria, have also been described.[50]

Based on the analysis of the transcriptome of the Peldon population,[21] together with RNA-Seq data derived from NTSR black-grass populations that had been selected for resistance to pendimethalin or fenoxaprop-*P*-ethyl, respectively,[33] CYPs from 11 families were found to be differentially expressed genes (DEGs), as compared with an HS population (Rothamsted). The CYP DEGs identified in these studies were derived from clan 71 (families 71, 73, 76, 81, 89, and 99), clan 72 (709 and 72), clan 84 (86 and 704), and clan 711 (711). These transcriptomic studies have the potential to shed new light on EMR mechanisms. While Peldon- and pendimethalin-selected NTSR populations had an increased capacity to detoxify herbicides metabolized by CYPs and GSTs, the fenoxaprop-selected population showed no such enhancement.[33] Furthermore, the spectrum of NTSR in Peldon extended over a wider range of herbicides than was observed in pendimethalin-selected population. As such, each population was ascribed an NTSR phenotype as being either Peldon (NTSR1), pendimethalin-selected (NTSR2), or fenoxaprop-*P*-ethyl-resistant plants (NTSR3). These three NTSR phenotypes can be used to explore the diverse functions of large gene families, such as CYPs, in NTSR and EMR. Based on the NTSR type, CYPs that are upregulated in both NTSR1 and NTSR2 are likely to be involved in EMR, while this is unlikely to be the case with any upregulated CYPs in the NTSR3 plants.

Intriguingly, in contrast to NTSR1 and NTSR2 populations, the majority of CYPs in NTSR3 were actually downregulated rather than enhanced when compared with those in HS plants.[33] This observation further confirms that the NTSR3 phenotype is not classically linked to EMR.

To establish the link between CYP-DEGs and EMR, a phylogenetic analysis was carried out on members of CYP families 81, 72, 709, and 71. These families are known to possess orthologs that function in herbicide detoxification in plants or have been shown to be highly expressed in NTSR populations in other weed species.[51–56] Several members of the CYP81 family are known to be involved in the metabolism of multiple herbicides in both crops and weeds.[57,58] The upregulated CYPs from family 81 identified in NTSR1 (*Am*CYP81A4 and *Am*CYP81A2) and NTSR2 populations (*Am*CYP81A1, *Am*CYP81A3, and *Am*CYP81A4) were included in the phylogenetic tree (Fig. 4.3). Among these, *Am*CYP81A4, a CYP that was upregulated in NTSR1 and NTSR2, was part of a monophyletic group that included *Hv*CYP81A63 from barley (*Hordeum vulgare*), a CYP that conferred high resistance to diclofop-methyl and pinoxaden when expressed in rice.[56] In addition, the related *Lr*CYP81B1 (BAD27506) from ryegrass (*L. rigidum*) was constitutively upregulated in populations resistant to ALS inhibitors. Similarly, the expression of the *Am*CYP81A4 ortholog in wild oat (*Avena fatua*=*Af*CYP81 5954.7) was also enhanced in NTSR biotypes.[52,59] The fact that *Am*CYP81A4

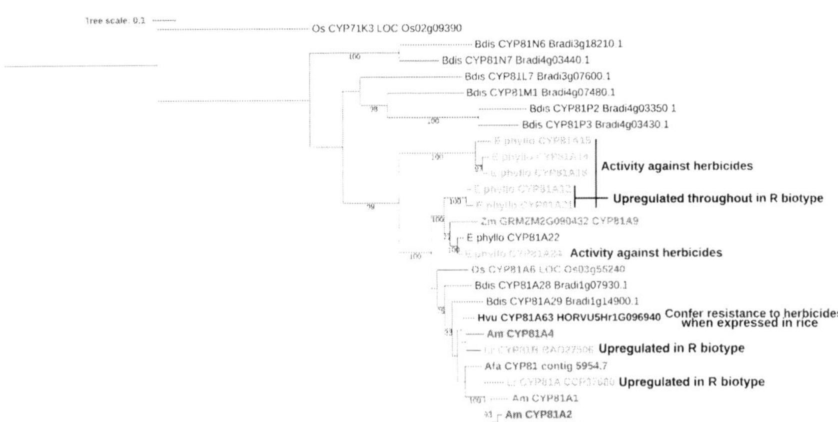

Figure 4.3 Phylogenetic analysis of CYP proteins from family 81. Amino acid sequences of CYPs from black grass (Am, *Alopecurus myosuroides*), brachypodium (Bdis, *Brachypodium distachyon*), wild oat (Afa, *Avena fatua*), maize (Zm, *Zea mays*), annual ryegrass (Lr, *Lolium rigidum*), rice (Os, *Oryza sativa*), later water grass (E phyllo, *Echinochloa phyllogon*), and barley (Hv, *Hordeum vulgare*) were used for maximum likelihood alignment for phylogenic analysis. The number on the branch represents the bootstrap support values above 70%. The scale bar indicates the inferred number of substitutions per site. (*See insert for color representation of the figure.*)

was identified in NTSR1 and NTSR2 populations that exhibit EMR strongly suggests that this CYP is likely to play a key role in EMR in black grass.[33] Indeed, *Am*CYP81A4 is able to catalyzed the demethylation of pendimethalin and chlorotoluron as well as the hydroxylation of chlorotoluron and tralkoxydim when heterologously expressed in yeast.[117] Additionally, *Am*CYP81A2 (NTSR1) and *Am*CYP81A3 (NTSR2), that share 93% identity, are part of the same monophyletic group as *Am*CYP81A4 (Fig. 4.3). Hence, these two CYPs could potentially have functions in herbicide metabolism in black grass.[33] It is noteworthy that *Am*CYP81A2 and *Am*CYP81A3 formed a monophyletic group with maximum bootstrap support with *Am*CYP81A1, an upregulated CYP in NTSR2 plants (Fig. 4.3). This monophyletic group is related to *Lr*CYP81A (CCP37680), which is constitutively upregulated in NTSR ryegrass.[52] Surprisingly, *Am*CYP81A2 did not show activity towards pendimethalin, chlorotoluron, or tralkoxydim as reported for *Am*CYP81A4.[117] This suggests it is not possible to predict the activity of members of the CYP81A family based only on sequence sequence similarity. Hence, the functional characterization using in vitro or in vivo systems is required to confirm the activity range of these CYP81s. Intriguingly, none of these CYP81s were identified in the NTSR3 population, which shows no evidence of EMR.

CYP72s are another family that has been confirmed to function in herbicide metabolism in various crop and weed species. While *Am*CYP72A3, *Am*CYP7A4, and *Am*CYP72A5 were enhanced in NTSR1, no CYP72s were differentially expressed in the NTSR2 population. This suggests that the upregulated CYP72As could function in the detoxification of herbicide chemistries that are metabolized in NTSR1 (Peldon) plants, but not in NTSR2 population. Two CYPs upregulated in NTSR1, *Am*CYP72A4 and *Am*CYP72A5, were part of a monophyletic group that included *Os*CYP72A18,[60] an enzyme that has known activity toward the herbicide pelargonic acid in rice, as well as being related to the maize *Zm*CYP71A5[61] which is linked to increased tolerance to triasulfuron induced by the co-treatment with the safener naphthalic anhydride. Moreover, *Lr*CYP72 A2218 and *Lr*CYP72A 1604 from ryegrass populations resistant to diclofop-methyl[53] are also part of the same clade, with *Lr*CYP72A 2218 enhanced in ryegrass populations resistant to ALS inhibitors and are inducible after exposure to iodosulfuron + mesosulfuron, or pyroxsulam.[52] The two CYP72s upregulated in the NTSR3 population (*Am*CYP72A2 and *Am*CYP72A1) were divergent (51–57% identity) to those enhanced in NTSR1 plants. Although *Am*CYP72A2 and *Am*CYP72A1 formed a clade that contained *Os*CYP72A31 from rice involved in tolerance to bispyribac sodium, this is not supported by a strong bootstrap value (data not shown). Based on this information, *Am*CYP72A1 and *Am*CYP72A2 are speculated to function in biological processes unrelated to herbicide detoxification in black grass.

Within the family 71, three CYPs were upregulated in NTSR1 (*Am*CYP71X1, *Am*CYP71X2, and *Am*CYP71C3) but no CYP71s were differentially expressed in the NTSR2 population. Additionally, *Am*CYP71X1 can metabolise chlorotoluron and tralkoxydim when heterologously expressed in yeast.[117] *Am*CYP71T1, which sits in a distinct clade, was the only CYP71 family that showed signficant increased expression in the NTSR3 population (Fig. 4.3). It is interesting that the down-regulated CYP71s in NTSR3 population, namely, *Am*CYP71R1, *Am*CYP71R2, *Am*CYP71AD1, *Am*CYP71C1, *Am*CYP71C2, and *Am*CYP71Y1, are clustered with CYP71s that have been reported to be involved in herbicide metabolism in crops and weeds. For instance, *Am*CYP71C1 and *Am*CYP71C2 clustered with maximum support value with *Ta*CYP71C6v1 from wheat, a CYP that acts on sulfonylurea herbicides (chlorsulfuron, triasulfuron, metsulfuron-methyl, bensulfuron-methyl, and tribenuron-methyl) when heterologously expressed in yeast,[51] as well as *Zm*CYP71C3v2 from maize whose expression is induced by triasulfuron.[61] In addition, *Am*CYP71R1 is also clustered with *Lr*CYP71R4 from ryegrass, which is upregulated in populations resistant to ALS inhibitors.[52] Based on the phylogenetic analysis and yeast heterologous expression system, this suggests that unlike CYP81s, CYP71s and CYP72As are more likely to be involved in the detoxification of specific herbicide chemistries. Taken together, the upregulation of CYP81s, CYP71s, and CYP72As in NTSR1 (Peldon) contributes to the broad-ranging NTSR and EMR observed in this population as compared to the selective upregulation of CYP81 members in the NTSR2 population that exhibits selective resistant to ACCase-inhibiting herbicides.

Finally, the genes upregulated in the family 709 (*Am*CYP709J1 and *Am*CYP709E1 in NTSR2 and NTSR1, respectively) clustered in distinct clades supported by high bootstrap values (above 90%, data not shown). This is consistent with the low level of sequence similarity (47%) between them. It is interesting that *Am*CYP709E1 clustered on a monophyletic group with the wheat enzyme *Ta*CYP709C1, which was shown to hydroxylate fatty acids with a putative role in plant defense. While *Ta*CYP709C1 has not been shown to have roles in herbicide metabolism, its expression in wheat is induced by the safener naphthalic acid anhydride and by phenobarbital, suggesting a potential role in inducible chemical stress tolerance.[59] Therefore, the increased transcript expression of *Am*CYP709s in the NTSR1 and NTSR2 populations might be an indication that general enhanced chemical stress tolerance is another NTSR mechanism apart from EMR that protects NTSR black grass populations.[117]

4.5 Roles for Glutathione-*S*-Transferases in NTSR

Plant glutathione-*S*-transferases (GSTs (EC 2.5.1.18) are a superfamily of enzymes that utilize the tripeptide (gamma-glutamyl-cysteinyl-glycine) as a cofactor, or co-substrate, to catalyze a range of reactions, most typically linked to cytoprotection. In crops and weeds, GSTs are part of the four plant-specific classes, namely, Tau (U), Phi (F), Lambda (L), and dehydroascorbate reductase (DHAR) and two classes (i.e., Theta and Zeta) found in all eukaryotes.[57] GSTU and GSTF are the most abundant classes and most relevant to this chapter, as they catalyze the S-glutathionylation and detoxification of xenobiotics including herbicides.[58] Both classes have additional activities as glutathione-dependent peroxidases (GPOXs), reducing potentially toxic organic hydroperoxides to the respective monohydroxy alcohols and affording plants protection from oxidative stress.[60] The other GST classes do not catalyze the conjugation of xenobiotics, having roles in primary metabolism (GSTT, lipid peroxidation; GSTZ, tyrosine degradation) or antioxidant recycling (GSTL, DHAR). These non-detoxifying GSTs are not classically linked to herbicide tolerance in black grass, though GSTLs are both chemically responsive and upregulated in NTSR plants.[24]

GSTUs and GSTFs have been shown to catalyze the S-glutathionylation of multiple herbicide chemistries, including members of the triazine, chloroacetanilide, aryloxyphenoxypropionate, sulfonylurea, thiocarbamate, and diphenyl ether classes.[61] Typically, this conjugation achieves a total detoxification of the respective herbicide, and it has led to GSTs being implicated in controlling the selectivity of herbicides in cereal crops and soybean.[62,63] As a natural extension to this, GSTs have also been linked to herbicide resistance in annual ryegrass,[64] foxtail (*A. aequalis* Sobol.),[65] and wild oat.[59] In black grass, GST activities toward the herbicides metolachlor, fluorodifen, and fenoxaprop-*P*-ethyl were elevated in the NTSR Peldon and Lincs E1 populations as compared with HS plants, and this was associated with the accumulation of related 25, 27, and 28 kDa polypeptides.[66] It was subsequently concluded that the high concentration of glutathione and an enhanced GST activity in the NTSR populations were partially responsible for the resistance to fenoxaprop-*P*-ethyl in these populations.[67] Similarly, GST polypeptides of 27.5 kDa were found in HS and Peldon plants, with an additional 30 kDa subunit observed only in the NTSR plants.[68,69] GST activity in black grass was found to be regulated by both plant age and exposure to environmental stress,[70,71] as well as by exposure to safeners such as flurazole, fenclorim, and benoxacor.[67] Detailed biochemical studies on the types of GSTs in black grass revealed a tau class enzyme (termed GST1) present in both HS and NTSR plants and a phi GST2 that was enhanced in resistant populations.[23] Whereas GST1 showed activity toward herbicides, GST2 showed a limited ability to *S*-glutathionylate xenobiotics but instead had a high GPOX activity. It was proposed that GST2 functioned in NTSR by scavenging the hydroperoxides formed during herbicide (paraquat, chlorotoluron, or fluorodifen) injury, thereby preventing membrane oxidation and phytotoxicity.[23] Further evidence of a link between resistance and antioxidant responses was subsequently reported through the observation that, as compared with HS, NTSR plants accumulated thiol and flavonoid antioxidants.[37] GST2 was later cloned and formerly classified as *Am*GSTF1, with its importance in NTSR clearly demonstrated through its transformation of *Arabidopsis* conferring tolerance to multiple herbicides.[24] Similarly, the suppression of expression of *Am*GSTF1 in black grass using virus-induced gene silencing (VIGS) reversed the NTSR phenotype.[72] In a large 190 population study, combined genetic studies showed that the relationship between the levels of *Am*GSTF1 protein expression and the degree of NTSR shown toward herbicides was quantitative.[19] Therefore, there is an overwhelming case that *Am*GSTF1 is a key quantitative regulator of NTSR in black grass, with orthologs apparently having similar roles in annual ryegrass,[24] brome,[32] and wild oat.[73]

Further insight into the role of GSTs in NTSR in black grass has been provided through detailed transcriptomic and proteomic studies.[21,33] Proteomic data identified two isoforms of the *Am*GSTF1 in the Peldon (NTSR1) population, as well as in the black grass populations experimentally selected to be resistant to pendimethalin (NTSR2) and fenoxaprop-*P*-ethyl (NTSR3). The significance of these multiple forms of *Am*GSTF1 is not obvious, as all isoenzymes tested to date have identical GST activities toward xenobiotics and endogenous metabolites.[74] In addition, the additional phi enzyme *Am*GSTF2 and a tau protein termed *Am*GSTU2 were also identified by proteomics in the NTSR1 and NTSR2 plants, but not in NTSR3 biotype.[21] *Am*GSTU2 was subsequently shown to be active in detoxifying fenoxaprop-*P*-ethyl and the chloroacetanilide metolachlor, suggesting that unlike the GSTFs, this tau enzyme was directly involved in NTSR through its functions in herbicide detoxification.[75] The only differentially expressed GSTU in NTSR3, when compared with the HS population, was GSTU11 that was downregulated and in a phylogenetic framework clustered with *Zm*br2 from maize reported as a key protein involved in the transfer of the natural product cyanidin-3-glucoside into the vacuole (Fig. 4.4).[76]

*Am*GSTU6 (upregulated in both NTSR1 and NTSR2) was part of a monophyletic group supported with maximum bootstrap that includes genes from ryegrass upregulated in populations resistant to diclofop-methyl.[53] The second NTSR-linked monophyletic group included *Am*GSTU8, *Am*GSTU10 (upregulated in NTSR2), *Am*GSTU2b, and *Am*GSTU2a (upregulated in NTSR1), which clustered with wheat *Ta*GSTU4. *Ta*GSTU4 was previously reported to exhibit detoxifying activity against multiple herbicides.[77] Within this clade, *Am*GSTU2a, *Am*GSTU2b, *Am*GSTU8, and *Am*GSTU10 were most likely derived from a lineage-specific duplication, as they clustered together in the phylogenetic tree and shared high level of similarity (91–95% identity).

Overall, there were clear differences in the number of GSTUs upregulated between the two EMR populations (NTSR1 and NTSR2) and NTSR3 that might explain the differences in herbicide resistance profiles,[33] with the phylogenies confirming a closer evolutionary relationship of the proteins identified in NTSR1 and NTSR2 with homologs from other species involving in herbicide detoxification.

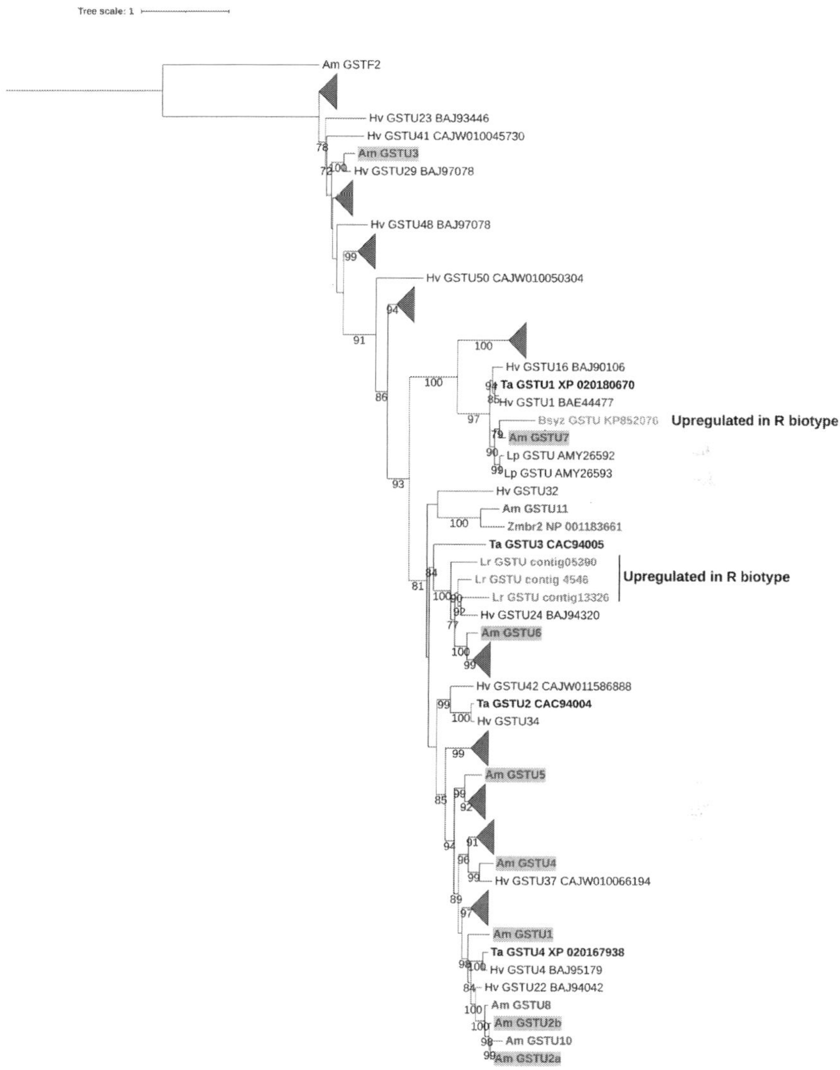

Figure 4.4 Phylogenetic analysis of GSTUs proteins in grass weeds and crop plants. Amino acid sequences of GSTUs from black grass (Am, *Alopecurus myosuroides*), barley (Hv, *Hordeum vulgare*), wheat (Ta, *Triticum aestivum*), maize (Zm, *Zea mays*), annual ryegrass (Lr, *Lolium rigidum*), perennial ryegrass (Lp, *Lolium perenne*), and American slough grass (Bsyz, *Beckmannia syzigachne*) were used for maximum likelihood alignment for phylogenic analysis. The number on the branch represents the bootstrap support values above 70%. The scale bar indicates the inferred number of substitutions per site. Clades comprising exclusively barley sequences were collapsed into triangles. (*See insert for color representation of the figure.*)

4.6 Roles for UDP-Glucose-Dependent Glycosyltransferases in NTSR

UDP-glycosides, notably UDP-glucose-dependent glycosyltransferases (EC 2.4.1), catalyze the conjugation of sugars to a wide range of aglycone acceptors. This conjugation generally reduces the toxicity and enhances the solubility of both plant secondary metabolites and xenobiotics.[78] In terms of herbicide metabolism, UGTs catalyze the glycosylation of a large number of phase 1 metabolites offering an alternative route to xenobiotic bioconjugation and detoxification to glutathione conjugation (Fig. 4.1B). The glycosylated products are then sequestered into the vacuole following the action of active transporters, notably by ABC transporters. UGTs are encoded by a diverse gene family, with more than 120 related genes identified in *Arabidopsis*,[46] and a great diversity determined between plants.[79] The phylogeny of the family is based on a family number that shows >45% similarity, a letter for the subfamily (sharing 60% or more), and final number on the individual gene identity.[75] UGTs catalyze the glycosylation of a variety of functional groups present in herbicide metabolites, including —COOH, —OH, —NH_2, and —SH moieties to give the respective carboxy ester and *O*-, *N*-, and *S*-ether conjugates.[80,81] Importantly, glycosylation cannot always be considered a final route of detoxification, with the formation of carboxy esters of herbicides such as diclofop-methyl, leading to labile conjugates that can then undergo hydrolysis re-releasing the phytotoxic acid.[82,83]

The role of glycosylation in EMR in black grass has received less attention than that directed to GSTs and CYPs. Intriguingly, a broader role for UGTs in NTSR has been inferred through the enhancement in flavonoid *O*- and *C*-glycosides observed in resistant black grass.[37,84] When UGT activity toward the flavonol quercetin and the xenobiotic 2,4,5-trichlorophenol was measured in crude extracts of HS and NTSR black grass, activities were determined to be two- to sixfold higher in the resistant population.[85] A UGT with activity toward 3-hydroxyflavone was subsequently identified that was enhanced in NTSR1 Peldon plants,[86] further suggesting that the expression of these enzymes is elevated in NTSR black grass.

To further investigate the types of UGTs enhanced by NTSR, the transcriptome data generated from studies with NTSR2 and NTSR3 plants were analyzed.[33] Overall, UGTs from the families 710C, 73B, 74F, 75E, 75K, 83A, 84C, 85A, 88B, 89B, and 90A as well as 7-deoxyloganetin glucosyltransferase; anthocyanidin 3-*O* glycosyltransferases; anthocyanidin 5,3-*O* glycosyltransferases; *O*-glycosyltransferases active toward *cis*-zeatin, crocetin, and scopoletin; and homogentisic acid (HGA) glucosyltransferases were all differentially expressed. While the majority of UGTs were downregulated, there were clear differences in the number of UGTs upregulated in NTSR2 (nine) and NTSR3 (three). The higher number of upregulated UGTs in NTSR2 population suggests a potential role for this enzyme family in

EMR. However, the function of individual enzymes in herbicide detoxification in NTSR populations remains largely unknown.

4.7 Potential Roles for ABC Transporters in NTSR

Within the context of NTSR in wild grasses, relatively little attention has been paid to the role of transporters responsible for the efflux or sequestration of herbicides and their metabolites in conferring resistance, despite their well-characterized role in antibiotic tolerance in prokaryotes and drug resistance in mammals. As such in this section of the chapter, additional background on the known biology of transporters linked to herbicide tolerance has been included, notably in the description of the ABC proteins. ABC proteins are encoded by a large gene family and transport a wide range of endogenous and xenobiotic ligands across membranes through a mechanism driven by ATP hydrolysis.[87] The ABC transporters are characterized by two transmembrane domains (TMDs) and two nucleotide-binding domains (NBDs).[83] Each TMD usually contains six transmembrane helices forming the translocation pathway across the membrane bilayer as well as the ligand-binding site, while the NBD binds and hydrolyzes ATP.[86,88,89] The full-size ABC transporters contain all four elements in a single polypeptide chain organized as TMD-NBD-TMD-NBD or NBD-TMD-NBD-TMD configurations, while half-size transporters combine two distinct TMD-NBD polypeptides.

NBDs are highly conserved in structure and sequence and contain motifs for the binding and hydrolysis of ATP, namely, the P-loop or Walker A motif (GXXGXGK(S/T)), a Walker B motif (ϕϕϕϕD, of which ϕ is a hydrophobic residue), a Q-loop that couple NBD to TMD events, and an ABC signature motif LSGGQ.[90,91] Unlike the NBDs, the TMDs generally share a similar topology within a transporter class but display no significant sequence conservation relating to the broad range of substrates that they transport.[92] ATP hydrolysis on the NBD drives conformational changes in the TMDs, resulting in alternating access to the two sides of the transporter driving unidirectional ligand movement across the lipid bilayer.[89] According to domain organizations and primary sequence, eukaryotic ABC transporters have been classified into eight subfamilies, from ABC-A to ABC-H, with an additional group proposed to include "prokaryotic-type" transporters.[93]

ABC transporters have a well-established role in drug elimination. In multidrug resistance (MDR) cancer cells in humans, the increased expression of members of this enzyme family, such as ABC-B, ABC-C, and ABC-G subfamilies, reduces the efficacy of cancer chemotherapy.[94,95] ABC transporters also play a key role in resistance to insecticides and antimicrobial drugs resistance in a broad range of insect pests and parasitic protozoa, including major vectors of disease in crops and animals.[92,96,97] Terrestrial plants have a large set of ABC proteins, for example, *Arabidopsis* and rice contain 130 and 133 open reading frames (ORFs), respectively, compared to 49 in humans.[97,98] This large number of transporters is believed to be a result of the diversity of plant secondary metabolism with its large numbers of natural products that need to be selectively transported for use in biological processes, including development, signaling, and defense against biotic and abiotic stress.[99] Plant ABC transporters that are localized in the plasma membrane and the membranes of tonoplasts, chloroplasts, mitochondria, and peroxisomes[97,100] are known to be involved in diverse functions, including phytohormone transport,[101] cuticle and pollen coat deposition,[102] transport of secondary metabolites,103 and defense against pathogens.[104,105]

Most ABC-C transporters in *Arabidopsis* localize to the vacuole membrane[103] with members shown to transport glutathione conjugates of *S*-(2,4-dinitrophenol) and the herbicide metolachlor.[106,107] Increasing evidence points to a key role for these transporters in herbicide detoxification in plants, notably through the transport of the respective phase 2 metabolites conjugated with glutathione, or glucose, from the cytosol into the vacuole.[108] Multiple substrate ligand sites have been putatively identified within the binding cavity of the ABC protein termed *At*MRP2 from *Arabidopsis*, explaining the underlying capacity to transport diverse ligands such as glucuronide substrates and glutathione conjugates.[108] With a role in herbicide detoxification established in *Arabidopsis*, there has been an increased interest in the roles of these transporters in NTSR. Importantly, transcripts encoding ABC proteins have been shown to be enhanced in NTSR populations of several weed species, notably *Conyza canadensis*,[109] *L. rigidum*,[53,110] *A. aequalis*,[71] *Myosoton aquaticum*,[111] and *Descurainia sophia*.[112] Moreover, the upregulation of ABC transporters induced by herbicide safeners, the chemical that enhanced herbicide detoxification, in cereal crops further indicates a role for this protein family in herbicide detoxification.[108,113]

In black grass, transcriptome analysis identified several contigs corresponding to ABC proteins that were significantly upregulated in NTSR1 (Peldon) compared with HS (Rothamsted) populations.[21] However, the ABC transporters identified as being associated with NTSR1 were not significantly upregulated in the NTSR2 plants when compared with HS population.[33] Furthermore, some 10 ABC proteins (in particular ABC-C, ABC-D, ABC-G, and MDR) were actively downregulated in the NTSR3 population, strongly suggesting that the ABCs were not associated with the singular resistance mechanism of NTSR3. Detailed subsequent analysis of the ABC transporters enhanced in the NTSR1 black-grass plants demonstrated the functional involvement of two of the induced ABC-C transporters, *Am*ABCC1 and *Am*ABCC2, in tolerance to the sulfonylurea herbicide mesosulfuron.[114] Thus, heterologous expression of these *AmABCCs* in yeast significantly enhanced tolerance to the recombinant cultures to mesosulfuron. However, the mechanism by which this protection was afforded was not definable, with no enhancement in the efflux of mesosulfuron metabolites from the yeast cells determined, suggesting that the transporters may have been excluding the herbicide itself. As such, further studies with these *Am*ABCCs in *planta* are needed to define their function in transporting parent herbicides, or their metabolites, when contributing to NTSR toward mesosulfuron or other herbicides.

4.8 Conclusions

This review brings together the latest understanding of the role of the enzymes and transporters responsible for xenobiotic detoxification in EMR which are linked to the NTSR phenotype in black grass. The sections above pointed to the major benefits to the subject of transcriptome studies, notably the identification of potential xenome gene family members as being enhanced in NTSR populations and therefore linked to resistance. The next major challenge will be the functional characterization of the individual proteins with respect to detoxification activity with different herbicide chemistries and their role in plant physiology and biochemistry. Overall, it seems clear that the upregulation of the xenome in NTSR black grass is part of a preprogrammed regulatory response to plant stress. As such, there is an intriguing link between herbicide safening in cereals and NTSR in weeds such as black grass. Many of the postemergence herbicide formulations used for black grass control contain both sulfonylureas, and FOPs are co-formulated with safeners for use in the field.[36,115] As NTSR in weeds and herbicide safening in cereal crops involve the enhanced expression of CYPs, GSTs, and ABC proteins, there has been interest in the link between safener use and the evolution of NTSR. It has been postulated that the continuous application of safeners in weeds could accelerate the evolution of NTSR through epigenetic effects controlling the expression of xenome proteins.[22] However, safeners have been shown to functionally induce very modest changes in the composition of GSTs, UGTs and CYPs in HS black grass,[21,117] which result in only small enhancements in the rate of metabolism of herbicides such as fenoxaprop-*P*-ethyl.[37] This suggests that the ability to directly upregulate detoxification capacity by safeners in black grass is inherently limited. Detailed studies have further demonstrated that while safeners can enhance herbicide detoxifying GSTs in HS plants, this is not the case in the NTSR1 (Peldon) population.[37] While the link between safening and NTSR is not as clear as might be anticipated, emerging evidence does suggest a close link between this type of herbicide resistance and tolerance to biotic and abiotic stress such as fungal infection or cold and osmotic stresses,[21,33] with even a potential joint association with transposable elements that can be activated by environmental stresses.[33] A further link to NTSR involving the infection of resistant populations of black grass with latent viruses suggested that while infection could not be directly correlated with resistance, it might provide a route for its enhanced evolution based on the de-regulation of protective gene networks.[116] Understanding this link between EMR and stress responses in black grass is now seen as a major focus for future research that will provide new insight into the mechanisms underpinning NTSR and the evolution of these resistance traits in the field.

Acknowledgments

The authors acknowledge funding from the Biotechnology and Biological Sciences Research Council (BBSRC grants BB/S005617/1 and BB/L001489/1 and CASE studentship awarded to SB).

References

1 Gwynne D, and Murray R, *Weed Biology and Control in agriculture and Horticulture*, Batsford Academic and Educational, London, UK, p. 258 (1985).
2 Langer RHM, and Hill GD, *Agricultural Plants*, Second edition, Cambridge University Press, Cambridge, UK, p. 387 (1991).
3 Moss SR, The agro-ecology and control of black-grass, *Alopecurus myosuroides* Huds., in modem cereal growing systems. *ADAS Quarterly Review* **38**:170–191 (1981).
4 Moss SR, Cocker KM, Brown AC, Hall L, and Field LM, Characterisation of target-site resistance to ACCase-inhibiting herbicides in the weed *Alopecurus myosuroides* (black-grass). *Pest Manag Sci* **59**:190–201 (2003).
5 Behrendt S, and Hanf M, *Grass Weeds in World Agriculture*, BASF Aktiengesellschaft, Ludwigshaten am Rhein, Germany (1979).
6 Chancellor RJ, Book review: weed biology and control in agriculture and horticulture. *Outlook Agr* **14**:150 (1985).
7 Moss SR, Competition between black-grass *(Alopecurus myosuroides)* and winter wheat. *Proc Brit Crop Prot Conf- Weeds* **2**:367–374 (1987).
8 Rieger MA, Lamond M, Preston C, Powles SB, and Roush RT, Pollen-mediated movement of herbicide resistance between commercial canola fields. *Science* **296**:2386–2388 (2002).
9 Busi R, Yu Q, Barrett-Lennard R, and Powles S, Long distance pollen-mediated flow of herbicide resistance genes in *Lolium rigidum*. *Theor Appl Genet* **117**:1281–1290 (2008).
10 Varah A, Ahodo K, Coutts SR, Hicks HL, Comont D, Crook L, *et al.*, The costs of human-induced evolution in an agricultural system. *Nat Sustain* **3**:63–71 (2020).

11 Hall LM, Moss SR, and Powles SB, Mechanisms of resistance to aryloxyphenoxypropionate herbicides in two resistant biotypes of *Alopecurus myosuroides* (blackgrass): herbicide metabolism as a cross-resistance mechanism. *Pestic Biochem Physiol* **57**:87–98 (1997).

12 Heap I, Global perspective of herbicide-resistant weeds. *Pest Manag Sci* **70**:1306–1315 (2014).

13 Hess M, Beffa R, Kaiser J, Laber B, Menne H, and Strek H, Status and development of ACCase and ALS resistant black- grass (Alopecurus myosuroides Huds.) in neighboring fields in Germany. *Julius Kühn Archiv* **434**:1–8 (2012).

14 Délye C, Michel S, Bérard A, Chauvel B, Brunel D, Guillemin JP, Dessaint F, and Le Corre V, Geographical variation in resistance to acetyl-coenzyme A carboxylase-inhibiting herbicides across the range of the arable weed *Alopecurus myosuroides* (black-grass). *New Phytol* **186**:1005–1017 (2010).

15 Gaines TA, Zhang W, Wang D, Bukur B, Chisholm ST, Shaner DL, *et al.*, Gene amplification confers glyphosate resistance in *Amaranthus palmeri. PNAS* **107**:1029–1034 (2010).

16 Salas RA, Dayan FE, Pan Z, Watson SB, Dickson JW, Scott RC, and Burgos NR, EPSPS gene amplification in glyphosate-resistant Italian ryegrass (*Lolium perenne* ssp. *multiflorum*) from Arkansas. *Pest Manag Sci* **68**:1223–1230 (2012).

17 Keshtkar E, Mathiassen SK, Moss SR, and Kudsk P, Resistance profile of herbicide-resistant *Alopecurus myosuroides* (black-grass) populations in Denmark. *Crop Prot* **69**:83–89 (2015).

18 Vrbničanin S, Pavlović D, and Božić D, Weed resistance to herbicides, In Pacanoski Z (Ed.), *Herbicide Resistance in Weeds and Crops*, IntechOpen, http://dx.doi.org/10.5772/65569 (2017).

19 Comont D, Lowe C, Hull R, Crook L, Hicks HL, Onkokesung N, *et al.*, Evolution of generalist resistance to herbicide mixtures reveals a trade-off in resistance management. *Nat Commun* **11**:3086 (2020).

20 Cocker KM, Moss SR, and Coleman JOD, Multiple mechanisms of resistance to Fenoxaprop-P-Ethyl in United Kingdom and other European populations of herbicide-resistant *Alopecurus myosuroides* (black-grass). *Pestic Biochem Physiol* **65**:169–180 (1999).

21 Tétard-Jones C, Sabbadin F, Moss S, Hull R, Neve P, and Edwards R, Changes in the proteome of the problem weed blackgrass correlating with multiple-herbicide resistance. *Plant J* **94**:709–720 (2018).

22 Délye C, Unravelling the genetic bases of non-target-site-based resistance (NTSR) to herbicides: a major challenge for weed science in the forthcoming decade. *Pest Manag Sci* **69**:176–187 (2013).

23 Cummins I, Cole DJ, and Edwards R, A role for glutathione transferases functioning as glutathione peroxidases in resistance to multiple herbicides in black-grass. *Plant J* **18**:285–292 (1999).

24 Cummins I, Wortley DJ, Sabbadin F, He Z, Coxon CR, Straker HE, *et al.*, Key role for a glutathione transferase in multiple-herbicide resistance in grass weeds. *PNAS* **110**:5812–5817 (2013).

25 Vila-Aiub MM, Balbi MC, Distéfano AJ, Fernández L, Hopp E, Yu Q, and Powles S, Glyphosate resistance in perennial *Sorghum halepense* (Johnsongrass), endowed by reduced glyphosate translocation and leaf uptake. *Pest Manag Sci* **68**:430–436 (2012).

26 Ge X, d'Avignon DA, Ackerman JJH, and Sammons RD, Rapid vacuolar sequestration: the horseweed glyphosate resistance mechanism. *Pest Manag Sci* **66**:345–348 (2010).

27 Menendez J, and de Prado R, Diclofop-methyl cross-resistance in a chlorotoluron-resistant biotype of *Alopecurus myosuroides. Pestic Biochem Physiol* **56**:123–133 (1996).

28 Hall LM, Moss SR, and Powles SB, Mechanism of resistance to chlorotoluron in two biotypes of the grass weed *Alopecurus myosuroides. Pestic Biochem Physiol* **53**:180–192 (1995).

29 Powles SB, and Yu Q, Evolution in action: plants resistant to herbicides. *Annu Rev Plant Biol* **61**:317–347 (2010).

30 Edwards R, and Dixon DP, Plant glutathione transferases. *Methods Enzymol* **401**:169–186 (2005).

31 Letouzé A, and Gasquez J, Enhanced activity of several herbicide-degrading enzymes: a suggested mechanism responsible for multiple resistance in blackgrass (*Alopecurus myosuroides* Huds.). *Agronomie* **23**:601–608 (2003).

32 Davies LR, Onkokesung N, Brazier-Hicks M, Edwards R, and Moss S, Detection and characterization of resistance to acetolactate synthase inhibiting herbicides in *Anisantha* and *Bromus* species in the United Kingdom. *Pest Manag Sci* **76**:2473–2482 (2020).

33 Franco-Ortega S, Goldberg-Cavalleri A, Walker A, Brazier-Hicks M, Onkokesung N and Edwards R, Non-target site herbicide resistance is conferred by two distinct mechanisms in black-grass (*Alopecurus myosuroides*). *Front Plant Sci* DOI: 10.3389/fpls.2021.636652 (2021).

34 Hyde RJ, Hallahan DL, and Bowyer JR, Chlorotoluron metabolism in leaves of resistant and susceptible biotypes of the grass weed *Alopecurus myosuroides. Pestic Sci* **47**:185–190 (1996).

35 Menendez J, and de Prado R, Detoxification of chlorotoluron in a chlorotoluron-resistant biotype of *Alopecurus myosuroides*. Comparison between cell cultures and whole plants. *Physiol Plant* **99**:97–104 (2006).

36 Tal A, Romano ML, Stephenson GR, Schwan AL, and Hall JC, Glutathione conjugation: a detoxification pathway for fenoxaprop-ethyl in barley, crabgrass, oat, and wheat. *Pestic Biochem Physiol* **46**:190–199 (1993).

37 Cummins I, Bryant DN, and Edwards R, Safener responsiveness and multiple herbicide resistance in the weed black-grass (*Alopecurus myosuroides*). *Plant Biotechnol J* **7**:807–820 (2009).

38 Kaiser YI, and Gerhards R, Degradation and metabolism of Fenoxaprop and Mesosulfuron + Iodosulfuron in multiple resistant blackgrass (*Alopecurus myosuroides*). *Gesunde Pflanzen* **67**:109–117 (2015).

39 Zhao N, Yan Y, Ge L, Zhu B, Liu W, and Wang J, Target site mutations and cytochrome P450s confer resistance to fenoxaprop-P-ethyl and mesosulfuron-methyl in *Alopecurus aequalis*. *Pest Manag Sci* **75**:204–214 (2019).

40 Dücker R, Zöllner P, Parcharidou E, Ries S, Lorentz L, and Beffa R, Enhanced metabolism causes reduced flufenacet sensitivity in black-grass (*Alopecurus myosuroides* Huds.) field populations. *Pest Manag Sci* **75**:2996–3004 (2019).

41 Dücker R, Parcharidou E, and Beffa R, Flufenacet activity is affected by GST inhibitors in blackgrass (*Alopecurus myosuroides*) populations with reduced flufenacet sensitivity and higher expression levels of GSTs. *Weed Sci* **68**:451–459 (2020).

42 Donald WW, and Shimabukuro RH, Selectivity of diclofop-methyl between wheat and wild oat: growth and herbicide metabolism. *Physiol Plant* **49**:459–464 (1980).

43 Kordi¢ S, Xenobiotic-hydrolysing esterases from crops and *Arabidopsis*: characterisation of a S-formylglutathione hydrolase, Durham theses, Durham University. Available at Durham E-Theses Online: http://etheses.dur.ac.uk/3182/ (2003).

44 Cummins I, Burnet M, and Edwards R, Biochemical characterisation of esterases active in hydrolysing xenobiotics in wheat and competing weeds. *Physiol Plant* **113**:477–485 (2001).

45 Cummins I, and Edwards R, Purification and cloning of an esterase from the weed black-grass (*Alopecurus myosuroides*), which bioactivates aryloxyphenoxypropionate herbicides. *Plant J* **39**:894–904 (2004).

46 Nelson D, and Werck-Reichhart D, A P450-centric view of plant evolution. *Plant J* **66**:194–211 (2011).

47 Werck-Reichhart D, and Feyereisen R, Cytochromes P450: a success story. *Genome Biol* **1**:reviews3003.1 (2000).

48 Bak S, Beisson F, Bishop G, Hamberger B, Höfer R, Paquette S, and Werck-Reichhart D, Cytochromes P450 in the arabidopsis book. *Am Soc Plant Biol* **9**:e0144 (2011).

49 Nelson DR, Cytochrome P450 nomenclature, 2004. *Methods Mol Biol* **320**:1–10 (2006).

50 Schuler MA, Duan H, Bilgin M, and Ali S, *Arabidopsis* cytochrome P450s through the looking glass: a window on plant biochemistry. *Phytochem Rev* **5**:205–237 (2006).

51 Xiang WS, Wang XJ, Ren TR, and Ju XL, Expression of a wheat cytochrome P450 monooxygenase cDNA in yeast catalyzes the metabolism of sulfonylurea. *Pestic Biochem Physiol* **85**:402–426 (2006).

52 Duhoux A, and Délye C, Reference genes to study herbicide stress response in *Lolium* sp.: up-regulation of P450 genes in plants resistant to acetolactate-synthase inhibitors. *PLoS ONE* **8**:e63576 (2013).

53 Pan G, Zhang X, Liu K, Zhang J, Wu X, Zhu J, and Tu J, Map-based cloning of a novel rice cytochrome P450 gene CYP81A6 that confers resistance to two different classes of herbicides. *Plant Mol Biol* **61**:933–943 (2006).

54 Dimaano NG, Yamaguchi T, Fukunishi K, Tominaga T, and Iwakami S, Functional characterization of cytochrome P450 CYP81A subfamily to disclose the pattern of cross-resistance in *Echinochloa phyllopogon*. *Plant Mol Biol* **102**:403–416 (2020).

55 Imaishi H, and Matumoto S, Isolation and functional characterization in yeast of CYP72A18, a rice cytochrome P450 that catalyzes (−1)–hydroxylation of the herbicide pelargonic acid. *Pestic Biochem Physiol* **88**:71–77 (2007).

56 Persans MW, Wang J, and Schuler A, Characterization of maize cytochrome P450 monooxygenases induced in response to safeners and bacterial pathogens. *Plant Physiol* **125**:1126–1138 (2001).

57 Liu YJ, Han XM, Ren LL, Yang HL, and Zeng QY, Functional divergence of the glutathione S-transferase supergene family in *Physcomitrella patens* reveals complex patterns of large gene family evolution in land plants. *Plant Physiol* **161**:773–786 (2013).

58 Frova C, Glutathione transferases in the genomics era: new insights and perspectives. *Biomol Eng* **23**:149–169 (2006).

59 Kandel S, Morant M, Benveniste I, Blée E, Werck-Reichhart D, and Pinot F, Cloning, functional expression, and characterization of CYP709C1, the first sub-terminal hydroxylase of long chain fatty acid in plants. Induction by chemicals and methyl jasmonate. *J Biol Chem* **280**:35881–35889 (2005).

60 Dixon DP, Hawkins T, Hussey PJ, and Edwards R, Enzyme activities and subcellular localization of members of the arabidopsis glutathione transferase superfamily. *J Exp Bot* **60**:1207–1218 (2009).

61 Cole DJ, Detoxification and activation of agrochemicals in plants. *Pestic Sci* **42**:209–222 (1994).

62 Hatton PJ, Dixon D, Cole DJ, and Edwards R, Glutathione transferase activities and herbicide selectivity in maize and associated weed species. *Pestic Sci* **46**:267–275 (1996).

63 Andrews CJ, Cummins I, Skipsey M, Grundy MN, Jepson I, Townson J, and Edwards R, Purification and characterisation of a family of glutathione transferases with roles in herbicide detoxification in soybean (*Glycine max* L.); selective enhancement by herbicides and herbicide safeners. *Pestic Biochem Physiol* **82**:205–219 (2005).

64 Duhoux A, Pernin F, Desserre D, and Délye C, Herbicide safeners decrease sensitivity to herbicides inhibiting acetolactate-synthase and likely activate non-target-site-based resistance pathways in the major grass weed Lolium sp. (Rye-grass). *Front Plant Sci* **8**:1310 (2017).

65 Zhao N, Li W, Bai S, Guo W, Yuan G, Wang F, Liu W, and Wang J, Transcriptome profiling to identify genes involved in mesosulfuron-methyl resistance in *Alopecurus aequalis*. *Front Plant Sci* **8**:1391 (2017).

66 Cummins I, Moss S, Cole DJ, and Edwards R, Glutathione transferases in herbicide-resistant and herbicide-susceptible black-grass (*Alopecurus myosuroides*). *Pestic Sci* **51**:244–250 (1997).

67 Sharples CR, Hull MR, and Cobb AH, The effect of herbicide safener on chlorotoluron susceptible and resistant black-grass (of *Alopecurus myosuroides*). *Proc Brit Crop Prot Conf-Weeds* **1**:359–360 (1995).

68 Reade JPH, and Cobb AH, Purification, characterisation and comparison of glutathione S-transferases from black-grass (*Alopecurus myosuroides* Huds) biotypes. *Pestic Sci* **55**:993–999 (1999).

69 Reade JP, and Cobb AH, New, quick tests for herbicide resistance in black-grass (*Alopecurus myosuroides* Huds) based on increased glutathione S-transferase activity and abundance. *Pest Manag Sci* **58**:26–32 (2002).

70 Milner LJ, Reade JP, and Cobb AH, Developmental changes in glutathione S-transferase activity in herbicide-resistant populations of *Alopecurus myosuroides* Huds (black-grass) in the field. *Pest Manag Sci* **57**:1100–1106 (2001).

71 Milner LJ, Reade JP, and Cobb AH, The effect of temperature on glutathione S-transferase activity and glutathione content in *Alopecurus myosuroides* (black grass) biotypes susceptible and resistant to herbicides. *Weed Res* **47**:106–112 (2007).

72 Mellado- Sánchez M, McDiarmid F, Cardoso V, Kanyuka K, and DR MG, Virus-mediated transient expression techniques enable gene function studies in black-grass. *Plant Phisiol* **183**:455–489 (2020).

73 Burns EE, Keith BK, Refai MY, Bothner B, and Dyer WE, Proteomic and biochemical assays of glutathione-related proteins in susceptible and multiple herbicide resistant *Avena fatua* L. *Pestic Biochem Physiol* **140**:69–78 (2017).

74 NandulaV RD, Ferhatoglu Y, Barrett M, Duke S, Dayan G-CA, Tétard-Jones C, *et al.*, Herbicide metabolism: crop selectivity, bioactivation, weed resistance, and regulation. *Weed Sci* **67**:149–175 (2019).

75 Li Y, Baldauf S, Lim EK, and Bowles DJ, Phylogenetic analysis of the UDP-glycosyltransferase multigene family of *Arabidopsis thaliana. J Biol Chem* **276**:4338–4343 (2001).

76 Marrs K, Alfenito M, Lloyd A, and Walbot V, A glutathione S-transferase involved in vacuolar transfer encoded by the maize gene bronze-2. *Nature* **375**:397–400 (1995).

77 Thom R, Cummins I, Dixon DP, Edwards R, Cole DJ, and Lapthorn AJ, Structure of a Tau class glutathione S-tranferase from wheat active in herbicdie detoxification. *Biochemistry* **41**:7008–7020 (2002).

78 Bowles D, Isayenkova J, Lim EK, and Poppenberger B, Glycosyltransferases: managers of small molecules. *Curr Opin Plant Biol* **8**:254–263 (2005).

79 Vogt T, Zimmermann E, Grimm R, Meyer M, and Strack D, Are the characteristics of betanidin glucosyltransferases from cell-suspension cultures of *Dorotheanthus bellidiformis* indicative of their phylogenetic relationship with flavonoid glucosyltransferases? *Planta* **203**:349–361 (1997).

80 Davies J, and Caseley JC, Herbicide safeners: a review. *Pestic Sci* **55**:1043–1058 (1999).

81 Owen WJ, Herbicide metabolism as a basis for selectivity, In Roberts T (Ed.), *Metabolism of Agrochemicals in Plants*, John Wiley and Sons, Chichester, UK, pp. 211–258 (2000).

82 Shimabukuro RH, Walsh WC, and Hoerauf RA, Metabolism and selectivity of diclofop-methyl in wild oat and wheat. *J Agric Food Chem* **27**:615–623 (1979).

83 Higgins CF, ABC transporters: from microorganisms to man. *Annu Rev Cell Biol* **8**:67–113 (1992).

84 Brazier MC, O-glucosyltransferases in wheat (Triticum aestivum) and the competing weed black-grass (Alopecurus myosuroides). Doctoral thesis, Durham University, UK (2003).

85 Brazier M, Cole DJ, and Edwards R, O-glucosyltransferase activities toward phenolic natural products and xenobiotics in wheat and herbicide-resistant and herbicide-susceptible black-grass (*Alopecurus myosuroides*). *Phytochemistry* **59**:149–156 (2002).

86 Beis K, Structural basis for the mechanism of ABC transporters. *Biochemical Soc T* **43**:889–893 (2015).

87 Henikoff S, Greene EA, Pietrokovski S, Bork P, Attwood TK, and Hood L, Gene families: the taxonomy of protein paralogs and chimeras. *Science* **278**:609–614 (1997).

88 Biemans-Oldehinkel E, Doeven MK, and Poolman B, ABC transporter architecture and regulatory roles of accessory domains. *FEBS Lett* **580**:1023–1035 (2006).

89 Gyimesi G, Ramachandran S, Kota P, Dokholyan NK, Sarkadi B, and Hegedűs T, ATP hydrolysis at one of the two sites in ABC transporters initiates transport related conformational transitions. *Biochim Biophys Acta* **1808**:2954–2964 (2013).

90 Schmitt L, Benabdelhak H, Blight MA, Holland IB, and Stubbs MT, Crystal structure of the nucleotide-binding domain of the ABC-transporter haemolysin B: identification of a variable region within ABC helical domains. *J Mol Biol* **330**:333–342 (2003).

91 Wilkens S, Structure and mechanism of ABC transporters. *F1000Prime Rep* **7**:14 (2015).

92 Bariami V, Jones CM, Poupardin R, Vontas J, and Ranson H, Gene amplification, abc transporters and cytochrome P450s: unraveling the molecular basis of pyrethroid resistance in the dengue vector, Aedes aegypti. *PLoS Neglect Trop D* **6**:e1692 (2012).

93 Verrier PJ, Bird D, Burla B, Dassa E, Forestier C, Geisler M, *et al.*, Plant ABC proteins – a unified nomenclature and updated inventory. *Trends Plant Sci* **13**:151–159 (2008).

94 Cole SPC, Bhardwaj G, Gerlach JH, Mackie JE, Grant CE, Almquist KC, Stewart AJ, Kurz EU, Duncan AM, and Deeley RG, Overexpression of a transporter gene in a multidrug-resistant human lung cancer cell line. *Science* **258**:1650–1654 (1992).

95 Robey RW, Polgar O, Deeken J, To KW, and Bates SE, ABCG2: determining its relevance in clinical drug resistance. *Cancer Metastasis Rev* **26**:39–57 (2007).

96 Park Y, González-Martínez RM, Navarro-Cerrillo G, Chakroun M, Kim Y, Ziarsolo P, Blanca J, Cañizares J, Ferré J, and Herrero S, ABCC transporters mediate insect resistance to multiple Bt toxins revealed by bulk segregant analysis. *BMC Biol* **12**:46 (2014).

97 Pramanik PK, Alam MN, Roy Chowdhury D, and Chakraborti T, Drug resistance in protozoan parasites: an incessant wrestle for survival. *J Glob Antimicrob Re* **18**:1–11 (2019).

98 Sánchez-Fernández R, Davies TGE, Coleman JOD, and Rea PA, The *Arabidopsis thaliana* ABC protein superfamily, a complete inventory. *J Biol Chem* **276**:30231–30244 (2001).

99 Hwang JU, Song WY, Hong D, Ko D, Yamaoka Y, Jand S, *et al.*, Plant ABC transporters enable many unique aspects of a terrestrial plant's lifestyle. *Mol Plant* **9**:338–355 (2016).

100 Francisco RM, Regalado A, Ageorges A, Burla BJ, Bassin B, Eisenach C, *et al.*, ABCC1, an ATP binding cassette protein from grape berry, transports anthocyanidin 3-O-glucosides. *Plant Cell* **25**:1840–1854 (2013).

101 Kuromori T, Miyaji T, Yabuuchi H, Shimizu H, Sugimoto E, Kamiya A, Moriyama Y, and Shinozaki K, ABC transporter AtABCG25 is involved in abscisic acid transport and responses. *PNAS* **107**:2361–2366 (2010).

102 Choi H, Ohyama K, Kim YY, Jin J-Y, Lee SB, Yamaoka Y, Muranaka T, Suh MC, Fujioka S, and Lee Y, The role of *Arabidopsis* ABCG9 and ABCG31 ATP binding cassette transporters in pollen fitness and the deposition of steryl glycosides on the pollen coat. *Plant Cell* **26**:310–324 (2014).

103 Nagy R, Grob H, Weder B, Green P, Klein M, Frelet-Barrand SJK, Brearley C, and Martinoia E, The *Arabidopsis* ATP-binding cassette protein AtMRP5/AtABCC5 is a high affinity inositol hexakisphosphate transporterinvolved in guard cell signaling and phytate storage. *J Biol Chem* **284**:33614–33622 (2009).

104 Krattinger SG, Lagudah ES, Spielmeyer W, Singh RP, Huerta-Espino J, McFadden H, Bossolini E, Selter LL, and Keller B, A putative ABC transporter confers durable resistance to multiple fungal pathogens in wheat. *Science* **323**:1360–1363 (2009).

105 Bienert MD, Siegmund SEG, Drozak A, Trombik T, Bultreys A, Baldwin IT, and Boutry M, A pleiotropic drug resistance transporter in *Nicotiana tabacum* is involved in defense against the herbivore *Manduca sexta*. *Plant J* **72**:745–757 (2012).

106 Lu YP, Li ZS, Drozdowicz YM, Hörtensteiner S, Martinoia E, and Rea PA, AtMRP2, an *Arabidopsis* ATP binding cassette transporter able to transport glutathione S-conjugates and chlorophyll catabolites: functional comparisons with AtMRP1. *Plant Cell* **10**:267–282 (1998).

107 Klein M, Burla B, and Martinoia E, The multidrug resistance-associated protein (MRP/ABCC) subfamily of ATP-binding cassette transporters in plants. *FEBS Lett* **580**:1112–1122 (2006).

108 Theodoulou FL, Clark IM, He XL, Pallett KE, Cole DJ, and Hallahan DL, Co-induction of glutathione-S-transferases and multidrug resistance associated protein by xenobiotics in wheat. *Pest Manag Sci* **59**:202–214 (2003).

109 Moretti ML, Alárcon-Reverte R, Pearce S, Morran S, and Hanson BD, Transcription of putative tonoplast transporters in response to glyphosate and paraquat stress in *Conyza bonariensis* and *Conyza canadensis* and selection of reference genes for qRT-PCR. *PLoS ONE* **12**:e0180794 (2017).

110 Duhoux A, Carrère S, Gouzy J, Bonin L, and Délye C, RNA-Seq analysis of rye-grass transcriptomic response to an herbicide inhibiting acetolactate-synthase identifies transcripts linked to non-target-site-based resistance. *Plant Mol Biol* **87**:473–487 (2015).

111 Liu W, Bai S, Zhao N, Jia S, Li W, Zhang L, and Wang J, Non-target site-based resistance to tribenuron-methyl and essential involved genes in *Myosoton aquaticum* (L.). *BMC Plant Biol* **18**:225 (2018).

112 Yang Q, Deng W, Li X, Yu Q, Bai L, and Zheng M, Target-site and non-target-site based resistance to the herbicide tribenuron-methyl in flixweed (*Descurainia sophia* L.). *BMC Genomics* **17**:551 (2016).

113 Zhang Q, Xu F, Lambert KN, and Riechers DE, Safeners coordinately induce the expression of multiple proteins and MRP transcripts involved in herbicide metabolism and detoxification in *Triticum tauschii* seedling tissues. *Proteomics* **7**:1261–1278 (2007).

114 Goldber-Cavalleri A, Onkokesung N, Franco-Ortega S, and Edwards R, ABC transporters linked to multiple herbicide resistance in blackgrass (*Alopecurus myosuroides*). *Front Plant Sci* **14**:1082761 (2023).

115 King SR, Effect of mesosulfuron rate and formulation on wild oat (*Avena Fatua*) control and malt barley tolerance. *Weed Technol* **21**:771–776 (2007).

116 Sabbadin F, Glover R, Stafford R, Rozado-Aguirre Z, Boonham N, Adams I, Mumford R, and Edwards R, Transcriptome sequencing identifies novel persistent viruses in herbicide resistant wild-grasses. *Sci Rep* **7**:41987 (2017).

117 Goldberg-Cavalleri A, Franco-Ortega S, Brown S, Walker A, Rougemont B, Sinclair J, Brazier-Hicks M, Dale R, Onkokesung N, and Edwards R, Functional characterization of cytochromes P450 linked to herbicide detoxification and selectivity in winter wheat and the problem competing weed blackgrass. *ACS Omega* **12**:12270–12287 (2025).

5

Herbicide Detoxification in Black Grass

Roland Beffa

Senior Scientist Consultant, Liederbach, Hesse, Germany

5.1 Introduction

The grass weed black grass (*Alopecurus myosuroides* Huds. or ALOMY), formerly known also as *Alopecurus agrestis* L., was first described by Linné and Salvius in 1762.[1] Practices in modern agriculture, in particular high frequency of cereals in the rotation, provided black grass the opportunity to extend its development as it is a winter annual weed well-adapted to winter cereals, especially wheat. Black-grass populations generally have two distinct emerging cohorts, a major one during autumn and a minor one in spring,[2] and therefore occur in spring crops but to a less extent. As a Poaceae,[3] black grass is wind-pollinated and characterized by protogyny. Thus, most of the times, it is cross-fertilized and classified as an obligate outcrosser and can produce hundreds of seeds.[4,5] Black grass is growing predominantly in heavy and wet soils.[6,7] Its seedling can emerge before the crop or a couple of weeks after crop emergence.

Black grass has become one of the most difficult weed species to handle in Western Europe, especially in cereals[8,9] whereas *Alopecurus aequalis* and *Apostichopus japonicus* are growing mainly in Asia (Table 5.1). In winter, wheat with already 12 black-grass plants per m^2 can cause a 5% yield loss causing a loss of 75 euros per hectare.[11] For a long time black grass was mainly controlled using herbicides, but in a few decades it has evolved resistance against up to five different herbicide modes of action[10] as for the other *Alopecurus* spp. (Tables 5.1 and 5.2). As the mechanism(s) of resistance were not characterized in most of the populations analyzed, poly-resistance was used instead of multiresistance for populations resistant to several herbicides acting with different modes of action (Tables 5.1 and 5.2). This poly-resistance can occur in different individuals in a given population or even in the same individuals or a mix of both situations.

Resistance in Weeds from Herbicide Metabolism, First Edition.
Edited by Vijay K. Nandula and Roland Beffa.

Table 5.1 *Alopecurus* spp. unique resistance cases.

Species	Number of unique cases	HRAC group resistance	Poly-resistance	Number of countries
Alopecurus myosuroides	36	1, 2, 3, 5, 15	Yes	15 (Europe)
Alopecurus aequalis	5	1, 2, 3	Yes	2 (China, Japan)
Apostichopus japonicus	4	1, 2, 5, 22	Yes	1 (China)

Source: Adapted from Heap.[10]

Table 5.2 *Alopecurus myosuroides* unique resistance cases.

Country	Number of unique cases	HRAC group resistance	Poly-resistance
Belgium	3	1, 5, 3	Yes
Czech Republic	1	2	–
Denmark	2	1,3	Yes
France	3	1, 2, 5	Yes
Germany	5	1, 2, 5, 15	Yes
Ireland	1	5	–
Israel	1	5	–
Italy	1	1	–
Netherland	4	1, 2, 5	Yes
Poland	3	1, 2	Yes
Spain	2	1, 5	Yes
Sweden	3	1, 2, 15	Yes
Switzerland	2	1, 2, 5	Yes
Turkey	1	1	–
United Kingdom	4	1, 2, 3, 5	Yes

Source: Adapted from Heap.[10]

As resistance to the postemergent herbicides, in particular those inhibiting acetyl-CoA carboxylase (ACCase, HRAC Group 1) and acetolactate synthase (ALS, HRAC Group 2), is becoming more widespread,[6,7] agronomic practices have been adapted to this situation and black-grass control gradually shifted to autumn treatments (preemergence herbicides) in winter crops. Herbicides, such as flufenacet or prosulfocarb, which are classified as inhibitors of the synthesis of very-long-chain fatty acids (VLCFAs, HRAC Group 15) are most relied on for that purpose, either alone or in mixtures.[9,12] In parallel, the use of integrated weed management (IWM), which combines chemical and nonchemical control measures, was promoted.[6,7] The most effective nonchemical tactics to control black grass and mitigate herbicide resistance evolution are ploughing, false seedbed, crop rotation with the introduction of spring crops, including spring cereals, and delayed autumn sowing.[6,7,13–15] Nevertheless, herbicide use is still the most cost-effective practice and mode of action has to be combined to mitigate resistance evolution.[16]

Resistance to herbicides is an adaptive trait and can be caused by different mechanisms occurring alone or in combination in different populations or individuals in a given population.[17] Its selection in weeds in the fields depends on many factors, e.g., gene mutation, initial frequency of resistance alleles, inheritance, weed fitness in the presence and absence of herbicides, mating system, and gene flow,[18] as well as the herbicide dose rate.[19] It can arise in weed population from mutations or pre-existing resistant genes can spread under selection pressure[20] as exemplified by the detection of a resistant-conferring mutation of the ACCase codon 1781 in a black grass present in an herbarium and collected in 1888.[21] Several resistance mechanisms have been described and can be categorized into specialist target-site resistance (TSR) and generalist non-target-site resistance (NTSR) (Table 5.3).[17,23] TSR includes mutations that alter the expression or the structure of the gene encoding the protein target of the herbicides. This is well-described for the mutations altering the coding sequence of, e.g., the ALS or the ACCase genes[23] or the increase of the copy number of the target gene like for the EPSPs gene[24] inhibited by glyphosate (HRAC Group 9). Even if it is a concern, the presence of TSR can be overcome by using herbicides acting with an alternative mode of action. NTSR mechanisms range from inhibition of the uptake or translocation of the herbicide and its increased vacuole sequestration[25,26] to enhanced herbicide detoxification[27] or other mechanisms including rapid cell death response.[28] Among these NTRS mechanisms, enhanced herbicide detoxification is a very strong concern since it can confer resistance to a broad spectrum of herbicides acting with several modes of action.[29] It can also cause unpredictable cross-resistance pattern across different herbicide modes of action including herbicides not yet marketed.[30–33] Although in certain cases metabolism-based resistance was linked with a single nucleotide polymorphisms,[32,34] the regulation of NTSR is generally considered polygenic traits.[31] Herbicide detoxification involved several pathways (Fig. 5.1) and is very complex since many genes involved are part of big gene families.[40–42] The detoxification of xenobiotics involved several steps starting with their oxidation or their activation by hydrolyses (phase I). It involves mainly cytochrome P450 (CYP) monooxygenases encoded by a gene superfamily.

Table 5.3 Number of the annotated black-grass genes belonging to gene families related to non-target-site resistance.[22]

Gene families	Number of genes
Cytochrome P450s (CYPs)	595
UDP-glycosyl transferases/UDP-glucosyl transferases (UGTs)	298
Glutathione-*S*-transferases (GSTs)	115
ATB-binding cassette (ABC) transporters	212
NADPH-dependent FMM reductases (FMMs)	30
Aldo/keto reductases (AKRs)	46
Peroxidases (PODs)	243
Peptidase family A1 (PEP)	111
Protein kinases (PKs)	427
Protein phosphatases (PPs)	130
APETALA2/ethylene-responsive factors (AP2/ERFs)	156
Myc-type/basic helix-loop-helix (bHLH)	179
3-Keto-acyl-CoA-synthase (KCS)	47

While direct biochemical activity measurement is quite difficult,[43] CYPs activity has frequently been linked with herbicide resistance, e.g., diclofop-methyl resistance in *Lolium* spp. populations.[29,35,44] The activated xenobiotics can then follow different pathways depending on their chemical characteristics (e.g., electrophilic sites) and may be detoxified by conjugation to glutathione (GSH) or glycosides (phase II).[36,37] Direct GSH conjugation without previous activation by CYPs has previously been described for several herbicides like flufenacet, *S*-metolachlor, atrazine, or S-éthyl-dipropylthiocarbamate (EPTC).[45–48] This generally leads to more hydrophilic and less toxic compounds.[36] Two main enzymes, glutathione-*S*-transferase (GST)[49,50] and glycosyl-transferase (UGT),[35,51] both encoded by a gene superfamily, are involved in phase II. Finally, in the third step, these conjugates are transported via ATP-binding cassette (ABC) transporters, encoded also by a gene superfamily, into the vacuole,[52,53] where further degradation like hydrolysis occurs via vacuolar peptidases.[54,55] The regulation and the functions of these processes as well as the role of individual isoforms are poorly understood. Nevertheless, the recent progress in weed genome sequencing[56] is given the tools to unravel in the different resistant populations about genes contributing to herbicide detoxification as well as their regulation in the resistant plants compared with the sensitive ones. This knowledge will offer to perform diagnostic and choose the best active herbicide or herbicide combinations to overcome this resistance.

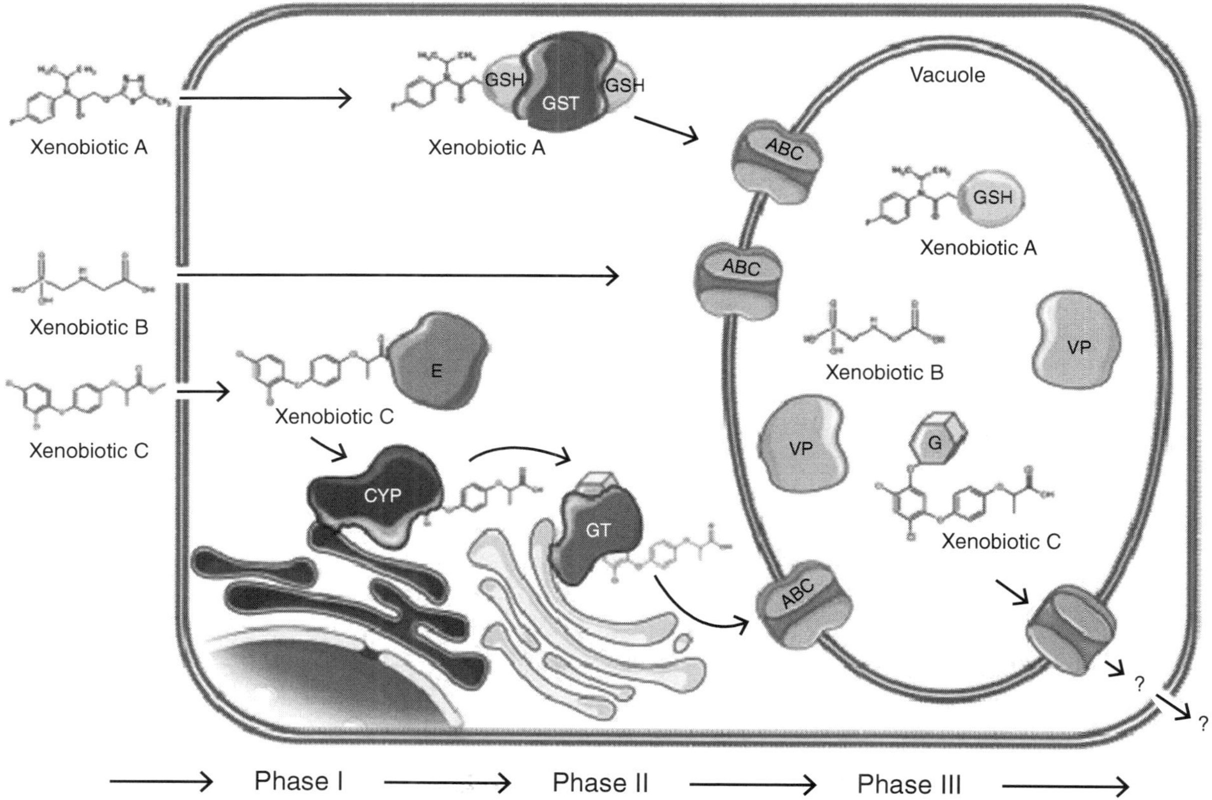

Figure 5.1 Summary of the main resistance mechanisms. Detoxification of xenobiotics in plants using the examples of flufenacet (xenobiotic A), glyphosate (xenobiotic B), and diclofop-methyl (xenobiotic C). The pathways include hydrolysis and hydroxylation catalyzed by esterases (E) and cytochrome P450 (CYP) monooxygenases in phase I; conjugation with glutathione (GSH) and glycosyl residues catalyzed by glutathione-*S*-transferases (GSTs) and glycosyl-transferases (GTs), respectively, in phase II. Phase III includes the transport into the vacuole, e.g., via ATP-binding cassette (ABC) transporters and degradation by vacuolar enzymes (VPs) as well as catabolism and compartmentalization. Source: Adapted from Refs.[35–39] (*See insert for color representation of the figure.*)

5.2 Molecular Mechanisms Involved in Herbicide Detoxification in Black Grass

Recent progress in genome sequencing and annotation[56] is an important step to find and characterize the genes involved in herbicide detoxification as well as their differences in structure and regulation between resistant and sensitive populations. Two independent black-grass sequencing projects using different populations[41,42] are sources of even complementary information. While the biochemical validation of CYP gene products are still poor in black grass, good progress has been recently done in characterizing GSTs, especially those, like flufenacet, involved in Group 15 herbicide detoxification.

5.2.1 Black-Grass GSTome

The widespread resistance of black grass to postemergent herbicides, mainly to HRAC Groups 1 and 2, has promoted an increased use of preemergent herbicides, such as flufenacet, diflufenican, pendimethalin, prosulfocarb, and aclonifen, applied in autumn.[12,57–59] Recently, cinmethylin acting with a new mode of action (HRAC Group 30) was added in the black-grass control tool box.[60] Since the use of flufenacet has gradually increased since the market introduction in 1998 in Germany,[61] selection pressure by this herbicide is also enhanced. Although this mode of action is classified as less prone to evolve resistance,[62] cases of resistance have been reported worldwide to two dicotyledons, namely, tall water hemp (*Amaranthus tuberculatus* [Moq.] JD Sauer)[63] and palmer amaranth (*Amaranthus palmeri* S. Watson)[64] and various monocotyledonous species.[10] The affected grass weeds include wild oat (*Avena fatua* L.),[65] barnyard grass (*Echinochloa crus-galli* [L.] P. Beauv.),[66] Italian ryegrass (*Lolium multiflorum* (Lam.)[67,68] and rigid ryegrass (*Lolium rigidum* Gaud.).[69,70] In parallel, it was observed that black grass evolved resistance against several VLCFA inhibitors (HRAC Group 15) including the oxyacetamide flufenacet.[39,71] TSR has so far not been described as a cause of resistance and is unlikely for the following two main reasons: (1) the putative binding site of flufenacet, the highly conserved cysteine residue in the active centers of β-keto-acyl-CoA synthases (KCSs), is crucial for the catalytic function[72,73] and (2) redundancy was observed among herbicide target enzyme family.[74,75] Thus, a single mutation in the gene(s) encoding the target is unlikely to have a significant effect on the resistance level. Instead, reduced herbicide efficacy of inhibitors of VLCFAs was linked to metabolic herbicide resistance, mostly in multiresistant populations.

Analytical studies in black grass revealed that GSTs (EC 2.5.1.18) are the enzymes catalyzing the first step of flufenacet detoxification *in vivo*.[39] This is not surprising since flufenacet acts by a nucleophilic attack of the thiol group of the

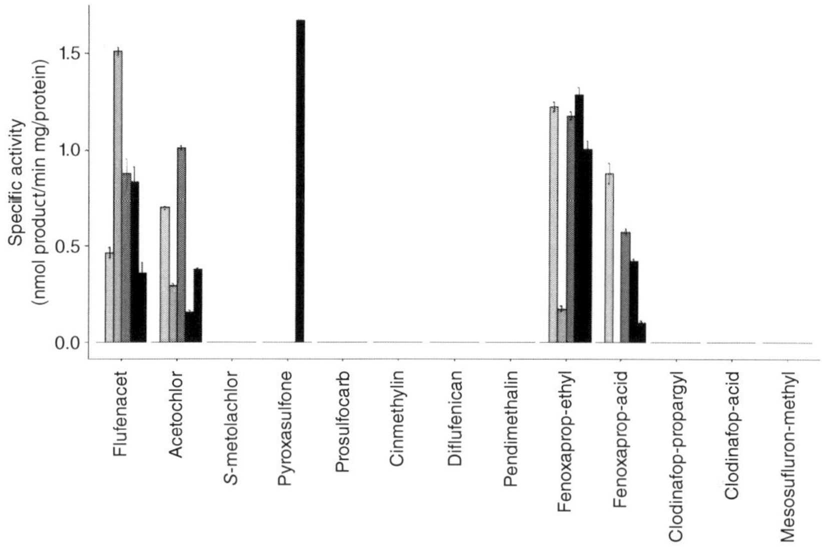

Figure 5.2 Enzymatic activity of recombinant EmGFP (□), ALOMY3G13667 (▩), ALOMY3G13668 (▩), ALOMY3G13670 (▩), ALOMY5G35766 (▩), and AmGSTF1/ALOMY3G11300 (■) on different substrates (flufenacet, acetochlor, *S*-metolachlor, pyroxasulfone, prosulfocarb, cinmethylin, diflufenican, pendimethalin, fenoxaprop-*P*-ethyl, fenoxaprop-acid, clodinafop-propargyl, clodinafop-acid, and mesosulfuron-methyl). The error bars represent the standard error of the mean ($n = 4$). (*See insert for color representation of the figure.*)

conserved cysteine in the active center of the target-condensing enzyme, resulting in a separation of the acetanilide and trifluoromethyl-thiadiazol groups.[72] Several upregulated GST genes in resistant black grass to flufenacet[76] were cloned and expressed as recombinant protein in *Escherichia coli*.[33] The biochemical activity of the respective recombinant GSTs was tested on several herbicides *in vitro* (Fig. 5.2).

All of the GSTs tested showed activity on the VLCFA inhibitor flufenacet with ALOMY3G13668 having the highest activity (1.50 nmol product/min mg/protein). Among the other herbicides of the same mode of action, acetochlor was degraded by all tested GSTs, pyroxasulfone only by the phi class GST ALOMY5G35766 (≥1.67 nmol product/min mg/protein, 100% of the substrate), and *S*-metolachlor and prosulforcarb by none of the GSTs. None of the other preemergent herbicides tested (cinmethylin, diflufenican, and pendimethalin) were degraded by the recombinant GSTs tested in the *in vitro* reactions. Among the postemergent herbicides, the ALS inhibitor mesosulfuron-methyl and the ACCase inhibitor clodinafop-propargyl and its active metabolite clodinafop-acid were not degraded by any of the GSTs tested. However, the ACCase inhibitor fenoxaprop-*P*-ethyl was detoxified by all the GSTs. The degradation rates of the active metabolite fenoxaprop-acid were in all cases lower than for the prodrug fenoxaprop-*P*-ethyl. ALOMY3G13668, which had the lowest activity on fenoxaprop-*P*-ethyl did not detoxify fenoxaprop-acid. This work showed that (1) several isoforms can detoxify a given herbicides; (2) the spectrum of each isoform can be narrow (one herbicide) or broad; (3) a given isoform does not detoxify all herbicides belonging to a given HRAC group (e.g., Group 15); (4) a given isoform can detoxify herbicides belonging to different HRAC groups (e.g., Groups 1 and 15). In addition, this study showed that if most of the GSTs catalyze the formation of the GH conjugate of the herbicide(s), one isoform ALOMY3G13668 produced flufenacet-alcohol and this was not the consequence of a mutation ASP68Glu at a position important in the active site for GSH binding.[33] This study showed that three tau and two phi class GSTs, which are differentially expressed in flufenacet-resistant black grass, were tested *in vitro* on various herbicide substrates. Thus, metabolic resistance is complex and does not necessarily confer strong resistance to a wide spectrum of herbicides. It was shown that the same enzyme can confer cross-resistance with other modes of action, while other active ingredients of the same mode of action or even chemical class may not be affected. In that regard, herbicides such as diflufenican (HRAC Group 12), pendimethalin (HRAC Group 3), prosulfocarb (HRAC Group 15), cinmethylin (HRAC Group 30), or mesosulfuron-methyl (HRAC Group 2) are particularly suitable compounds for an herbicide management to delay flufenacet resistance evolution in black-grass population showing such resistance pattern. Moreover, alternation of compounds of the same class that are not affected by cross-resistance, such as flufenacet and *S*-metolachlor, is preferable to repetitive use of the same compound. A sustainable alternation of active ingredients to slow down resistance evolution can therefore not only be based on the classification of herbicides by their mode of action but rather on a classification by their resistance mechanisms, detoxification pathways, and observed resistance patterns. In a further study,[33] the analysis of the whole black-grass GSTome showed that several GSTs active on herbicide resistance are localized in clusters in the genome suggesting a possible co-regulation. It was also shown in that study that a repressor active in the sensitive plants could be the cause of the constitutive upregulation of these GSTs in the flufenacet-resistant plants. It was suggested to use overexpressed GSTs, like AmGSTF1, to perform rapid diagnostics to detect herbicide resistance in black grass.[77,78] Nevertheless, AmGSTF1 is not overexpressed in all resistant populations, even multiresistant populations to mesosulfuron-methyl, fenoxaprop-*P*-ethyl, and flufenacet, as shown by Parcharidou et al.[22] Rather to rely on one genetic marker, it would be more reliable to define a set of markers, including GSTs, CYPs, and UGTs, especially since herbicide resistance can evolve in a nonparallel way.[41] A better knowledge of the weed genomes will greatly help progress in that direction.[56]

5.2.2 Black-Grass CYPs, ABC Transporters, and UGTs

The black-grass genome sequencing has allowed to annotate the genes belonging to the different gene families thought to be related to NTSR.[22,41] CYPs superfamily was represented by 595 genes, ABC transporters by 212 genes, and UGTs by 298 genes.[22] So far, indirect evidence or analytical data[34] have suggested that CYPs might be involved in herbicide detoxification in black grass, in particular for sulfonylurea herbicides (HRAC Group 2). So far, no biochemical data on the activity of a given isoform are available for black grass. Other species are more advanced like for *L. rigidum*,[29] *Echinochloa phyllopogon*,[43,79] or *A. palmeri*.[80,81] Recent studies related to CYPs[82] or ABC transporters[83] tried to show indirect correlation between expression and resistance by herbicide detoxification. Even less biochemical evidence are available for UGT isoforms.

5.3 Perspectives

The progress in genome sequencing and knowledge[41,42,56] will allow to study the biochemical activity of each enzyme isoforms of the different genes involved in herbicide detoxification in black grass. It is not obvious to transfer data from one species to the other. In addition, between two resistant populations to the same herbicides, different genes might be involved.[22] The knowledge of the genome structure and the gene promoters will allow to study their regulation in herbicide-resistant populations compared with the susceptible ones. In addition, a better knowledge on the biochemical function of the different GSTs, CYPs, UGTs, and ABC transporters will allow (1) to better understand their involvement, if any, in plant fitness,[84] (2) to develop better diagnostics using molecular markers to improve the choice of strategies to mitigate and/or overcome herbicide resistance,[16] (3) to improve resistance evolution predicting models[85] (4) to develop new herbicides,[86] and (5) to refine the definitions of cross-resistance and multiresistance in order so that it can be applied to the generalist herbicide metabolic resistance.[87] This can led to complete the herbicide mode of action classification valid for the specialist TSR by an additional classification according to the generalist metabolic degradation pathway involved in the different herbicide weed populations.[17]

References

1 Linné CV, Salvius L, and Linneai C, Species plantarum: exhibentes plantas rite cognitas, ad genera relates, cum differentiis specificis, nominibus trivialibus, synonymis selectis, locis natalibus, secundum systema sexuale digestas. *Impensis Direct Laurentii Salvii.* DOI: 10.5962/bhl.title.11179 (1762).

2 Barralis G, La biologie du vulpin des champs (*Alopecurus agrestis* L.). I. Dormance primaire et faculté germinative. *Rev Gen Bot* **77**:429–433 (1970).

3 Naylor REL, *Alopecurus myosuroides* Huds. (*A. Agrestis* L.). *J Ecol* **60**:611 (1972).

4 Moss SR, The production and shedding of *Alopecurus myosuroides* Huds. Seeds in winter cereals crops. *Weed Res* **23**:45–51 (1983).

5 Johnsson H, Meiotic aberrations and sterility in *Alopecurus myosuroides* Huds. *Hereditas* **30**:469–566 (1944).

6 Moss SR, Herbicide Resistance in weeds, In Hatcher PE, and Froud-Williams RJ (Eds), *Weed Research: Expending Horizons*, West Sussex John Wiley and Sons Ltd, Chichester, pp. 181–214 (2017).

7 Moss SR, Black-grass (*Alopecurus myosuroides*): why has this weed become such a problem in Western Europe and what are the solutions. *Outlook Pest Manag* **28**:207–212 (2017).

8 Keshtkar E, Mathiassen SK, Moss SR, and Kudsk P, Resistance profile of herbicide-resistant *Alopecurus myosuroides* (black-grass) populations in Denmark. *Crop Prot* **69**:83–89 (2015).

9 Moss SR, Perryman SAM, and Tatnell LV, Managing herbicide resistant blackgrass (*Alopecurus myosuroides*): theory and practice. *Weed Technol* **21**:300–339 (2007).

10 Heap I, The international herbicide-resistant weed database. Online May 25th 2024 (2024).

11 Staples J, Counting the cost of controlling blackgrass. Today's Farm May–June 2021: 18–19. (2021).

12 Bailly CG, Dale RP, Archer SA, Right DJ, and Kaundun SS, Role of residual herbicides for the management of multiple herbicide resistance to ACCase and ALS inhibitors in a black-grass population. *Crop Prot* **34**:96–103 (2012).

13 Lutman PJW, Moss SR, Cook S, and Welham SJ, A review of the effects of crop agronomy on the management of *Alopecurus myosuroides*. *Weed Res* **53**:299–313 (2013).

14 Moss SR and Lutman P, Black-grass: the potential of non-chemical control. A Rothamsted technical publication, Rothamsted Research, Harpenden, Herts AL5 2JQ, UK (2013).

15 Herrmann J, Analysis of the spatial and temporal dynamics of herbicide resistance to 446 ACCase- and ALS-Inhibitors in *Alopecurus myosuroides* Huds. and their causes. PhD thesis, 447 TU Braunschweig. Available at: https://doi.org/10.24355/dbbs.084-201701240925-0 (2016).

16 Soteres J, and Peterson M, Industry views of monitoring and mitigation of herbicide resistance. *Weed Sci* **63**:972–975 (2015).

17 Comont D, Lowe C, Hull R, Crook L, Hicks HL, *et al.*, Evolution of generalist resistance to herbicide mixtures reveals a trade-off in resistance management. *Nat Commun* **11**(3086):1–9 (2020).

18 Jaseniuk M, Brûlé-Babel AL, and Morrison IN, The evolution and genetics of herbicide resistance in weeds. *Weed Sci* **42**:176–193 (1996).
19 Neve P, and Powles S, Recurrent selection with reduced herbicide rates results in the rapid evolution of herbicide resistance in *Lolium rigidum. Theor Appl Genet* **110**:1154–1166 (2005).
20 Maxwell BD, and Mortimer AM, Selection for herbicide resistance, In Powles SB, and JAM H (Eds), *Herbicide Resistance in Plants, Biology and Biochemistry*, CRS Press, Boca Rato, pp. 1–26 (1994).
21 Délye C, Deulvot C, and Chauvel B, DNA analysis of herbarium specimens of the grass weed *Alopecuros myosuroides* reveals herbicide resistance pre-dated herbicides. *PLoS One* **8**:e75117 (2013).
22 Parcharidou E, Dücker R, and Beffa R, Genome-wide study of glutathione transferases and their regulation in flufenacet susceptible and resistant black-grass (*Alopecurus myosuroides* Huds.). *Pest Manag Sci* **80**:3035–3046 (2024).
23 Powles SB, and Yu Q, Evolution in action: plants resistant to herbicides. *Annu Rev Plant Biol* **61**:317–347 (2010).
24 Gaines TA, Zhang W, Wang D, Bukun B, Chisholm ST, *et al.*, Gene amplification confers glyphosate resistance in *Amaranthus palmeri. Proc Natl Acad Sci USA* **107**:1029–1034 (2010).
25 Shaner DL, Role of translocation as a mechanism of resistance to glyphosate. *Weed Sci* **57**:118–123 (2009).
26 Ge X, d'Avignon DA, Ackerman JJ, and Sammons RD, Rapid vacuolar sequestration: the horseweed glyphosate resistance mechanism. *Pest Manag Sci* **66**:345–348 (2010).
27 Délye C, Gardin JAC, Boucansauld K, Chauvel B, and Petit C, Non-target-site-based resistance should be the center of attention for herbicide resistance research: *Alopecurus myosuroides* as an illustration. *Weed Res* **51**:433–437 (2011).
28 Van Horn CR, Moretti ML, Robertson RR, Segobye K, Weller SC, *et al.*, Glyphosate resistance in *Ambrosia trifida*: part 1. Novel rapid cell death response for glyphosate. *Pest Manage Sci* **74**:1071–1078 (2018).
29 Han H, Yu Q, Beffa R, Gonzalez S, and Maiwald F, Cytochrome P450 CYP81A10v7 in *Lolium rigidum* confers metabolic resistance to herbicides across at least five modes of action. *Plant J* **105**:79–92 (2021).
30 Beckie HJ, and Tardif FJ, Herbicide cross resistance in weeds. *Crop Prot* **35**:15–28 (2012).
31 Délye C, Unravelling the genetic basis of non-target-site-based resistance (NTSR) to herbicides: a major challenge for weed science in the forthcoming decade. *Pest Manag Sci* **69**:176–187 (2012).
32 Busi R, Resistance to herbicides inhibiting the biosynthesis of very-long-chain fatty acids. *Pest Manag Sci* **70**:378–384 (2014).

33 Parcharidou E, Dücker R, Zöllner P, Ries S, Orru R, and Beffa R, Recombinant glutathione transferases from flufenacet resistant black-grass (*Alopecurus myosuroides* Huds.) form different flufenacet metabolites and differ in their interaction with pre- and post-emergence herbicides. *Pest Manag Sci* **79**:3376–3386 (2023).

34 Beffa R, Figge A, Lorentz L, Hess M, Laber B, Ruiz-Santaella JP, and Strek H, Weed resistance diagnostic technologies to detect herbicide resistance in cereal-growing areas. A review. *Julius-Kühn-Archiv* **434**:75–80 (2012).

35 Gaines TA, Lorentz L, Figge A, Herrmann J, Maiwald F, *et al.*, RNAseq transcriptome analysis to identify genes involved in metabolism-based diclofop resistance in *Lolium rigidum*. *Plant J* **78**:865–876 (2014).

36 Coleman J, Blake-Calff M, and Davies E, Detoxification of xenobiotics by plants: chemical modification and vacuolar compartmentation. *Trends Plant Sci* **2**:144–151 (1997).

37 Yuan JS, Tranel PJ, and Stewart CN, Non-target-site herbicide resistance: a family business. *Trends Plant Sci* **12**:6–13 (2007).

38 Sammons RD, and Gaines TA, Glyphosate resistance: state of knowledge. *Pest Manag Sci* **70**:1367–1377 (2014).

39 Dücker R, Zöllnen P, Lümmen P, Riess S, Collavo A, and Beffa R, Glutathione transferase plays a major role in flufenacet resistance of ryegrass (*Lolium* spp.) field populations. *Pest Manag Sci* **75**:3084–3092 (2019).

40 Dücker R, Molecular mechanisms of flufenacet resistance in grass weeds. University of Göttingen. Thesis (2018).

41 Cai L, Comont D, MacGregor D, Lowe C, Beffa R, *et al.*, The blackgrass genome reveals pattern of non-parallel evolution of polygenic herbicide resistance. *New Phytol* **237**:1891–1907 (2023).

42 Kersten S, Chang G, Huber CD, Voichek Y, Lanz C, *et al.*, Standing genetic variation fuels rapid evolution of herbicide resistance in blackgrass. *Proc Natl Acad Sci USA* **120**:e2206808120 (2023).

43 Dimaano NG, and Iwakami S, Cytochrome P450-mediated herbicide metabolism in plants: current understanding and prospects. *Pest Mang Sci* **77**:22–32 (2021).

44 Yu Q, and Powles SB, Metabolism based herbicide resistance and cross-resistance in crop weeds: a threat to herbicide sustainability and crop production. *Plant Physiol* **166**:1106–1118 (2014).

45 Lamoureux GL, Shimabukuro RH, Swanson HR, and Frear DS, Metabolism of 2-chloro-4-ethylamino-6-isopropylamino-s-triazine (atrazine) in excised sorghum leaf sections. *J Agric Food Chem* **18**:81–86 (1970).

46 Ezra G, and Stephenson GR, Comparative metabolism of atrazine and EPTC in Proso millet (*Panicum millaceum* L.) and corn. *Pestic Biochem Physiol* **24**:207–212 (1985).

47 Bieseler B, Fedtke C, Neuefeind W, Etzel W, Prade L, and Reinemer P, Maize selectivity of FOE 5043: breakdown of active ingredient by glutathione-S-transferases. *Pflanzenschutz-Nachrichten Bayer* **50**:117–142 (1997).

48 Dixon D, Cole DJ, and Edwards R, Characterization of multiple glutathione transferases containing the GST I subunit with activities toward herbicide substrates in maize (Zea mays). *Pestic Sci* **50**:72–82 (1997).

49 Cummins I, Dixon DP, Freitag-Pohl S, Skipsey M, and Edwards R, Multiple roles for plant glutathione transferases in xenobiotic detoxification. *Drug Metab Rev* **43**:266–280 (2011).

50 Cummins I, Wortley DJ, Sabbadin F, He Z, Coxon CR, *et al.*, Key role for a glutathione transferase in multiple-herbicide resistance in grass weeds. *Proc Natl Acad Sci* **110**:5812–5817 (2013).

51 Tal A, Romano ML, Stephenson GR, Schwan AL, and Hall JC, Glutathione conjugation: a detoxification pathway for fenoxaprop-ethyl in barley, crabgrass, oat, and wheat. *Pestic Biochem Physiol* **46**:190–199 (1993).

52 Bartholomew DM, Van Dyk DE, Law SMC, O'Keefe DP, Rea PA, and Viitanen PV, Alternate energy-dependent pathways for the vacuolar uptake of glucose and glutathione conjugates. *Plant Physiol* **130**:1562–1572 (2002).

53 Klein M, Burla B, and Martinoia E, The multidrug resistance-associated protein (MRP/ABCC) subfamily of ATP-binding-cassette transporters in plants. *FEBS Lett* **580**:1112–1122 (2006).

54 Wolf AE, Dietz KJ, and Schroder P, Degradation of glutathione S-conjugates by a carboxypeptidase in the plant vacuole. *FEBS Lett* **384**:31–34 (1996).

55 Chronopoulou E, Ataya FS, Pouliou F, Perperopoulou F, Georgakis N, Nianiou-Obeidat I, *et al.*, Structure, evolution and functional roles of plant glutathione transferases, In Hossain M, Mostafa M, Diaz-Vivancos T, Burritt D, Fugita M, and Tran LS (Eds), *Glutathione in Plant Growth, Development, and Stress Tolerance*, Springer, Cham, Switzeland, pp. 195–213 (2017).

56 Montgomery J, Morran S, MacGregor DR, McElroy JS, Neve P, *et al.*, Current status of community resources and priorities for weed genomics research. *Genome Biol* **25**. DOI: 10.1186/s13059-024-03274-y (2024).

57 Bayer (b), 2022 online, Available on https://cropscience.bayer.co.uk/aclonifeneffect [Accessed December 27, 2022] (2023).

58 Bayer (c), 2022 online, Available on https.//cropscience.bayer.co.uk/our-products/herbicides/liberator/ [Accessed December 27, 2022] (2023).

59 Klingenhagen G, Comparison of different black-grass populations (*Alopecurus myosuroides* Huds.) in their susceptibility to herbicides under field conditions. *Julius Kuhn Archiv* **434**:81–87 (2012).

60 Messelhäuser MH, Linn AI, Mathes A, Sievernich B, and Gerhards R, Development of an agar bioassay sensitivity test in *Alopecurus myosuroides f*or the pre-emergence herbicides cinmethylin and flufenacet. *Agronomy* **11**:1408 (2021).

61 BVL https://www.bvl.bund.de/DE/Arbeitsbereiche/04_Pflanzenschutzmittel/01_Aufgaben/02_ZulassungPSM/03_PSMInlandsabsatzAusfuhr/psm_PSMInlandsabsatzAusfuhr_node.html (2023).

62 Moss SR, Ulber L, and den Hoed I, A herbicide resistance risk matrix. *Crop Prot* **115**:13–19 (2019).

63 Strom SA, Gonzini LC, Mitsdarfer C, Davis AS, Riechers DE, and Hager AG, Characterization of multiple herbicide resistant waterhemp (*Amaranthus tuberculatus*) populations from Illinois to VLCFA-inhibiting herbicides. *Weed Sci* **67**:369–379 (2019).

64 Brabham C, Norsworthy JK, Houston MM, Varanasi VK, and Barber T, Confirmation of S-metolachlor resistance in palmer amaranth (*Amaranthus palmeri*). *Weed Technol* **33**:720–726 (2019).

65 Mangin AR, Hall L, and Beckie H, Triallate resistant wild oat (*Avena fatua* L.): unexpected resistance to pyroxasulfone and sulfentrazone. *Can J Plant Sci* **97**:1–6 (2017).

66 Juliano LM, Casimero MC, and Liewellyn R, Multiple herbicide resistance in barnyardgrass (*Echinochloa cruss-galli*) in direct-seeded rice in the Philippines. *Int J Pest Manag* **56**:299–307 (2010).

67 Dücker R, Zöllner P, Parcharidou E, Riess S, Lorentz L, and Beffa R, Enhanced metabolism causes reduced flufenacet sensitivity in black-grass (*Alopecurus myosuroides* Huds.) field populations. *Pest Manag Sci* **75**:2996–3004 (2019).

68 Rauch TA, Thill DC, Gersdorf SA, and Price WJ, Whitespread occurrence of herbicide resistant Italian ryegrass (*Lolium multiflorum*) in northern Idaho and eastern Washington. *Weed Tech* **24**:281–288 (2010).

69 Busi R, and Powles SB, Cross-resistance to prosulfocarb and triallate in pyroxasulfone-resistant *Lolium rigidum*: evolved cross-resistance in *L. rigidum*. *Pest Manag Sci* **69**:1378–1384 (2013).

70 Busi R, Porri A, Gaines TA, and Powles SB, Pyroxasulfone resistance in *Lolium rigidum* is metabolism-based. *Pestic Biochem Physiol* **148**:74–80 (2018).

71 Rosenhäuer M, and Petersen J, Bioassay development for the identification of pre-emergence herbicide resistance in *Alopecurus myosuroides* (Huds.) populations. *Gesunde Pflanzen* **67**:141–150 (2015).

72 Böger P, Matthes B, and Schmalfuss J, Towards the primary target of chloroacetamides – new findings pave the way. *Pest Manag Sci* **56**:497–508 (2000).

73 Krähmer H, Dücker R, and Beffa R, Inhibitors of the synthesis of very long-chain fatty acid (VLCFAs), In *Modern Plant Protection Compounds*, Wiley, pp. 351–363 (2018).

74 Tanetani Y, Ikeda M, Kaku K, Shimizu T, and Matsumoto H, Role of metabolism in the selectivity of a herbicide, pyroxasulfone, between wheat and rigid ryegrass seedlings. *J Pestic Sci* **38**:152–156 (2013).

75 Todd J, Post-Beittenmiller D, and Jaworski JG, KCS1 encodes a fatty acid elongase 3-ketoacyl-CoA synthase affecting wax biosynthesis in *Arabidopsis thaliana*. *Plant J* **17**:119–130 (1999).

76 Dücker R, Parcharidou E, and Beffa R, Flufenacet activity is affected by GST inhibitors in blackgrass (*Alopecurus myosuroides*) populations with reduced flufenacet sensitivity and higher expression levels of GSTs. *Weed Sci* **68**:451–459 (2020).

77 Onkokesung N, Brazier-Hicks M, Tetard-Jones C, Bentham H, and Edwards R, Molecular diagnostics fofr real-time determination of herbicide resistance in wild grasses. *J Biotechnol* **358**:64–66 (2022).

78 Lowe C, Onkokesung N, Goldberg A, Beffa R, Neve P, Edwards R, and Comont D, RNA and protein biomarkers for detecting enhanced metabolic resistance to herbicides mesosulfuron-methyl and fenoxaprop-ethyl in black-grass (*Alopecurus myosuroides*). *Pest Mang Sci* **80**:2539–2551 (2024).

79 Suda H, Kubo T, Yoshimto Y, Tanaka K, Tanaka S, *et al.*, Transcriptionally linked simultaneous overexpression of P450 genes for broad-spectrum herbicide resistance. *Plant Physiol* **192**:3017–3029 (2023).

80 Küpper A, Peter F, Zöllner P, Lorentz L, Tranel PJ, Beffa R, and Gaines TA, Tembotrione detoxification in 4-hydroxyphenylpyruvate dioxygenase (HPPD) inhibitor-resistant palmer amaranth (*Amaranthus palmeri* S. Wats.). *Pest Manag Sci* **74**:2325–2334 (2018).

81 Rigon CA, Küpper A, Sparks C, Montgomery J, Peter F, *et al.*, Cytochrome P450 CYP72A219 is involved in evolution of metabolic resistance to tembotrione in *Amaranthus palmeri* populations. DOI: 10.1101/2023.12.13.571468 (2023).

82 Franco-Ortega S, Goldberg-Cavalleri A, Walker A, Brazier-Hicks M, Onkokesung N, and Edwards R, Non-target site herbicide resistance is conferred by two distinct mechanisms in black-grass (*Alopecurus myosuroides*). *Front Plant Sci* **12**:636652 (2021).

83 Goldberg-Cavalleri A, Onkokesung N, Franco-Ortega S, and Ewards R, ABC transporters linked to multiple herbicide resistance in blackgrass (*Alopecurus myosuroides*). *Front Plant Sci* **14**:1082761 (2023).

84 Keshtkar E, Abdolshayi R, Sasanfar H, Zand E, Beffa R, Dayan FE, and Kudsk P, Assessing fitness costs from a herbicide-resistance management perspective: a review and insight. *Weed Sci* **67**:137–148 (2019).

85 Lepke J, Herrmann J, Remy N, Beffa R, and Richter O, Weed resistance prediction. A random Forest analysis based on field histories. *Front Agron* **6**. DOI: 10.3389/fagro.2024.1407422 (2024).

86 Powles SB, Herbicide discovery through innovation and diversity. *Adv Weed Sci* **40**. DOI: 10.51694/AdvWeedSci/2022;40:seventy-five012 (2022).

87 Bobadilla LK, and Tranel PJ, Predicting the unpredictable: the regulatory nature and promiscuity of herbicide cross resistance. *Pest Manag Sci* **80**:235–244 (2024).

6

Metabolism-Based Resistance in *Lolium rigidum* and *Raphanus raphanistrum*

Hugh J. Beckie, Heping Han, and Qin Yu

Australian Herbicide Resistance Initiative, School of Agriculture and Environment, The University of Western Australia, Perth, Western Australia, Australia

6.1 Introduction

Lolium rigidum Gaud. (annual ryegrass) is an adaptable annual grass weed found in the Mediterranean area of southern Europe and North Africa.[1] It is a genetically diverse, obligate, outcrossing diploid species. Highly productive varieties of *L. rigidum* were selected and introduced into southern Australia in the late 1800s as a pasture species.[2] With a shift to continuous grain cropping in the 1980s, it became an abundant, troublesome crop weed.

Lolium rigidum is a prolific seed producer, and field infestations of the species can produce up to 45,000 seeds/m^2.[3,4] Seeds have a short period (about two months) of primary dormancy.[5] On average, 15–30% of the seed bank can persist into the following year,[6] and studies estimate seed bank longevity to a maximum of four years.[7]

The biological characteristics, widespread adoption of no-tillage, and almost exclusive reliance on herbicidal control of *L. rigidum* have collectively resulted in rapid evolution of herbicide resistance in the species since first reported in 1982.[8] Most *L. rigidum* populations across the Australian grain-growing region have evolved resistance to herbicides of different sites of action (SOA).[9–11] In Western Australia, overreliance on postemergence herbicides, such as acetyl-CoA carboxylase (ACCase) and acetolactate synthase (ALS) inhibitors, has led to the rapid selection of high levels of multiple resistance in >95% of populations as documented by periodic random surveys (Fig. 6.1).[12]

Resistance in Weeds from Herbicide Metabolism, First Edition.
Edited by Vijay K. Nandula and Roland Beffa.

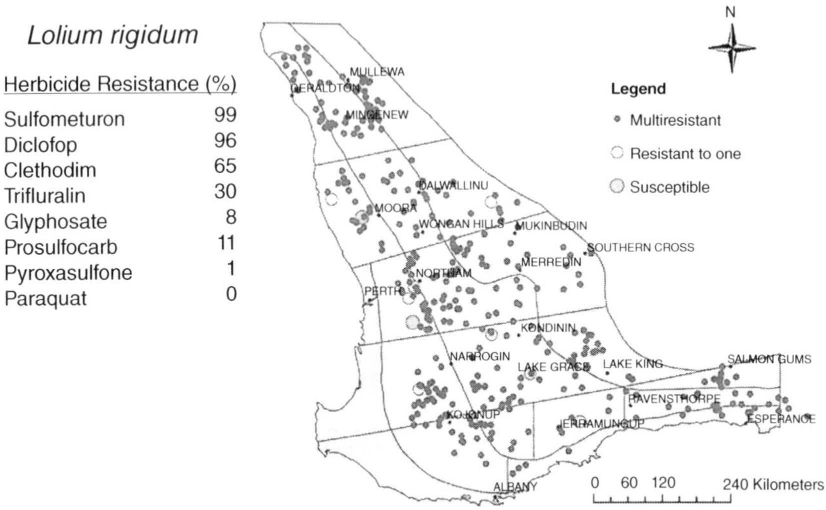

Figure 6.1 Distribution and abundance of multiple-resistant *Lolium rigidum* across Western Australia in 2015. Source: Adapted from Owen and Beckie.[11] (*See insert for color representation of the figure.*)

Raphanus raphanistrum L. (wild radish) is an important annual dicot weed infesting crops in many parts of the world.[13,14] Even low plant densities can result in significant crop yield loss.[15] In Australia, it is the most economically damaging dicot weed of the major grain crops. Plants surviving herbicidal control can produce a prolific amount of seed to rapidly replenish the soil seed bank. In a wheat crop, *R. raphanistrum* density of 50 plants/m^2 produced 17,000 seeds/m^2.[16] In addition to having a protracted germination period, seed can remain dormant in the soil seed bank for up to seven years[17] with as much as 90% of freshly shed, pod-enclosed seed not germinating the following season.[18]

Significant genetic variation is exhibited both within and among *R. raphanistrum* populations.[19] This genetic diversity is expressed by high morphological and phenological variation.[13,20] Herbicide resistance in populations of this outcrossing species is widespread in Western Australia. In a random survey in the Western Australian grain belt in 2015, 88% of wild radish populations were ALS inhibitor-resistant, 65% were phytoene desaturase (PDS) inhibitor (diflufenican)-resistant, 61% were auxinic (2,4-D)-resistant, and 14% were photosystem II (PS-II) inhibitor (atrazine)-resistant.[11] Of concern is the widespread occurrence of multiple-resistant wild radish populations (combinations of ALS inhibitor, synthetic auxin, PDS inhibitor, and PS-II inhibitor) (Fig. 6.2). In a 2003 random survey across the state, 58% of populations were resistant to two SOA, 18% to three SOA, and 7% to four SOA.[21] The most prevalent multiple resistance pattern was ALS inhibitor plus synthetic auxin.

Raphanus raphanistrum

Herbicide Resistance (%)	
Chlorsulfuron	88
Imazamox + imazapyr	70
Diflufenican	65
2,4-D	61
Atrazine	14
Glyphosate	0

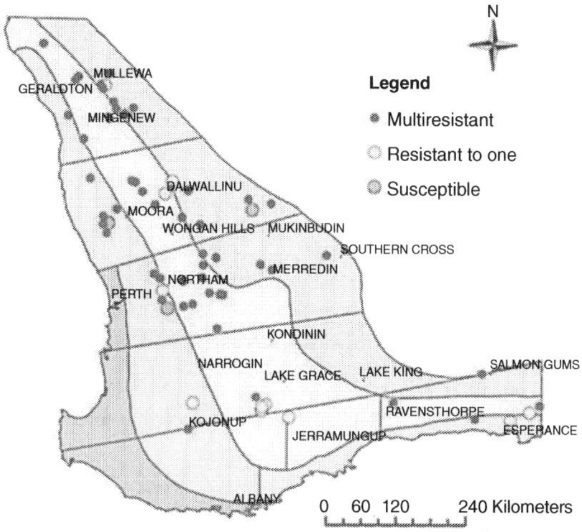

Figure 6.2 Distribution and abundance of multiple-resistant *Raphanus raphanistrum* across Western Australia in 2015. Source: Adapted from Owen and Beckie.[11] (*See insert for color representation of the figure.*)

Multiple resistance in *L. rigidum* or *R. raphanistrum* populations in Australia can result from gene flow and/or sequential herbicide selection. When determining why a weed population exhibits herbicide resistance, most studies focus, at least initially, on target-site resistance because it is generally easier to investigate target-site resistance than non-target-site resistance. While target-site resistance is specific to one SOA, non-target-site resistance can confer cross-resistance to two or more SOA. In the following section, we describe recent studies mainly from Australia documenting populations of these two weed species with non-target-site metabolic resistance, which can compromise the efficacy of herbicides of different SOA. The final section outlines some future directions for research.

6.2 Metabolic Resistance in *Lolium rigidum*

Lolium rigidum has evolved widespread metabolic resistance to ACCase- and ALS-inhibiting herbicides in Australia.[22–27] The first report by Christopher et al.[22] described increased metabolism of chlorsulfuron, an ALS-inhibiting herbicide, in an Australian population that was cross-resistant to diclofop-methyl, an ACCase-inhibiting herbicide. Enhanced diclofop metabolism in another Australian population was also documented.[28] From 33 diclofop-methyl-resistant *L. rigidum* populations collected in a 2010 field survey in Western Australia, the HPLC (high pressure liquid chromatography) profile of [^{14}C]-diclofop-methyl *in vivo* metabolism revealed that 79% of populations exhibited enhanced diclofop-acid metabolism. ACCase gene mutations conferring resistance were also found in 70% of the populations.[26]

It has long been established that metabolic resistance in *L. rigidum* is likely due to enhanced activity of cytochrome P450 monooxygenase enzymes.[22,28] Resistance to both chlorsulfuron and diclofop-methyl via enhanced metabolism by different cytochrome P450 enzymes was inferred in two Australian *L. rigidum* populations; the P450 inhibitors malathion and amitrole reversed chlorsulfuron and diclofop-methyl resistance, respectively.[29]

Metabolism pathways of diclofop-methyl and chlorsulfuron have been documented. Diclofop-methyl metabolism in *L. rigidum* involves hydrolysis of diclofop-acid and then aryl hydroxylation, followed by glucose conjugation.[30–32] A wheat-like metabolism of chlorsulfuron was reported in *L. rigidum*, in which chlorsulfuron was hydroxylated at the phenyl ring, followed by glucose conjugation.[22,25] In addition, the cleavage of chlorsulfuron into chlorobenzene sulfonamide and triazine amine was also reported in *L. rigidum*.[25]

A number of studies have investigated metabolic resistance in *L. rigidum* to other herbicide SOA. Studies by Burnet et al.[33,34] investigated enhanced metabolism via cytochrome P450s of the PS-II inhibitors simazine and chlortoluron in Australian populations of *L. rigidum*. More recently, metribuzin resistance was confirmed in an Australian *L. rigidum* population.[35] [^{14}C]-Metribuzin metabolism was 2.3-fold greater in resistant than susceptible plants. The cytochrome P450 monooxygenase inhibitor piperonyl butoxide enhanced plant mortality response to metribuzin and atrazine in both resistant and susceptible populations, suggesting the likely involvement of this enzyme family in enhanced detoxification of these PS-II inhibitors in resistant plants.[35]

Evolved metabolic resistance to the microtubule assembly-inhibiting herbicide trifluralin in a weed species was not reported until 2018. Previous studies indicated that trifluralin was not readily metabolized by plants. However, enhanced metabolism of another dinitroaniline herbicide, pendimethalin, was previously described in *L. rigidum*.[36] Enhanced metabolism of trifluralin in three *L. rigidum* populations originating from Australian grain fields was confirmed by measuring the amount of [^{14}C] label partitioning into the polar phase of a hexane:methanol system.[37] As the P450 inhibitor phorate can partially reverse trifluralin resistance in one of the populations, cytochrome P450 monooxygenases were implicated. In a *L. rigidum* population from Western Australia, both target-site mutation (Val202Phe) and enhanced metabolism conferred resistance to trifluralin.[38]

Resistance to pyroxasulfone, a very-long-chain fatty acid (VLCFA) inhibitor, in an Australian population was metabolism-based; similar to wheat, almost 90%

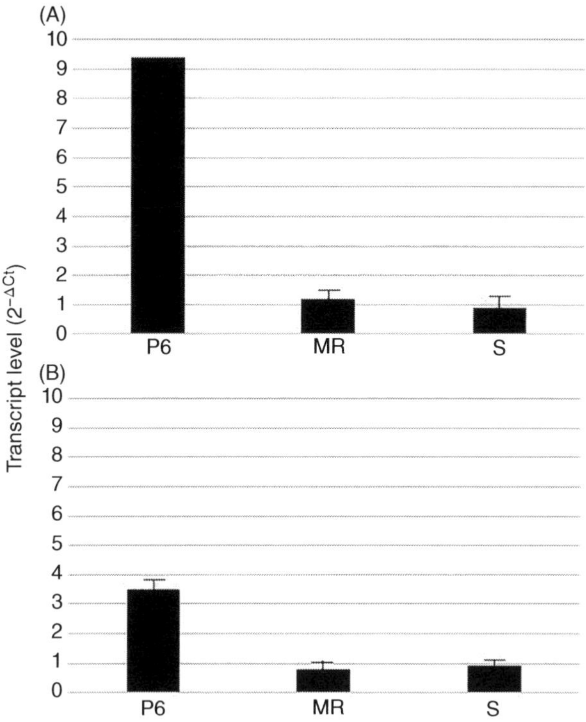

Figure 6.3 Transcript levels of GST-1 (A) and GST-2 (B) genes in pyroxasulfone-treated (P6, progeny of a multiple-resistant [MR] population after six generations of recurrent selection with pyroxasulfone), MR (but susceptible to pyroxasulfone), and susceptible (S) *Lolium rigidum* (bars denote standard error). Source: Adapted from Busi et al.[39]

of parental [^{14}C]-labeled pyroxasulfone was metabolized 24 hours after herbicide treatment.[39] Several metabolites were formed via a glutathione-*S*-transferase (GST)-catalyzed glutathione conjugation pathway. The constitutive overexpression from six- to nine-fold of two resistance-endowing *GST* genes was associated with the resistance phenotype. Both *GST-1* and *GST-2* (Tau class) had significantly higher transcription in resistant than susceptible plants (Fig. 6.3). Additionally, enhanced GST activity was implicated in the resistance of some European populations of *Lolium* spp. (*Lolium perenne* ssp. *multiflorum* and *L. rigidum*) to flufenacet, another VLCFA inhibitor.[40]

As outlined above, metabolic resistance in *L. rigidum* is well known, but a molecular understanding has been lacking. Most studies on metabolic herbicide resistance in *L. rigidum* were conducted at the biochemistry level.[11] Precise biochemical and molecular genetic elucidation of metabolic resistance had been stalled due to a number of factors such as complex cytochrome P450 superfamilies, high genetic diversity in *Lolium* or other cross-pollinated species, and the complexity of genetic control of metabolic resistance.[27] However, progress is being made on the P450 genes responsible for resistance in *L. rigidum*. RNA sequencing (RNA-Seq) transcriptome analysis implicated involvement of P450, glucosyl-transferase (GT), and GST genes in metabolic resistance in two Australian and eight French *L. rigidum* populations.[42] Identification, cloning, and functional characterization of a *L. rigidum* P450 gene endowing resistance to herbicides across several SOA and chemistry groups were recently reported.[43] A metabolic-resistant subset from a field-evolved resistant *L. rigidum* population (SLR31), a susceptible and derived F_2 population, was investigated for candidate P450 gene discovery by RNA-Seq. A P450 gene, *CYP81A10v7*, was identified with higher expression in herbicide-resistant versus susceptible plants. Transgenic rice overexpressing this *CYP81A10v7* gene became highly resistant to the ACCase-inhibiting herbicides diclofop-methyl and tralkoxydim and the ALS-inhibiting herbicide chlorsulfuron, and moderately resistant to the 4-hydroxyphenylpyruvate dioxygenase (HPPD)-inhibiting herbicide mesotrione, PS-II-inhibiting herbicides atrazine and chlorotoluron, and the microtubule assembly-inhibiting herbicide trifluralin (Fig. 6.4). This broad herbicide cross-resistance profile in *CYP81A10v7* transgenic rice generally reflected that of the resistant *L. rigidum* subset containing CYP81A10v7. Furthermore, transgenic rice overexpressing the *CYP81A10v7* gene displayed enhanced rates of metabolism of diclofop-methyl and mesotrione. This study clearly demonstrated that a single P450 gene in a weed species delivers cross-resistance to at least seven herbicides from five SOA. Most recently, the *CYP81A10v7* transgenic rice was found to be resistant to bixlozone, a newly registered herbicide in Australia (data not shown). This ability of a single P450 gene to regulate the metabolism of multiple herbicides is a major threat to the efficacy of currently available and yet-to-be-commercialized herbicides.

Metabolic cross-resistance to ACCase-, ALS-, PS-II-, and HPPD-inhibiting herbicides due to enhanced expression of CYP81A subfamily genes has independently evolved in the unrelated weedy species *Echinochloa phyllopogon*[44–47] and *L. rigidium*,[43] indicating molecular convergence at least for the CYP81A subfamily conferring resistance to these four herbicide SOA. However, due to differences in CYP81A sequences in *L. rigidum* and *E. phyllopogon* (72–73% identity between CYP81A10v7 and CYP81A12/21), herbicide-metabolizing activities of CYP81As against an herbicide vary in the two species. For example, CYP81A10v7 in

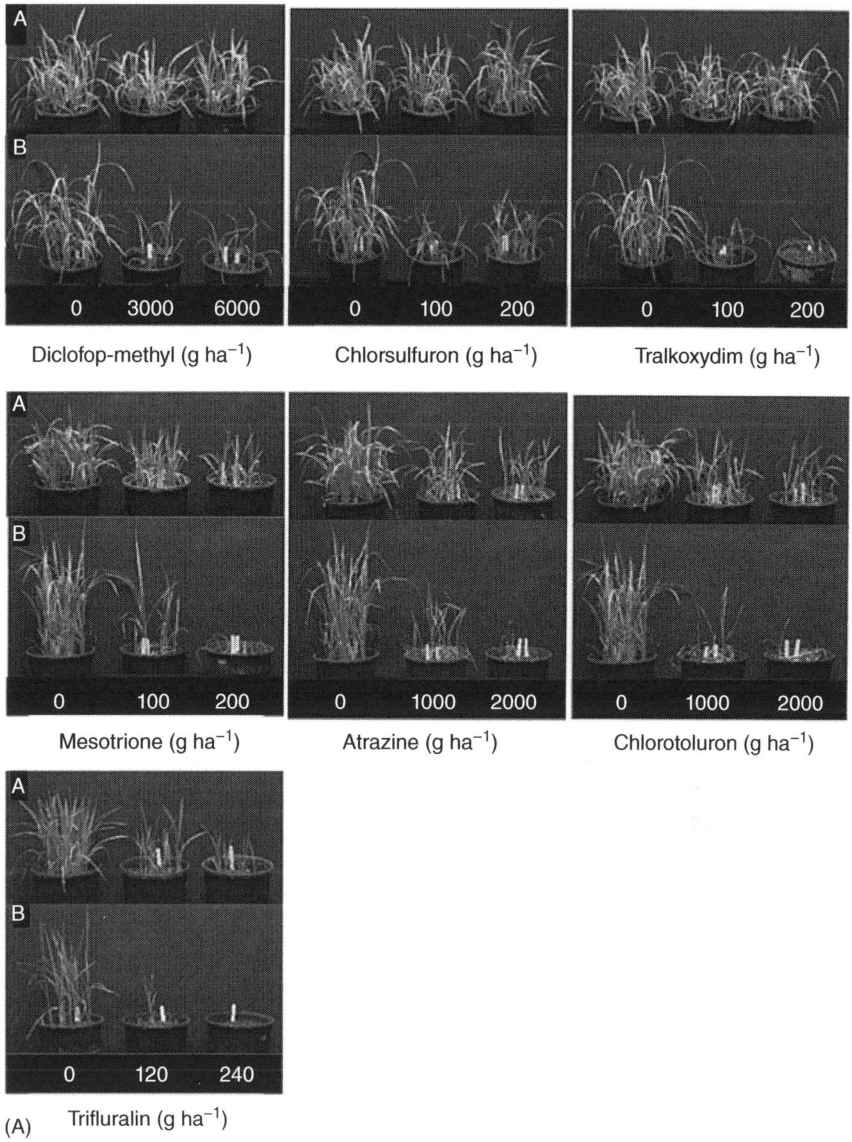

Figure 6.4 (A) Sensitivity of transgenic rice seedlings expressing CYP81A10v7 and green fluorescent protein (GFP) (negative control). The four leaf stage rice seedlings were treated with the respective herbicide, and photos were taken three weeks after treatment. A: CYP81A10v7, B: GFP. Source: Adapted from Han et al.[43] (*See insert for color representation of the figure.*) (B) Mortality response of transgenic rice seedlings expressing CYP81A10v7 (○) and green fluorescent protein (GFP) (●) (negative control) to herbicides of different sites of action. The *Lolium rigidum* seedlings were treated with diclofop-methyl, chlorsulfuron, tralkoxydim, mesotrione, atrazine, chlorotoluron, or trifluralin.

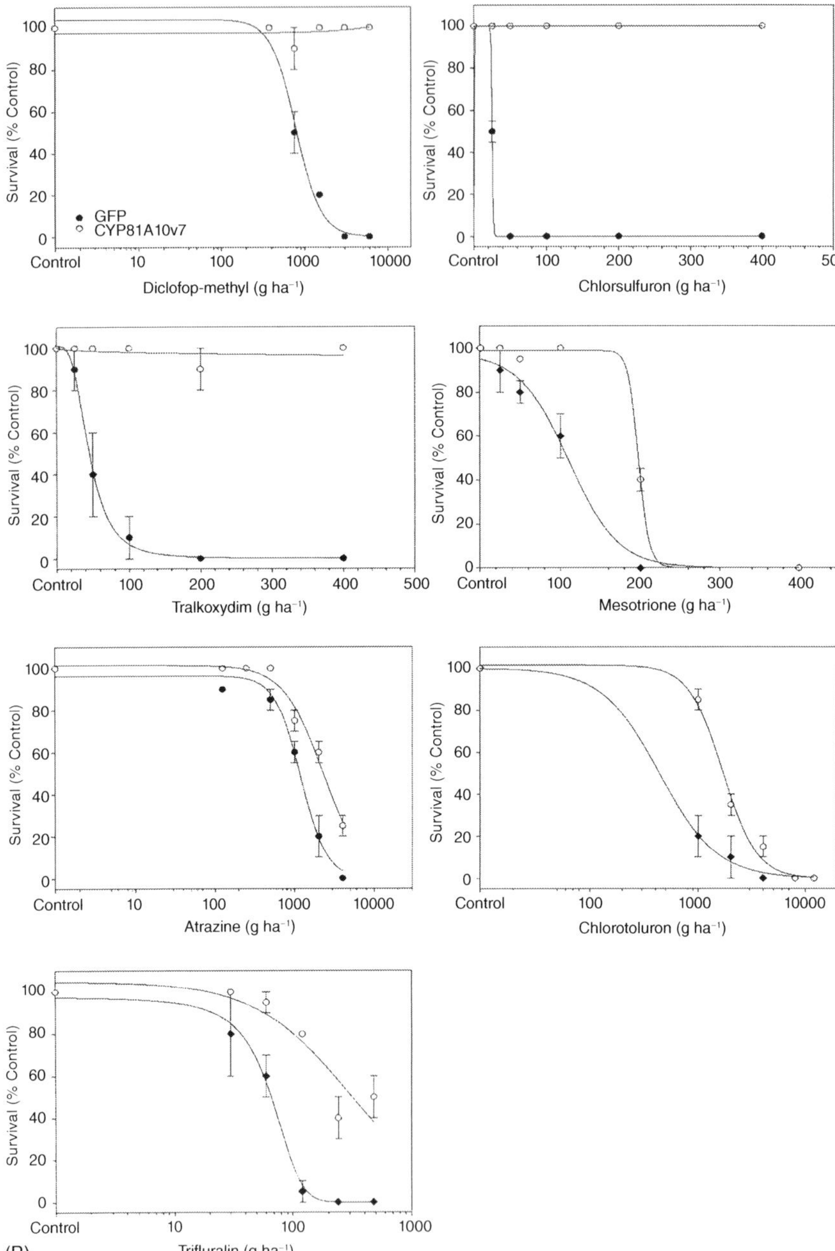

Figure 6.4 (*Continued*)

L. rigidium but not CYP81A12/21 in *E. phyllopogon* (see Dimaano et al.[47]) confers resistance to the microtubule assembly-inhibiting herbicide trifluralin.

Identification of a single CYP81A10v7 gene endowing cross-resistance to chemically dissimilar herbicides with a major effect on diclofop-methyl and chlorsulfuron explains why metabolic resistance to chlorsulfuron in *L. rigidum* can be selected by diclofop-methyl and resistance to both herbicides are often linked.[24] However, this P450 species may not be the sole gene responsible for metabolic herbicide resistance in the *L. rigidum* population. For instance, the P450 inhibitor malathion can reverse resistance to chlorsulfuron but not to diclofop-methyl.[48] Therefore, among other possibilities, these inhibitor studies suggest involvement of additional unidentified P450s and/or other metabolic enzymes at the population level. Indeed, genetic control of metabolic resistance to ACCase- and ALS-inhibiting herbicides in *L. rigidum* involves one to multiple loci.[49,50] Previous studies have shown that a herbicide can be metabolized by several P450 enzymes. For example, the ALS-inhibiting herbicide bensulfuron-methyl can be metabolized by four P450s (CYP81A12, CYP81A21, CYP71C6v1, and CYP72A31), and the PS-II-inhibiting herbicide chlorotoluron by multiple P450s (e.g., CYP71A10, CYP71A11, CYP73A1, CYP76B1, CYP81A10v7, CYP81B1, and CYP81B2) (Table 6.1), with different catalytic efficiency. In addition, GSTF1 in *Alopecurus myosuroides* Huds.[63] and GSTF2 in *Amaranthus tuberculatus* Moq. J. D. Sauer.[64] have been demonstrated to be responsible for non-target-site resistance to atrazine and/or chlorotoluron.

Although the role of CYP81A10v7 in herbicide metabolic resistance was clearly identified in the resistant subset from *L. rigidum* population SLR31, it was not consistently manifested among other herbicide-resistant populations.[43] This result is not unexpected, as different metabolic genes or mechanisms can be selected within individuals and among populations especially in a genetically diverse, cross-pollinated weed species like *L. rigidum*. This may also reflect flexibility of P450 (and other metabolic enzyme families) in regulating adaptive evolution of metabolic resistance in *L. rigidum*. Although overexpression of CYP72A is often associated with metabolic resistance to ACCase- and ALS-inhibiting herbicides and with 2,4-D-induced protection against these herbicides, in *Lolium* species,[42,65,66] transgenic rice overexpressing *CYP72A7v1* did not endow resistance to the ACCase- and ALS-inhibiting herbicides tested.[43] Hence, the specific role of the CYP72A subfamily in herbicide resistance remains elusive. One possibility is that CYP72As have close linkage to resistance-endowing CYP81As in a gene cluster and are coordinately upregulated with the latter. Indeed, P450s normally cluster together on the same chromosome in plants,[67,68] and co-regulated gene clusters have been reported for the biosynthesis of allelochemicals such as DIMBOA and momilactone (see Giacomini et al.[69]; Mao et al.[70]).

Table 6.1 Herbicide-metabolizing P450 genes identified from weedy or domestic plants.

P450s	Herbicide group	Target herbicides	Plants	Citation
CYP706A3	Tubulin	Benefin, ethalfluralin, fluchloralin, pendimethalin, trifluralin	*Arabidopsis*	Abdollahi et al.[51]
CYP71A10	PS-II	Fluometuron, linuron, diuron, chlorotoluron	Soybean	Siminszky et al.[52]
CYP71A11	PS-II	Chlorotoluron	Tobacco	Yamada et al.[53]
CYP71C6v1	ALS	Chlorsulfuron, triasulfuron, metsulfuron-methyl, bensulfuron-methyl, tribenuron-methyl	Wheat	Xiang et al.[54]
CYP72A31	ALS	Bispyribac sodium, bensulfuron-methyl	Rice	Saika et al.[55]
CYP72A18	Other	Pelargonic acid	Rice	Imaishi and Matumoto[56]
CYP73A1	PS-II	Chlorotoluron	Jerusalem artichoke	Pierrel et al.[57]
CYP76B1	PS-II	Linuron, chlorotoluron, isoproturon, diuron	Jerusalem artichoke	Didierjean et al.[58]
CYP81A6	PS-II ALS	Bentazon Bensulfuron-methyl	Rice	Pan et al.[59]
CYP81A10v7	ALS ACCase PS-II HPPD Tubulin	Chlorsulfuron Diclofop-methyl, tralkoxydim Atrazine, chlorotoluron Mesotrione Trifluralin		Han et al.[43]
CYP81A12	ALS ACCase	Bensulfuron-methyl penoxsulam, diclofop-methyl, tralkoxydim, pinoxaden	Water grass[a)]	Iwakami et al.[44,45]; Guo et al.[46]
	Carotenoid synthesis	Clomazone		

CYP81A15	ACCase	Pinoxaden	Water grass	Iwakami et al.[45]; Guo et al.[46]
	Carotenoid synthesis	Clomazone		
CYP81A21	ALS	Bensulfuron-methyl, penoxsulam	Water grass	Iwakami et al.[44,45]; Guo et al.[46]
	ACCase	Diclofop-methyl, tralkoxydim, pinoxaden		
	Carotenoid synthesis	Clomazone		
CYP81A24	ACCase	Pinoxaden, tralkoxydim	Water grass	Iwakami et al.[45]; Guo et al.[46]
	Carotenoid synthesis	Clomazone		
CYP81A63	ACCase	Diclofop-methyl, pinoxaden	Water grass	Iwakami et al.[45]
CYP81B1	PS-II	Chlorotoluron	Jerusalem artichoke	Cabello-Hurtado et al.[60]
CYP81B2	PS-II	Chlorotoluron	Tobacco	Yamada et al.[53]
CYP749A16	ALS	Trifloxysulfuron	Cotton	Thyssen et al.[61]
Nsf1	ALS	Rimsulfuron, nicosulfuron	Maize	Dam et al.[62]
	PS-II	Diuron		
	HPPD	Mesotrione, isoxaflutole		
	PPO	Acifluorfen, flumioxazin, sulfentrazone		
	Synthetic auxin	Dicamba		

a) *Echinochloa phyllopogon.*

6.3 Metabolic Resistance in *Raphanus raphanistrum*

Metabolic herbicide resistance is less reported in dicot weeds, limited to only a few species.[71,72] One of the earliest reports of enhanced metabolism in a herbicide-resistant dicot weed was that of *Stellaria media* (L.) Vill. (common chickweed) resistant to mecoprop.[73] A susceptible *R. raphanistrum* population subjected to recurrent (four generations) low-dose 2,4-D selection evolved resistance to not only 2,4-D but also the ALS-inhibiting herbicides chlorsulfuron and metosulam.[74] Pretreatment of the resistant population with the cytochrome P450 inhibitor malathion restored chlorsulfuron to full efficacy, indicating that cross-resistance was likely due to P450-mediated enhanced herbicide metabolism.

A *R. raphanistrum* population from Western Australia (H1/16) was determined to be resistant to metribuzin, a PS-II-inhibiting herbicide, via both *psbA* gene mutation (Ser264Gly) and enhanced metabolism.[75] The resistant plants metabolized metribuzin at least twice as fast as herbicide-susceptible plants, although target-site resistance was the major resistance mechanism in this population (whole-plant LD_{50} resistant:susceptible ratio of 74). This was the first study to implicate enhanced herbicide metabolism as a field-evolved resistance mechanism in this weed species.

Another population from Western Australia (H2/10) was confirmed to be resistant to HPPD-inhibiting herbicides due to non-target-site-based mechanisms, even though it was not selected with this SOA herbicide group.[76] Dose–response experiments with mesotrione, tembotrione, and isoxaflutole indicated that the resistance factor (based on whole-plant LD_{50} resistant:susceptible ratio) ranged from 4 to 6.5. This population was also resistant to ALS, PS-II, and PDS inhibitors, as well as to 2,4-D herbicides. Only two other weed species from North America have been previously documented as HPPD-inhibitor–resistant, namely, *Amaranthus palmeri* S. Watson and *A. tuberculatus*.[77] The time required for resistant *R. raphanistrum* plants to metabolize 50% [^{14}C]-mesotrione was eight-fold faster than that for susceptible plants (Fig. 6.5). The accumulation rate of the major metabolites was at least five-fold greater in the resistant than susceptible plants. The resistant *R. raphanistrum* population and maize (naturally tolerant) share similar but not identical metabolic pathways for HPPD-inhibiting herbicides.[76] The major metabolite observed at 120 hours after treatment was likely 4-hydroxy-mesotrione, as deduced in metabolic studies involving maize or *Amaranthus* spp.[78] Tembotrione resistance but not mesotrione or isoxaflutole resistance was reversed by malathion, a cytochrome P450 inhibitor, suggesting that P450s may be involved in *R. raphanistrum* resistance to HPPD-inhibiting herbicides.[76] A recent study[79] further identified one 2-oxoglutarate/Fe(II)-dependent dioxygenase gene (*Rr2ODD1*) and two P450 genes (*RrCYP704C1* and *RrCYP709B1*), which were significantly highly

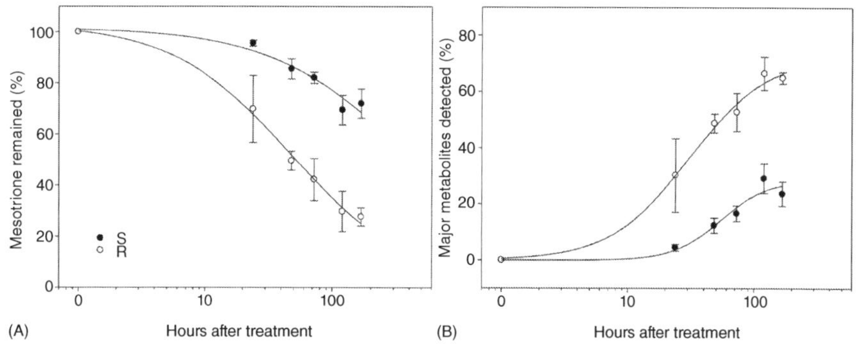

Figure 6.5 Percentage of mesotrione remaining (intact mesotrione) in resistant (solid circle) and susceptible (open circle) *Raphanus raphanistrum* plants as a function of time after herbicide treatment (A) and accumulation of its major metabolites (B). Source: From Reference[76]/ John Wiley & Sons.

expressed in resistant (R) compared to susceptible (S) plants. Gene functional characterization using *Arabidopsis* transformation showed overexpression of RrCYP709B1 conferred a modest level of resistance to mesotrione. UPLC-MS/MS analysis showed that tissue mesotrione levels in RrCYP709B1 transgenic *Arabidopsis* plants were significantly lower than that in the wild type. In addition, overexpression of Rr2ODD1 or RrCYP704C1 in *Arabidopsis* endowed resistance to tembotrione and isoxaflutole. Although each gene confers a modest level of resistance, overexpression of the multiple herbicide-metabolizing genes could contribute to HPPD-inhibiting herbicide resistance in this wild radish population.

The same *R. raphanistrum* population (H2/10) also displayed non-target-site resistance (resistance factor of five) to the PDS-inhibitor herbicide diflufenican.[80] Malathion reversed diflufenican resistance in this H2/10 population, further implicating enhanced herbicide metabolism involving cytochrome P450s. Discovery of P450 genes responsible for metabolic resistance in this weed is challenging due to generally low-level resistance plus genetic heterogeneity and polymorphisms. For example, the absence/presence of a certain target-site *HPPD* and *PDS* transcript vary among populations.

6.4 Future Research

Gene discovery for metabolic herbicide resistance is challenging, as plants possess hundreds of P450s with varying substrate specificities.[47,67] In *L. rigidum*, P450-mediated metabolic resistance has been known at the whole-plant level for decades, but there has been less progress until recently in resistance gene discovery because of the large genome size (2.44 Gb), lack of quality genome sequences and genetic linkage maps, great genetic heterogeneity, and difficulty in obtaining true breeding lines due to self-incompatibility.

Thus far, we have cloned at least 12 CYP81A10 variants in resistant and susceptible *L. rigidum* plants.[43] In addition to genes from different families or different members from the same family, it remains to be determined whether and how these variants affect the level and pattern of herbicide metabolism and thus resistance. For the well-characterized CYP81A10v7,[43] future research will investigate how the gene is regulated. Cloning and analysis of the promoter sequence and its methylation status, as well as transcription factors, may reveal genetic and epigenetic regulation dynamics of the P450 gene. Genome-based work is needed to reveal the genetic architecture of P450-driven evolution of metabolic resistance in *L. rigidum*. Metabolomic studies will help understand the endogenous function of CYP81A10v7 in secondary metabolism in plants, in addition to its significant role in herbicide metabolic resistance.

Virus-induced gene silencing can be explored in *L. rigidum* to speed up the process of functional screening of candidate resistance genes. A *L. rigidum* genetic transformation system compatible with CRISPR/Cas9-based gene editing will be established to knock out the responsible P450 genes, potentially for proof of concept of genetic sensitizing/reversal of resistance. In addition, a yeast transformation system will be explored for identification of major herbicide metabolites produced by P450s owing to the fast growth of the yeast cells and hence the ease of purifying the microsomes containing P450 enzyme activities.

As HPPD-inhibiting herbicides continue to enter the marketplace and the overall use of this SOA herbicide group in major agronomic crops continues to increase worldwide, it is imperative to document and investigate the mechanistic basis of target-site and non-target-site resistance in resistant weed biotypes. To date, enhanced metabolism is a prominent resistance mechanism in herbicide-resistant biotypes of grass weeds.[81] The availability of genome information for *R. raphanistrum* and its close relativè *Raphanus sativus* (radish) will facilitate metabolic resistance gene discovery in this weed species in the future. In particular, future research is required in the following areas: (1) identification and validation of candidate genes involved in mesotrione (and the PDS inhibitor diflufenican) metabolic resistance in *R. raphanistrum* and (2) *Arabidopsis* genetic transformation and functional characterization of the candidate genes, including herbicide resistance screening and HPPD herbicide metabolism analysis.

References

1 Bennett SJ, Hayward MD, and Marshall DF, Morphological differentiation in four species of the genus *Lolium*. *Genet Resour Crop Evol* **47**:247–255 (2000).

2 Goggin DE, Powles SB, and Steadman KJ, Understanding *Lolium rigidum* seeds: the key to managing a problem weed? *Agronomy* **2**:222–239 (2012).

3 Gill GS, Why annual ryegrass is a problem in Australian agriculture. *Plant Prot Q* **11**:193–195 (1996).

4 Stanton R, Piltz J, Pratley J, Kaiser A, Hudson D, and Gill G, Annual ryegrass (*Lolium rigidum*) seed survival and digestibility in cattle and sheep. *Aust J Exp Agric* **42**:111–115 (2002).

5 Steadman KJ, Dormancy release during hydrated storage in *Lolium rigidum* seeds is dependent on temperature, light quality, and hydration status. *J Exp Bot* **55**:929–937 (2004).

6 Chauhan BS, Gill G, and Preston C, Influence of tillage systems on vertical distribution, seedling recruitment and persistence of rigid ryegrass (*Lolium rigidum*) seed bank. *Weed Sci* **54**:669–676 (2006).

7 Narwal S, Sindel BM, and Jessop RS, Dormancy and longevity of annual ryegrass (*Lolium rigidum*) as affected by soil type, depth, rainfall, and duration of burial. *Plant Soil* **310**:225–234 (2008).

8 Heap I, and Knight R, A population of ryegrass tolerant to the herbicide diclofop-methyl. *J Aust Inst Agric Sci* **48**:156–157 (1982).

9 Broster J, Koetz E, and Wu H, Herbicide resistance levels in annual ryegrass (*Lolium rigidum* gaud.) in southern New South Wales. *Plant Prot Q* **26**:22–28 (2011).

10 Boutsalis P, Gill G, and Preston C, Incidence of herbicide resistance in rigid ryegrass (*Lolium rigidum*) accross southeastern Australia. *Weed Technol* **26**:391–398 (2012).

11 Owen M and Beckie H, Update on herbicide resistance status in the Western Australian grain belt (2020). Grains Research Update, Perth, Australia. Available: http://grdc.com.au/resources-and-publications/grdc-update-papers [1 March 2020].

12 Owen MJ, Martinez NJ, and Powles SB, Multiple herbicide-resistant *Lolium rigidum* (annual ryegrass) now dominates across the Western Australian grain belt. *Weed Res* **54**:314–324 (2014).

13 Cheam AH, and Code GR, The biology of Australian weeds 24. *Raphanus raphanistrum* L. *Plant Prot Q* **10**:2–13 (1995).

14 Blackshaw RE, Lemerle D, Mailer R, and Young KR, Influence of wild radish on yield and quality of canola. *Weed Sci* **50**:344–349 (2002).

15 Code GR, and Donaldson TW, Effect of cultivation, sowing methods and herbicides on wild radish populations in wheat crops. *Aust J Exp Agric* **36**:437–442 (1996).

16 Reeves TG, Code GR, and Piggin CM, Seed production and longevity, seasonalemergence and phenology of wild radish (*Rahanus raphanistrum*). *Aust J Exp Agric* **21**:524–530 (1981).

17 Young KR, Germination and emergence of wild radish (*Raphanus raphanistrum* L.). PhD dissertation, University of Melbourne (2001).
18 Cheam AH, Seed production and seed dormancy in wild radish (*Raphanus raphanistrum*) and some possabilities for improving control. *Weed Res* **26**:405–413 (1986).
19 Madhou P, Wells A, Pang E, and Stevenson T, Genetic variation in populations of Western Australian wild radish. *Aust J Agric Res* **56**:1079–1087 (2005).
20 Conner J, and Via S, Patterns of phenotypic and genetic correlations among morphological and life history traits in wild radish, *Raphanus raphanistrum*. *Evolution* **47**:704–711 (1993).
21 Walsh MJ, Owen MJ, and Powles SB, Frequency and distribution of herbicide resistance in *Raphanus raphanistrum* populations randomly collected across the Western Australian wheatbelt. *Weed Res* **47**:542–550 (2007).
22 Christopher JT, Powles SB, Liljegren DR, and Holtum JAM, Cross-resistance to herbicides in annual ryegrass (*Lolium rigidum*). II. Chlorsulfuron resistance involves a wheat-like detoxification system. *Plant Physiol* **95**:1036–1043 (1991).
23 Christopher JT, Powles SB, and Holtum JAM, Resistance to acetolactate synthase-inhibiting herbicides in annual ryegrass (*Lolium rigidum*) involves at least two mechanisms. *Plant Physiol* **100**:1909–1913 (1992).
24 Holtum JAM, Matthews JM, Hausler RE, Liljegren DR, and Powles SB, Cross-resistance to herbicides in annual ryegrass (*Lolium rigidum*). 3. On the mechanism of resistance to diclofop-methyl. *Plant Physiol* **97**:1026–1034 (1991).
25 Cotterman JC, and Saari LL, Rapid metabolic inactivation is the basis for cross-resistance to chlorsulfuron in diclofop-methyl-resistant rigid ryegrass (*Lolium rigidum*) biotype SR484. *Pestic Biochem Physiol* **43**:182–192 (1992).
26 Han H, Yu Q, Owen MJ, Cawthray GR, and Powles SB, Widespread occurrence of both metabolic and target-site herbicide resistance mechanisms in *Lolium rigidum* populations. *Pest Manag Sci* **72**:255–263 (2016).
27 Yu Q, and Powles S, Metabolism-based herbicide resistance and cross-resistance in crop weeds: a threat to herbicide sustainabiity and global crop production. *Plant Physiol* **166**:1106–1118 (2014).
28 Preston C, Tardif FJ, Christopher JT, and Powles SB, Multiple resistance to dissimilar herbicide chemistries in a biotype of *Lolium rigidum* due to enhanced activity of several herbicide degrading enzymes. *Pestic Biochem Physiol* **54**:123–134 (1996).
29 Yu Q, Abdallah B, Han H, Owen M, and Powles S, Distinct non-target site mechanisms endow resistance to glyphosate, ACCase and ALS-inhibiting herbicides in multiple herbicide-resistant *Lolium rigidum*. *Planta* **230**:713–723 (2009).
30 Shimabukuro RH, Walsh WC, and Hoerauf RA, Metabolism and selectivity of diclofop-methyl in wild oat and wheat. *J Agric Food Chem* **27**:615–623 (1979).

31 Donald WW, and Shimabukuro RH, Selectivity of diclofop-methyl between wheat and wild oat: growth and herbicide metabolism. *Physiol Plant* **49**:459–464 (1980).

32 Shimabukuro RH, and Hoffer BL, Metabolism of diclofop-methyl in susceptible and resistant biotypes of *Lolium rigidum*. *Pestic Biochem Physiol* **39**:251–260 (1991).

33 Burnet MWM, Loveys BR, Holtum JAM, and Powles SB, Increased detoxification is a mechanism of simazine resistance in *Lolium rigidum*. *Pestic Biochem Physiol* **46**:207–218 (1993a).

34 Burnet MWM, Loveys BR, Holtum JAM, and Powles SB, A mechanism of chlorotoluron resistance in *Lolium rigidum*. *Planta* **190**:182–189 (1993b).

35 Ma H, Lu H, Han H, Yu Q, and Powles S, Metribuzin resistance via enhanced metabolism in a multiple herbicide resistant *Lolium rigidum* population. *Pest Manag Sci* **76**:3785–3791 (2020).

36 Tardif FJ, and Powles SB, Effect of malathion on resistance to soil-applied herbicides in a population of rigid ryegrass (*Lolium rigidum*). *Weed Sci* **47**:258–261 (1999).

37 Chen J, Goggin D, Han H, Busi R, Yu Q, and Powles S, Enhanced trifluralin metabolism can confer resistance in *Lolium rigidum*. *J Agric Food Chem* **66**:7589–7596 (2018).

38 Chen J, Chu Z, Han H, Goggin DE, Yu Q, Sayer C, and Powles SB, A Val-202-Phe α–tubulin mutation and enhanced metabolism confer dinitroaniline resistance in a single *Lolium rigidum* population. *Pest Manag Sci* **76**:645–652 (2020).

39 Busi R, Porri A, Gaines TA, and Powles SB, Pyroxasulfone resistance in *Lolium rigidum* is metabolism-based. *Pestic Biochem Physiol* **148**:74–80 (2018).

40 Dücker R, Zöllner P, Lümmen P, Ries S, Collavo A, and Beffa R, Glutathione transferase plays a major role in flufenacet resistance in ryegrass (*Lolium* spp.) field populations. *Pest Manag Sci* **75**:3084–3092 (2019).

41 Fischer TC, Klattig JT, and Gierl A, A general cloning strategy for divergent plant cytochrome P450 genes and its application in *Lolium rigidum* and *Ocimum basilicum*. *Theor Appl Genet* **103**:1014–1021 (2001).

42 Gaines T, Lorentz L, Figge A, Herrmann J, Maiwald F, Ott M-C, *et al.*, RNA-Seq transcriptome analysis to identify genes involved in metabolism-based diclofop resistance in *Lolium rigidum*. *Plant J* **78**:865–876 (2014).

43 Han H, Yu Q, Beffa R, González S, Maiwald F, Wang J, and Powles SB, Cytochrome P450 CYP81A10v7 in *Lolium rigidum* confers metabolic resistance to herbicides across at least five modes of action. *Plants J* **105**:79–92 (2021).

44 Iwakami S, Endo M, Saika H, Okuno J, Nakamura N, Yokoyama M, *et al.*, Cytochrome P450 CYP81A12 and CYP81A21 are associated with resistance to two acetolactate synthase inhibitors in *Echinochloa phyllopogon. Plant Physiol* **165**:618–629 (2014).

45 Iwakami S, Kamidate Y, Yamaguchi T, Ishizaka M, Endo M, Suda H, *et al.*, CYP81A P450s are involved in concomitant cross-resistance to acetolactate synthase and acetyl-CoA carboxylase herbicides in *Echinochloa phyllopogon. New Phytol* **221**:e15552 (2019).

46 Guo F, Iwakami S, Yamaguchi T, Uchino A, Sunohara Y, and Matsumoto H, Role of CYP81A cytochrome P450s in clomazone metabolism in *Echinochloa phyllopogon. Plant Sci* **283**:321–328 (2019).

47 Dimaano NG, and Iwakami S, Cytochrome P450-mediated herbicide metabolism in plants: current understanding and prospects. *Pest Manag Sci* **77**:22–32 (2021).

48 Christopher JT, Preston C, and Powles SB, Malathion antagonizes metabolism-based chlorsulfuron resistance in *Lolium rigidum. Pestic Biochem Physiol* **49**:172–182 (1994).

49 Busi R, Vila-Aiub MM, and Powles SB, Genetic control of a cytochrome P450 metabolism-based herbicide resistance mechanism in *Lolium rigidum. Heredity* **106**:817–824 (2011).

50 Han H, Yu Q, Vila-Aiub M, and Powles S, Genetic inheritance of cytochrome P450- mediated metabolic resistance to chlorsulfuron in a multiple herbicide resistant *Lolium rigidum* population. *Crop Prot* **65**:57–63 (2014).

51 Abdollahi F, Taghi Alebrahim M, Ngov C, Lallemand E, Zheng Y, Villette C, *et al.*, Innate promiscuity of the CYP706 family of P450 enzymes provides a suitable context for the evolution of dinitroaniline resistance in weed. *New Phytol* **229**:3253–3268 (2021).

52 Siminszky B, Corbin FT, Ward ER, Fleischmann TJ, and Dewey RE, Expression of a soybean cytochrome P450 monooxygenase cDNA in yeast and tobacco enhances the metabolism of phenylurea herbicides. *PNAS* **96**:1750–1755 (1999).

53 Yamada T, Kambara Y, Imaishi H, and Ohkawa H, Molecular cloning of novel cytochrome P450 species induced by chemical treatments in cultured tobacco cells. *Pestic Biochem Physiol* **68**:11–25 (2000).

54 Xiang W, Wang X, and Ten T, Expression of a wheat cytochrome P450 monooxygenase cDNA in yeast catalyzes the metabolism of sulfonylurea herbicides. *Pestic Biochem Physiol* **85**:1–6 (2006).

55 Saika H, Horita J, Taguchi-Shiobara F, Nonaka S, Nishizawa-Ykoi A, Iwakami S, *et al.*, A novel rice cytochrome P450 gene, CYP72A31, confers tolerance to acetolactate synthase-inhibiting herbicides in rice and Arabidopsis. *Plant Physiol* **166**:1232–1240 (2014).

56 Imaishi H, and Matumoto S, Isolation and functional characterization in yeast of CYP72A18, a rice cytochrome P450 that catalyzes (w-1)-hydroxylation of the herbicide pelargonic acid. *Pestic Biochem Physiol* **88**:71–77 (2007).

57 Pierrel MA, Batard Y, Kazmaier M, Mignotte-Vieux C, Durst F, and Werck-Reichhart D, Catalytic properties of the plant cytochrome P450 CYP73 expressed in yeast. *Eur J Biochem* **224**:835–844 (1994).

58 Didierjean L, Gondet L, Perkins R, Lau SM, Schaller H, O'Keefe DP, and Werck- Reichhart D, Engineering herbicide metabolism in tobacco and Arabidopsis with CYP76B1, a cytochrome P450 enzyme from Jerusalem artichoke. *Plant Physiol* **130**:179–189 (2002).

59 Pan G, Zhang X, Liu K, Zhang J, Wu X, Zhu J, and Tu J, Map-based cloning of a novel rice cytochrome P450 gene CYP81A6 that confers resistance to two different classes of herbicides. *Plant Mol Biol* **61**:933–943 (2006).

60 Cabello-Hurtado F, Batard Y, Salaun JP, Durst F, Pinot F, and Werck-Reichhart D, Cloning, expression in yeast and functional characterization of CYP81B1, a plant P450 which catalyzes in-chain hydroxylation of fatty acids. *J Biol Chem* **272**:7260–7267 (1998).

61 Thyssen GN, Naoumkina M, McCarty JC, Jenkins JN, Florane C, Li P, and Fang DD, The P450 gene CYP749A16 is required for tolerance to the sulfonylurea herbicide trifloxysulfuron sodium in cotton (*Gossypium hirsutum* L.). *BMC Plant Biol* **18**:e186 (2018).

62 Dam, T, Guida Jr, AD, Hazel CB, Bailin LI, Williams ME, Polynucleotide encoding a maize herbicide resistance gene and methods for use. US 2008/0096218 A0096211 (2008).

63 Cummins I, Wortley DJ, Sabbadin F, He Z, Coxon CR, Straker HE, *et al.*, Key role for a glutathione transferase in multiple-herbicide resistance in grass weeds. *PNAS* **110**:5812–5817 (2013).

64 Evans AF Jr, O'Brien SR, Ma R, Hager AG, Riggins CW, Lambert KN, and Riechers DE, Biochemical characterization of metabolism-based atrazine resistance in *Amaranthus tuberculatus* and identification of an expressed GST associated with resistance. *Plant Biotechnol J* **15**:1238–1249 (2017).

65 Han HP, Yu Q, Cawthray G, and Powles SB, Enhanced herbicide metabolism induced by 2,4-D in herbicide susceptible *Lolium rigidum* provides protection against herbicide injury. *Pest Manag Sci* **69**:996–1000 (2013).

66 Duhoux A, Carrère S, and Délye C, Transcriptional markers enable identification of rye- grass (*Lolium* sp.) plants with non-target-site-based resistance to herbicides inhibiting acetolactate synthase. *Plant Sci* **257**:22–36 (2017).

67 Nelson D, and Werck-Reichhart D, A P450-centric view of plant evolution. *Plant J* **66**:194–211 (2011).

68 Liu Z, Tavares R, Forsythe ES, Andre F, Lugan R, Jonasson G, *et al.*, Evolutionary interplay between sister cytochrome P450 genes shapes plasticity in plant metabolism. *Nat Commun* **7**:e13026 (2016).

69 Giacomini DA, Patterson EL, Küpper A, Beffa R, Gaines TA, and Tranel PJ, Coexpression clusters and allele-specific expression in metabolism-based herbicide resistance. *Genome Biol Evol* **12**:2267–2278 (2020).

70 Mao L, Kawaide H, Higuchi T, Chen M, Miyamoto K, Hirata Y, *et al.*, Genomic evidence for convergent evolution of gene clusters for momilactone biosynthesis in land plants. *PNAS* **117**:12472–12380 (2020).

71 Werck-Reichhart D, Hehn A, and Didierjean L, Cytochromes P450 for engineering herbicide tolerance. *Trends Plant Sci* **5**:116–123 (2000).

72 Beckie HJ, Herbicide-resistant weeds: management tactics and practices. *Weed Technol* **20**:793–814 (2006).

73 Coupland D, Lutman PJW, and Heath C, Uptake, translocation and metabolism of mecaprop in a sensitive and resistant biotype of *Stellaria media*. *Pestic Biochem Physiol* **36**:61–67 (1990).

74 Ashworth MB, Walsh MJ, Flower KC, and Powles SB, Recurrent selection with reduced 2,4-D amine doses results in the rapid evolution of 2,4-D herbicide resistance in wild radish (*Raphanus raphanistrum* L.). *Pest Manag Sci* **72**:2091–2098 (2016).

75 Lu H, Yu Q, Han H, Owen MJ, and Powles SB, Metribuzin resistance in a wild radish (*Raphanus raphanistrum*) population via both *psbA* gene mutation and enhanced metabolism. *J Agric Food Chem* **67**:1353–1359 (2019).

76 Lu H, Yu Q, Han H, Owen MJ, and Powles SB, Evolution of resistance to HPPD-inhibiting herbicides in a wild radish population via enhanced herbicide metabolism. *Pest Manag Sci* **76**:1929–1937 (2020a).

77 Heap IM, *International Herbicide-Resistant Weed Database* (2021) Available: www.weedscience.org [15 March 2021].

78 Ma R, Kaundun SA, Tranel PJ, Riggins CW, McGinness DL, Hager AG, *et al.*, Distinct detoxification mechanisms confer resistance to mesotrione and atrazine in a population of waterhemp. *Plant Physiol* **163**:363–377 (2013).

79 Lu H, Liu YZ, Li MS, Han H, Zhou FY, Nyporko A, Yu Q, Qiang S, and Powles SB, Multiple metabolic enzymes can be involved in cross-resistance to HPPD-inhibiting herbicides in wild radish. *J Agric Food Chem* **71**:9302–9313 (2023).

80 Lu H, Yu Q, Han H, Owen MJ, and Powles SB, Non-target-site resistance to PDS- inhibiting herbicides in a wild radish (*Raphanus raphanistrum*) population. *Pest Manag Sci* **76**:2015–2020 (2020b).

81 Nandula VK, Riechers DE, Ferhatoglu Y, Barrett M, Duke SO, Dayan FE, *et al.*, Herbicide metabolism: crop selectivity, bioactivation, weed resistance, and regulation. *Weed Sci* **67**:149–175 (2019).

7

Management of Metabolism-Based Resistance in *Lolium rigidum* in Australia

Christopher Preston

School of Agriculture, Food & Wine, University of Adelaide, Glen Osmond South Australia, Australia

7.1 Introduction: Complications of Metabolism-Based Herbicide Resistance

Where herbicide resistance is the result of a mutation in gene encoding the target enzyme (target-site resistance), resistance to one or more herbicides with the same mode of action can occur.[1] Target-site resistance allows rotation of herbicide modes of action to easily control the resistant weed species, provided the alternative herbicides are effective. The evolution of target-site resistance has resulted in activities to educate herbicide users about herbicide modes of action through mode of action labeling.[2–4]

In contrast, managing herbicide resistance due to enhanced metabolism of herbicides is more difficult due to the complex patterns of cross-resistance that occur across herbicide modes of action. While multiple enzymatic systems break down different herbicides in plants, some, such as aryl acylamidase and aldo-keto reductase, only act on single herbicides or herbicides from one mode of action.[5,6] Due to the specificity of these enzymes, resistance is limited and herbicide rotation remains an effective strategy. This is not true of other enzymatic systems that detoxify herbicides. For example, glutathione-*S*-transferases (GST) interact with electrophilic groups on herbicides, most notably on the triazine, chloroacetamine, thiocarbamate, and diphenylether herbicides.[7] Cytochrome P450 (CYP) oxygenases can metabolize some of the same herbicides, but also a wide range of other herbicides.[8] GST and CYP occur as gene families, with the CYP family being particularly large in plants.[9] To date, several different CYPs have been identified as contributing to the detoxification of one or more herbicides.[10–13] In contrast to target-site resistance, predicting resistance patterns where resistance is due to

Resistance in Weeds from Herbicide Metabolism, First Edition.
Edited by Vijay K. Nandula and Roland Beffa.

increased metabolism of herbicides has proved difficult as resistance is based on chemical features of the herbicide molecule, rather than its site of action.

The practices required for the management of metabolism-based herbicide resistance are different and considerably more complex than those required for target-site resistance.[14] Australian farmers have been managing metabolism-based herbicide resistance in *Lolium rigidum* since the 1980s, and a suite of practices have been employed with varying degrees of success.[15–17] This chapter will discuss the management of metabolism-based herbicide resistance in *L. rigidum* in Australia as a case study.

7.2 Problem of Herbicide Resistance in *L. rigidum* in Australia

Lolium rigidum is particularly challenging to manage due to its biology and adaptability. *L. rigidum* is an obligate outcrossing species as a result of a self-incompatibility gametic system.[18] In addition, *L. rigidum* can cross readily with two related species: *L. perenne* ssp. *perenne* and *L. perenne* ssp. *multiflorum*.[19] This has resulted in considerable genetic diversity within populations.[20] *L. rigidum* is widespread across the temperate world including southern Australia.[21] It was widely planted as a pasture species during the first half of the twentieth century in Australia. In addition, there were imports into Australia of *L. perenne* ssp. *perenne* and *L. perenne* ssp. *multiflorum* from South Africa, Algeria, the United States, and New Zealand.[22] *L. rigidum* was planted mainly in lower rainfall districts, with the other species mainly in the higher rainfall districts where perennial pastures are common.[22] However, there was significant overlap in distribution allowing considerable exchange of genes. Over time, *L. rigidum* became the most significant weed of grain crops in Australia.[19]

As a result of its obligate outcrossing nature, *L. rigidum* has become an accumulator of herbicide resistance mechanisms. Currently in Australia, *L. rigidum* is resistant to nine different herbicide modes of action and populations have been reported with resistance to seven of these modes of action.[23] Resistance to herbicides is extensive in this species being found in more than 50% of crop fields in every state where *L. rigidum* is common.[24–27] The extent of resistance varies between the states ranging from 55% of samples in Tasmania to >99% of samples in Western Australia, and overall more than 80% of all crop fields sampled contained herbicide-resistant *L. rigidum* (Table 7.1). The amount of resistance present tends to reflect the extent of continuous crop production fields in each state. In addition, resistance to multiple modes of action (two or more) is common in individual populations, ranging from 19% of samples from Tasmania to 96% of samples in Western Australia (Table 7.1).

Table 7.1 Extent of resistance to multiple herbicide modes of action in *L. rigidum* in Australia as determined from random surveys of crop fields.

	State				
	New South Wales	Victoria	Tasmania	South Australia	Western Australia
Modes of action	Samples with resistance (%)				
0	21.1	7.5	45.2	9.2	0.3
1	22.4	21.3	35.7	30.1	4.4
2	50.4	64.0	19.0	28.0	64.8
3	6.0	6.7	0	28.0	29.6
4	0	0.4	0	4.5	0.8
5	0	0	0	0.3	0
6	0	0	0	0	0
Samples tested	588	353	52	448	362

Note: Samples collected from crop fields were tested with herbicides from six modes of action.
Source: Adapted from Refs.[26,27]

The mechanisms responsible for resistance vary within and between populations of *L. rigidum*, with both target-site mutations and increased herbicide detoxification common in populations resistant to ACCase inhibitors including 70% of resistant populations containing both mechanisms.[28,29] This accumulation of herbicide resistance mechanisms in *L. rigidum* populations results in the complex resistance patterns seen across multiple modes of action.

7.3 Herbicide Resistance Testing to Understand Which Herbicides are Still Effective

Rotation of herbicide modes of action is a less effective strategy where metabolism-based resistance is present. Herbicide resistance testing is a method of understanding the herbicides that are still available for use. In Australia, herbicide resistance testing involves collecting seeds or plants from surviving weeds in fields and testing them in a laboratory for survival to herbicides of interest. Herbicide resistance testing of annual ryegrass has a long history in Australia,[30] with several commercial providers of tests. The commercial providers conduct tests by applying herbicides to whole plants, which means that resistance is identified regardless of the mechanism present.[31]

The Australian experience with herbicide resistance testing of weeds has identified several challenges. First, collecting the seed can be time-consuming while other activities, such as crop harvest, are underway. Second, it is important to collect samples from the area where the herbicide has failed to provide the most useful information. Finally, ensuring the correct herbicides are in the testing panel.[30] Historically, there was a tendency to test for the herbicide that had failed in the field. While this provided useful information, it is an unreliable indicator of the herbicide options still available for control. More recently, users have been encouraged to test for the herbicides they may wish to use over the next few years. This provides more useful information that can help decision-making.

Despite the wide use of herbicide resistance tests for *L. rigidum* in Australia, there is an argument that insufficient testing is conducted.[30] A test showing resistance to herbicides in one field may be used to infer that other fields on the same farm, or adjacent farms, have the identical problem when an herbicide fails. As discussed earlier, *L. rigidum* populations can vary between and within fields in the herbicides they are resistant to, as well as the mechanisms of resistance present.[28] Therefore, assuming resistance status from tests conducted elsewhere is risky.

7.4 Herbicide-Based Resistance Management Strategies

Despite the extensive resistance to herbicides present in *L. rigidum* in Australia, herbicides remain key components of *L. rigidum* management. In 2016, grain farmers reported spending AUD187 million per annum on herbicides to specifically control herbicide-resistant weeds, predominantly *L. rigidum*.[32] A number of different herbicide strategies have been promoted to help manage resistance to herbicides. The most obvious strategy has been rotation of modes of action.[33] However, the difficulty with metabolism-based resistance mechanisms is that unexpected cross-resistance occurs. This makes rotation of modes of action less useful than where target-site resistance mechanisms occur.[14] Some herbicide approaches that have been proposed, such as using all herbicides in mixtures,[34] are not widespread due to difficulties with crop selectivity.

7.4.1 Crop-Topping

One of the more useful strategies developed for the management of *L. rigidum* in crop production in Australia has been the adoption of late season herbicide applications to stop weed seed set. Collectively known as crop-topping, these practices were originally developed in the 1990s in pulse crops,[15] but registrations have extended to cover most winter crops grown.[35] Crop-topping initially involved the use of paraquat on pulse crops applied prior to harvest to reduce annual ryegrass seed set. Subsequently, glyphosate has been registered for the same use in pulse and other crops.[36] Crop-topping has proven particularly useful as it can reduce the amount of weed seed that returns to the seed bank, helping to delay the onset of resistance to other herbicides.[36]

7.4.2 Double Knock

Another herbicide strategy used is the double knock, which was originally developed to help manage resistance to glyphosate.[37] The double knock is the practice where paraquat is applied to follow a pre-seeding burn-down application of glyphosate to control any potential glyphosate-resistant survivors. This concept has since been expanded to include other herbicides.

7.4.3 Preemergent Herbicide Strategies

The most widespread current herbicide tactic primarily used for the management of *L. rigidum* in Australia is the adoption of preemergent herbicides applied immediately before seeding the crop. Following the widespread evolution of resistance to the postemergent herbicides for the control of *L. rigidum* in cereal crops in Australia, preemergent herbicides were adopted as a major control tactic.[38] The first to be widely used was trifluralin. At the same time, farmers were moving to no-till seeding systems. To protect wheat crops from trifluralin, Australian farmers adopted a narrow point soil opener that moved soil from the crop seed row into the inter-row, rather than use a disk-based soil opener with less soil movement.[39] In this system, herbicide is applied in a concentrated band to the weed seed sitting on the soil surface, followed by incorporation of the soil throw, as illustrated in Fig. 7.1. This very specific environment results in better performance of trifluralin against *L. rigidum* compared with previous systems where tillage had been used prior to herbicide application[40] and facilitated the widespread adoption of preemergent herbicides in no-till seeding systems.

Since the late 2000s, several additional preemergent herbicides have been developed for the control of *L. rigidum* in all grain crops grown in Australia.[38] There has also been increasing interest in improving preemergent herbicide performance, particularly in cereals, as often there are no further herbicide options available. One of the challenges of using preemergent herbicides as the main herbicide control is that they are subject to variable performance due to differences in soil type and rainfall pattern. Considerable effort has gone into investigating the use of preemergent herbicide mixtures for better control of *L. rigidum* and for delaying resistance.[34,38] The right choice of mixture can improve the overall reliability of weed control.[41] This is illustrated in Fig. 7.2 showing the effect

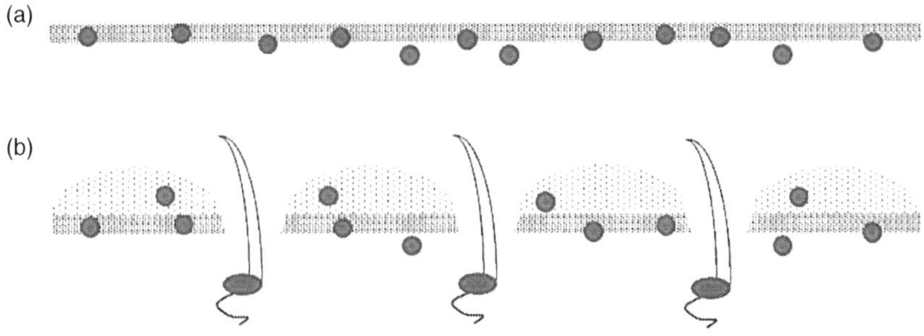

Figure 7.1 Diagrammatic representation of the use of preemergent herbicides in the no-till knife-point sowing system in Australia. As no pre-seeding tillage is conducted, weed seeds occur on or close to the soil surface. The herbicide is applied prior to sowing producing a concentrated band close to the weed seeds (a). The knife-point sowing system throws herbicide-treated soil out of the crop row and onto the inter-row, resulting in an herbicide-free area for the crop to emerge while most weed seeds germinate in herbicide-treated soil (b).

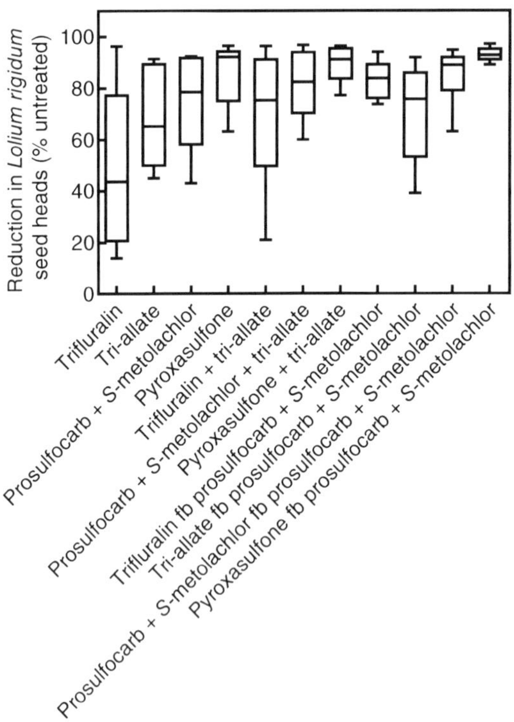

Figure 7.2 Box and whiskers plot of six trials conducted with various preemergent herbicide products, mixtures, and sequences for the control of annual ryegrass in wheat in south-eastern Australia in 2011 and 2012. Trials were conducted at Yarrawonga VIC, Lake Bolac VIC, Wagga Wagga NSW, Manoora SA, and Saddleworth SA. "+" indicates that herbicides were applied as a mixture (prosulfocarb + *S*-metolachlor is a single herbicide product in Australia). "fb" indicates that the second herbicide was applied early postemergent at the one-leaf *L. rigidum* growth stage. Line in the middle of the box is the mean of all six trials, and whiskers indicate the trials with the highest and lowest control. Source: C. Preston, H. Wu, S. Kleemann and G. Gill, unpublished data.

of preemergent herbicides alone and in mixtures in controlling *L. rigidum* measured as the reduction in weed seed set at the end of the crop-growing season. Several of these trial sites had *L. rigidum* with resistance to trifluralin, which reduced the performance of that herbicide. The mixtures and sequences of preemergent herbicides reduced the variability of performance compared to single products applied alone. Mixtures of trifluralin with other herbicides also tended to improve control in situations where trifluralin resistance was present.

7.5 Non-herbicide-Based Resistance Management Strategies

In addition to herbicide strategies, various non-herbicide strategies have been used for the control of herbicide-resistant *L. rigidum*. In general, non-herbicide strategies are not as effective or as easy to employ as herbicides, so they tend to be used in addition to herbicides, rather than replacing them. Crop rotation can allow the use of alternative herbicides or non-herbicide practices.[15] More than 25% of Australian crop farmers report that they have changed the crops grown to deal with herbicide-resistant weeds.[32]

7.5.1 Tillage

As previously mentioned, Australian crop production has mainly moved to a no-till farming system that employs no soil disturbance, except during the sowing operation. The main sowing equipment used is a narrow-point opener that moves soil from the crop seeding row onto the inter-row.[39] As a consequence, tillage has been greatly reduced as a management tactic for weeds of grain production.[32]

Some limited tillage is still used in specific circumstances, for example, moldboard plough operations to bury weed seed are used in a limited way.[42] They can be successful for a period of time in greatly reducing *L. rigidum* populations, but require that the field is not tilled again for a long period of time to keep the weed seed buried. Moldboard ploughing is expensive, so it is rarely conducted unless there is a requirement to amend soil problems, such as non-wetting sands. With non-wetting sands, burial of the non-wetting upper layer occurs, producing a surface where water can penetrate easily. At the same time, *L. rigidum* seeds are buried in-depth from which they cannot emerge.[43]

7.5.2 Crop Competition

Employing crop competition against weeds can reduce emergence of weeds and, more importantly, the amount of seed they set.[44] There are numerous ways that crop competition can be increased, and most of them have been promoted for use in managing herbicide-resistant *L. rigidum* in Australia. One of the difficulties encountered with improving crop competitiveness is that it can require changes to other agronomic management practices, such as increased fertilizer requirements and increased fungicide applications. This often results in compromises where crop competition is not fully exploited.

One important tactic to improve crop competition is increasing crop-seeding rates in cereals. This was shown to be useful in reducing the effect of *L. rigidum* on crop yield and reducing weed seed set. An extensive set of trials across Australia demonstrated that seeding rates in wheat-producing densities of at least 200 plants/m^2 were required to reduce *L. rigidum* biomass and hence seed set.[45] In these trials, *L. rigidum* biomass was reduced by 50% as the wheat density was increased from 100 to 200 plants/m^2. The benefits of higher seeding rates were smaller, with a further 50% reduction in biomass at a wheat density of 400 plants/m^2. As a consequence of this research, there has been an increase in the average crop-seeding rate in wheat across much of Australia. The same benefits occur for other cereal crops, but are more variable for canola[17] or pulse crops,[46] resulting in limited change to seeding rates in these crops. While crop type and crop cultivar can also increase crop competition against weeds,[47–49] there is limited exploitation of these factors, except for growing hybrid canola cultivars.[50]

Reducing row spacing was also promoted for improving crop competition against *L. rigidum* as wider crop rows leave a larger area between the rows where weeds can establish. However, the effects of row spacing on *L. rigidum* biomass and seed production are variable.[51,52] Reducing the crop row spacing from 36 or 27 cm to 9 or 18 cm resulted in a significant reduction in *L. rigidum* seed production; however, there was no additional benefit of 9 cm over 18 cm spacing.[52] This suggests that changes in row spacing are only effective over a limited range. In addition, the no-till seeding system with preemergent grass herbicides used across much of southern Australia leaves the inter-row relatively undisturbed and with high herbicide concentration, reducing the ability of *L. rigidum* to emerge in the inter-row.

Seeding cereal crops in an east to west direction rather than in a north to south direction has also been used as a means of favoring the crop over weeds. In winter, less light reaches the soil surface in the inter-row in east- to west-sown crops compared with crops sown in a north-to-south direction. Trials have shown *L. rigidum* biomass reductions of 51% with east to west seeding in wheat and of 37% reductions for barley crops.[53,54] As a result of differences in crop architecture, there is little value in changing seeding direction in canola or pulse crops.[53]

7.5.3 Hay Crops

Growing oats to cut for hay has proven to be an effective way of managing herbicide-resistant *L. rigidum.*[15] Oats are the most competitive cereal crop, and the crop is cut prior to seed maturation on *L. rigidum. L. rigidum* seed banks can be reduced by as much as 85% with a crop of oats cut for hay.[55] It is important that any regrowth of *L. rigidum* is controlled once the hay bales have been removed from the field. In addition to deliberately growing oats and cutting for hay, farmers have also opportunistically cut cereal or canola crops for hay if the crops have been damaged by frost or have excessive weed populations. Cutting such crops early allows for high levels of reduction in *L. rigidum* seed production. An alternative to hay is growing a crop specifically for weed control. In this practice, the crop is terminated prior to weed flowering by cultivation (green manure) or by herbicides (brown manure), resulting in no weed seed set.[56]

7.5.4 Harvest Weed Seed Control

Harvest weed seed control (HWSC) is a group of tactics invented in an attempt to reduce the amount of weed seed returning to the field during the harvest operation. Seed of *L. rigidum* is usually still present in the seed spike at harvest. As the seed is small, it often exits from the harvester in the chaff fraction and is spread across the field. HWSC practices take advantage of this to collect the chaff fraction and manage it, rather than return the weed seeds to the field.[57] There are several pieces of equipment that have been used to do this, ranging from chaff carts that collect and remove the chaff fraction, chutes that funnel the chaff and straw into narrow windrows for burning, chaff chutes that place the chaff on wheel tracks, and impact mills that process the chaff, destroying any weed seeds present.[58] All of these practices have similar efficacy at reducing *L. rigidum* populations in the next season, because they all rely on the weed seed being harvested with the crop to work.[59] If the weed seed has shed or the plants have lodged prior to harvest, the seed cannot be captured by the harvester. Overall, HWSC can reduce *L. rigidum* populations in the next year by about 50%.[59] Efficacy of HWSC can be increased by harvesting earlier, cutting the crop at a lower height, or swathing the crop prior to harvest.

7.6 Integrated Management of *L. rigidum* Populations

Perhaps, the most important change made in the management of herbicide-resistant *L. rigidum* in Australia has been the move from the use of individual tactics to control this species to combinations of tactics. Through the 1980s and early 1990s, considerable attention was paid to the single-season economic return

Table 7.2 Estimated costs of weeds in grain crop production in Australia from 1987 to 2016 (in 2016 AUD corrected for inflation).

	Costs (AUD million)		
	1987	**1999**	**2016**
Revenue loss	1240	648	745
Control costs	1862	1242	2573
Total	3102	1990	3318

Note: Revenue losses include yield losses due to residual weeds and losses as well as weed contamination of grain.
Source: Refs.[32,63,64]

of using herbicides to control *L. rigidum*.[60–62] With the widespread evolution of herbicide resistance, the approach moved to the employment of strategies to reduce weed numbers and keep them low over several years.[55] One result reveals that costs of weed control in grain crop production in Australia have increased considerably; however, revenue losses due to weeds have decreased as a proportion of total costs (Table 7.2). Between 1987 and 2016 estimates of revenue losses caused by weeds in grain crops in Australia decreased by 40% in real terms. At the same time, control costs for weeds increased by 40% in real terms. Much of this is a result of a focus on reducing *L. rigidum* numbers, rather than trying to maximize economic returns every year. Between 1987 and 1999, cost decreased as herbicides were substituted for more expensive multiple tillage operations. From 1999, control costs increased as additional herbicides and non-chemical tactics were introduced to manage herbicide-resistant weeds.

7.6.1 Synergy Between Tactics

Developing strategies that work and are profitable can be challenging. As many of the non-chemical tactics employed are less effective at controlling *L. rigidum* compared with herbicides, they are not generally used as a direct replacement for a failed herbicide.[35] These tactics are often added to the use of herbicides. There are opportunities to improve the efficacy by using tactics by identifying synergies between tactics. An example is the use of crop competition in combination with preemergent herbicides to control *L. rigidum*. In this situation, the preemergent herbicides control the early emerging *L. rigidum*, while the crop suppresses later emerging individuals.[65]

Earlier sowing of wheat, while the soil is still warm, allows the crop to be more competitive against later germinating cohorts of *L. rigidum*. Where this is coupled with effective preemergent herbicides that control the first flush of weeds, this can greatly reduce the amount of seed set compared with later sowing of the crop.[65] A similar benefit can be achieved with using more competitive hybrid cultivars of canola instead of open-pollinated cultivars.[51]

Increasing competition can also aid other tactics such as HWSC. *L. rigidum* seed heads are more likely to be higher in the crop canopy if the crop is more competitive. This ensures that more of the seed can be managed through HWSC.[66]

7.6.2 Stacking Tactics

Research investigating the efficacy of strategies has shown that in addition to crop rotations, multiple tactics are required every year to reduce *L. rigidum* weed populations. For example, employing a single tactic every year resulted in an increase in *L. rigidum* seed banks (Fig. 7.3).[55] Using two tactics every year resulted in seed banks remaining the same. However, employing additional tactics resulted in a decline in *L. rigidum* seed banks over time. Likewise, the use of narrow row spacing and HWSC every year was able to reduce *L. rigidum* populations.[52]

The exact tactics adopted by farmers to manage herbicide-resistant *L. rigidum* vary widely. Location, average seasonal rainfall, soil type, and farmer preference all influence the tactics adopted. In general, tactics that are relatively easy to use and are profitable are more likely to be adopted. The adoption of additional tactics

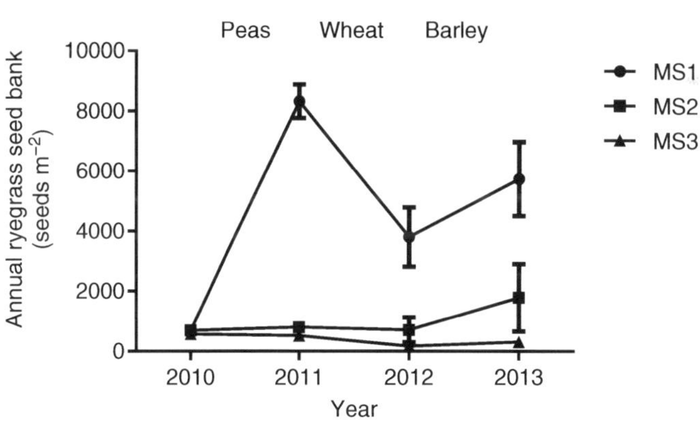

Figure 7.3 Changes in pre-seeding *L. rigidum* seed bank in response to three different weed management strategies in a trial at Roseworthy, South Australia. The trial had a crop rotation of peas followed by wheat and then barley with three management strategies (MS1 to 3) involved in increasing levels of intensity of management in each crop in the rotation. MS1 included a single preemergent herbicide in all crops; MS2 had an additional management strategy in the pea and wheat crops; and MS3 had two additional management strategies in the pea crop and one additional management strategy in the wheat crop. The symbols are means ± SEM for three replicates. Source: Data from Kleemann et al.[55] where further details on management strategies can also be found.

Table 7.3 Adoption of additional practices by farmers in Western Australia and south-eastern Australia (South Australia, Victoria, Tasmania, and New South Wales) for the control of weeds in small grain production.

	Farmers employing various weed management practices (% of farmers)	
Tactic	**Western Australia**	**South-eastern Australia**
Pre-seeding cultivation	15	30
Double knock	73	53
Crop-topping	52	47
Harvest weed seed control (HWSC)	62	39
Hay	78	50
Manure crop	26	28

Individual practices may be used on all or only some of the cropped land in any one year.
Source: Adapted from Llewellyn et al.[32]

for *L. rigidum* control in Australia is high. A high percentage of farmers employ various non-herbicide tactics for weed control (Table 7.3). The financial cost of these additional tactics is estimated at AUD152 million,[32] not including the cost of additional herbicides to specifically control herbicide-resistant weeds.

While several chemical and non-chemical tactics have been adopted for the management of herbicide-resistant *L. rigidum* in Australia, most of them are clustered at the beginning of the crop-growing season or at the end (Fig. 7.4). Only postemergent herbicides and crop competition are available as tactics

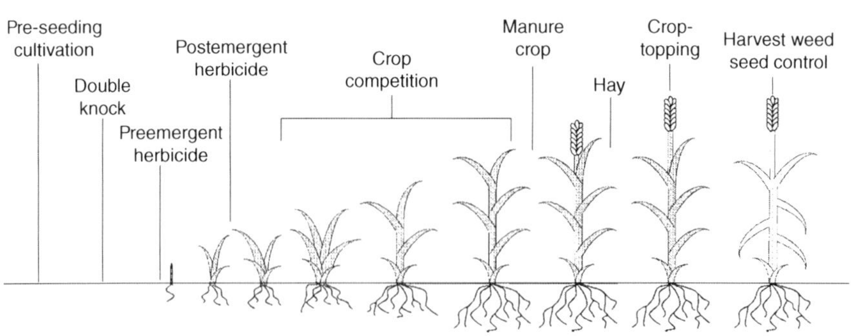

Figure 7.4 Diagrammatic representation of weed management tactics used against *L. rigidum* in Australia is employed during the life of a crop. While crop competition exerts its effects during the middle period of crop growth, decisions about its implementation need to be taken prior to seeding the crop.

during the main crop-growth period, and neither of these are available in all grain crops. Stacking tactics from different phases of cropping season is the most effective strategy. For example, the combination of preemergent herbicides, crop competition, and HWSC is highly effective.

7.7 Conclusions

The extent and complexity of herbicide resistance in *L. rigidum* in Australia has resulted in a complete change in the approach to small grains crop production in Australia. Weed management has now become a key driver of farmer decision-making, influencing crop choice as well as herbicide choice, seeding equipment, and the introduction of new weed management practices. This has resulted in a much more robust weed management program for *L. rigidum* that often includes multiple tactics in each cropping season. This has come at cost to farmers, not only from the additional tactics required, but also from the increased complexity of weed management. However, weed populations in grain crops tend to be smaller and reduce crop yield by less.

Surveys of herbicide-resistant *L. rigidum* in Australia have shown that resistance to herbicides is continuing to increase in *L. rigidum* populations. For example, in Western Australia, resistance has increased from 72% of samples collected in 1998 to 99% of samples collected in 2010.[22] In addition, there has been an increase in the number of herbicide modes of action affected by resistance (Table 7.1). This will likely mean that new herbicides introduced for the management of *L. rigidum* in recent years will fail.[67] The genetic diversity present in *L. rigidum* populations in Australia is also likely to result in evolution of avoidance of other control tactics.[19] Therefore, there will be a need to develop new tactics to manage *L. rigidum* in grain crops in the future and management will remain dynamic.

There are several lessons to be learned from the Australian experience. The employment of tactics that reduce *L. rigidum* seed production or reduce seed return to fields and the packaging of multiple tactics into strategies have been the two keys to effective management of this weed species. Similar approaches can be adopted for other weed species. Herbicides remain a key component of weed management systems and are augmented by other tactics, rather than being replaced. Finally, weed management practices need to be both profitable and adoptable to ensure their uptake by farmers.

References

1 Powles SB, and Yu Q, Evolution in action: plants resistant to herbicides. *Annu Rev Plant Biol* **61**:317–347 (2010).

2 Shaner DL, Howard S, and Chalmers I, Effectiveness of mode of action labelling for resistance management: a survey of Australian farmers, In *Brighton Crop Protection Conference – Weeds*, Vol. 3, pp. 797–804, (1999).
3 Retzinger EJ Jr, and Mallory-Smith C, Classification of herbicides by site of action for weed resistance management strategies. *Weed Technol* **11**:384–393 (1997).
4 Beckie HJ, Chang FY, and Stevenson FC, The effect of labeling herbicides with their site of action: a Canadian perspective. *Weed Technol* **13**:655–661 (1999).
5 Leah JM, Caseley JC, Riches CR, and Valverde B, Association between elevated activity of aryl acylamidase and propanil resistance in jungle-rice, *Echinochloa colona*. *Pestic Sci* **42**:281–289 (1994).
6 Pan L, Yu Q, Han H, Mao L, Nyporko A, Fan L, Bai L, and Powles S, Aldo-keto reductase metabolizes glyphosate and confers glyphosate resistance in *Echinochloa colona*. *Plant Physiol* **181**:1519–1534 (2019).
7 Cummins I, Dixon DP, Freitag-Pohl S, Skipsey M, and Edwards R, Multiple roles for plant glutathione transferases in xenobiotic detoxification. *Drug Metab Rev* **43**:266–280 (2011).
8 Dimaano NG, and Iwakami S, Cytochrome P450-mediated herbicide metabolism in plants: current understanding and prospects. *Pest Manag Sci* **77**:22–32 (2021).
9 Yuan JS, Tranel PJ, and Stewart CN Jr, Non-target-site herbicide resistance: a family business. *Trends Plant Sci* **12**:6–13 (2007).
10 Ohkawa H, Shiota N, Imaishi H, Yamada T, Inui H, and Ohkawa Y, Cytochrome P450 monooxygenases metabolizing herbicides. *Biotechnol Biotechnol Equip* **12**:17–22 (1998).
11 Iwakami S, Endo M, Saika H, Okuno J, Nakamura N, Yokoyama M, Watanabe H, Toki S, Uchino A, and Inamura T, Cytochrome P450 CYP81A12 and CYP81A21 are associated with resistance to two acetolactate synthase inhibitors in *Echinochloa phyllopogon*. *Plant Physiol* **165**:618–629 (2014).
12 Iwakami S, Kamidate Y, Yamaguchi T, Ishizaka M, Endo M, Suda H, *et al.*, CYP 81A P450s are involved in concomitant cross-resistance to acetolactate synthase and acetyl-CoA carboxylase herbicides in *Echinochloa phyllopogon*. *New Phytol* **221**:2112–2122 (2019).
13 Han H, Yu Q, Beffa R, González S, Maiwald F, Wang J, and Powles SB, Cytochrome P450 CYP81A10v7 in *Lolium rigidum* confers metabolic resistance to herbicides across at least five modes of action. *Plant J* **105**:79–92 (2020).
14 Preston C, Herbicide resistance in weeds endowed by enhanced detoxification: complications for management. *Weed Sci* **52**:448–453 (2004).
15 Gill GS, and Holmes JE, Efficacy of cultural control methods for combating herbicide-resistant *Lolium rigidum*. *Pestic Sci* **51**:352–358 (1997).

16 Llewellyn RS, Lindner RK, Pannell DJ, and Powles SB, Herbicide resistance and the adoption of integrated weed management by Western Australian grain growers. *Agric Econom* **36**:123–130 (2007).

17 Lemerle D, Luckett DJ, Wu H, and Widderick MJ, Agronomic interventions for weed management in canola (*Brassica napus* L.) – a review. *Crop Prot* **95**:69–73 (2017).

18 Klaas M, Yang B, Bosch M, Thorogood D, Manzanares C, Armstead IP, *et al.*, Progress towards elucidating the mechanisms of self-incompatibility in the grasses: further insights from studies in *Lolium*. *Ann Bot* **108**:677–685 (2011).

19 Matzrafi M, Preston C, and Brunharo CA, Evolutionary drivers of agricultural adaptation in *Lolium* spp. *Pest Manag Sci* **77**:2209–2218 (2021).

20 Guo ZH, Fu KX, Zhang XQ, Zhang CL, Sun M, Huang T, Peng Y, Huang LK, Yan YH, and Ma X, SSRs transferability and genetic diversity of three allogamous ryegrass species. *CR Biol* **339**:60–67 (2016).

21 Matzrafi M, Preston C, and Brunharo CA, Evolutionary drivers of agricultural adaptation in *Lolium* spp. *Pest Manag Sci* **77**:2209–2218 (2021).

22 Reed KFM, Perennial pasture grasses - an historical review of their introduction, use and development for southern Australia. *Crop Pasture Sci* **65**:691–712 (2014).

23 Heap I, The International Herbicide-Resistant Weed Database. www.weedscience.org [Accessed May 4, 2024] (2024).

24 Boutsalis P, Gill GS, and Preston C, Incidence of herbicide resistance in rigid ryegrass (*Lolium rigidum*) across southeastern Australia. *Weed Technol* **26**:391–398 (2012).

25 Broster JC, Koetz EA, and Wu H, Herbicide resistance levels in annual ryegrass (*Lolium rigidum* gaud.) in southern New South Wales. *Plant Prot Q* **26**:22–28 (2011).

26 Owen MJ, Martinez NJ, and Powles SB, Multiple herbicide-resistant *Lolium rigidum* (annual ryegrass) now dominates across the Western Australian grain belt. *Weed Res* **54**:314–324 (2014).

27 Broster J, Boutsalis P, Gill GS, and Preston C, The extent of herbicide resistance in *Lolium rigidum* gaud.(annual ryegrass) across South-Eastern Australia as determined from random surveys. *Crop Pasture Sci* **73**:1308–1317 (2022).

28 Malone JM, Boutsalis P, Baker J, and Preston C, Distribution of herbicide-resistant acetyl-coenzyme a carboxylase alleles in *Lolium rigidum* across grain cropping areas of South Australia. *Weed Res* **54**:78–86 (2014).

29 Han H, Yu Q, Owen MJ, Cawthray GR, and Powles SB, Widespread occurrence of both metabolic and target-site herbicide resistance mechanisms in *Lolium rigidum* populations. *Pest Manag Sci* **72**:255–263 (2016).

30 Broster JC, Pratley JE, Ip RHL, Ang L, and Seng KP, A quarter of a century of monitoring herbicide resistance in *Lolium rigidum* in Australia. *Crop Pasture Sci* **70**:283–293 (2019).

31 Boutsalis P, and Broster J, Herbicide resistance testing of *Lolium rigidum* by commercial institutions, In *Proceedings of the 15th Australian Weeds Conference*, ed. by Preston, C, Watts JH, and Crossman ND, pp. 488–490 (2006).

32 Llewellyn R, Ronning D, Clarke M, Mayfield A, Walker S, and Ouzman J, *Impact of Weeds in Australian Grain Production*, Grains Research and Development Corporation, Canberra, p. 112 (2016).

33 Powles SB, and Howat PD, Herbicide-resistant weeds in Australia. *Weed Technol* **4**:178–185 (1990).

34 Busi R, Powles SB, Beckie HJ, and Renton M, Rotations and mixtures of soil-applied herbicides delay resistance. *Pest Manag Sci* **76**:487–496 (2020).

35 Harries M, Flower KC, Scanlan CA, Rose MT, and Renton M, Interactions between crop sequences, weed populations and herbicide use in Western Australian broadacre farms: findings of a six-year survey. *Crop Pasture Sci* **71**:491–505 (2020).

36 Walsh MJ, and Powles SB, Management strategies for herbicide-resistant weed populations in Australian dryland crop production systems. *Weed Technol* **21**:332–338 (2007).

37 Neve P, Diggle AJ, Smith FP, and Powles SB, Simulating evolution of glyphosate resistance in *Lolium rigidum* II: past, present and future glyphosate use in Australian cropping. *Weed Res* **43**:418–427 (2003).

38 Boutsalis P, Gill GS, and Preston C, Control of rigid ryegrass in Australian wheat production with Pyroxasulfone. *Weed Technol* **28**:332–339 (2014).

39 Rainbow R, and Derpsch R, Advances in no-till farming technologies and soil compaction management in rainfed farming systems, In Tow P, Cooper I, Partridge I, and Birch C (Eds), *Rainfed Farming Systems*, Springer, Dordrecht, pp. 991–1014 (2011).

40 Chauhan BS, Gill GS, and Preston C, Effect of seeding systems and dinitroaniline herbicides on emergence and control of rigid ryegrass (*Lolium rigidum*) in wheat. *Weed Technol* **21**:53–58 (2007).

41 Brunton DJ, Boutsalis P, Gill G, and Preston C, Control of thiocarbamate-resistant rigid ryegrass (*Lolium rigidum*) in wheat in southern Australia. *Weed Technol* **34**:19–24 (2019).

42 Renton M, and Flower KC, Occasional mouldboard ploughing slows evolution of resistance and reduces long-term weed populations in no-till systems. *Agric Syst* **139**:66–75 (2015).

43 Hall DJ, Davies SL, Bell RW, and Edwards TJ, Soil management systems to overcome multiple constraints for dryland crops on deep sands in a water limited environment on the south coast of Western Australia. *Agronomy* **10**:1881. DOI: 10.3390/agronomy10121881 (2020).

44 Bajwa AA, Walsh M, and Chauhan BS, Weed management using crop competition in Australia. *Crop Prot* **95**:8–13 (2017).

45 Lemerle D, Cousens R, Gill G, Peltzer S, Moerkerk M, Murphy C, Collins D, and Cullis B, Reliability of higher seeding rates of wheat for increased competitiveness with weeds in low rainfall environments. *J Agric Sci* **142**:395–409 (2004).

46 Lemerle D, Verbeek B, and Diffey S, Influences of field pea (*Pisum sativum*) density on grain yield and competitiveness with annual ryegrass (*Lolium rigidum*) in South-Eastern Australia. *Aust J Exp Agric* **46**:1465–1472 (2006).

47 Lemerle D, Verbeek B, and Coombes N, Losses in grain yield of winter crops *from Lolium rigidum* competition depend on crop species, cultivar and season. *Weed Res* **35**:503–509 (1995).

48 Lemerle D, Verbeek B, and Orchard B, Ranking the ability of wheat varieties to compete with *Lolium rigidum*. *Weed Res* **41**:197–209 (2001).

49 Lemerle D, Luckett DJ, Lockley P, Koetz E, and Wu H, Competitive ability of Australian canola (*Brassica napus*) genotypes for weed management. *Crop Pasture Sci* **65**:1300–1310 (2014).

50 Mwendwa JM, Brown WB, Weston PA, Haque KS, Preston C, and Weston LA, Evaluation of selected commercial oilseed rape cultivars for early vigour, weed suppression and yield in southern New South Wales. *Weed Res* **60**:450–463 (2020).

51 Lemerle D, Lockley P, Koetz E, and Diffey S, Herbicide efficacy for control of annual ryegrass (*Lolium rigidum* gaud.) is influenced more by wheat seeding rate than row spacing. *Crop Pasture Sci* **64**:708–715 (2013).

52 Borger CPD, Riethmuller G, and D'Antuono M, Eleven years of integrated weed management: long-term impacts of row spacing and harvest weed seed destruction on *Lolium rigidum* control. *Weed Res* **56**:359–366 (2016).

53 Borger CP, Hashem A, and Pathan S, Manipulating crop row orientation to suppress weeds and increase crop yield. *Weed Sci* **58**:174–178 (2010).

54 Borger CPD, Hashem A, and Powles SB, Manipulating crop row orientation and crop density to suppress *Lolium rigidum*. *Weed Res* **56**:22–30 (2016).

55 Kleemann SG, Preston C, and Gill GS, Influence of management on long-term seedbank dynamics of rigid ryegrass (*Lolium rigidum*) in cropping systems of southern Australia. *Weed Sci* **64**:303–311 (2016).

56 Monjardino M, Pannell DJ, and Powles SB, The economic value of haying and green manuring in the integrated management of annual ryegrass and wild radish in a Western Australian farming system. *Aust J Exp Agric* **44**:1195–1203 (2005).

57 Walsh M, Newman P, and Powles S, Targeting weed seeds in-crop: a new weed control paradigm for global agriculture. *Weed Technol* **27**:431–436 (2013).

58 Walsh MJ, and Powles SB, Harvest weed seed control: impact on weed management in Australian grain production systems and potential role in global cropping systems. *Crop Past Sci* **73**:313–324 (2022).

59 Walsh MJ, Aves C, and Powles SB, Harvest weed seed control systems are similarly effective on rigid ryegrass. *Weed Technol* **31**:178–183 (2017).

60 Black ID, and Dyson CB, An economic threshold model for spraying herbicides in cereals. *Weed Res* **33**:279–290 (1993).

61 Streibig JC, Combellack JH, Pritchard GH, and Richardson RG, Estimation of thresholds for weed control in Australian cereals. *Weed Res* **29**:117–126 (1989).

62 Pannell DJ, An economic response model of herbicide application for weed control. *Aust J Agric Econ* **34**:223–241 (1990).

63 Combellack JH, Weed control pursuits in Australia. *Chem Ind* **8**:273–280 (1987).

64 Jones R, Vere DT, Alemseged Y, and Medd RW, Estimating the economic cost of weeds in Australian annual winter crops. *Agric Econ* **32**:253–265 (2005).

65 Preston C, Kleemann SGL, Noack S, Hooper P, and Gill GS, Using time of sowing of wheat and pre-emergent herbicides to control annual ryegrass (*Lolium rigidum*), In *Proceedings of the 18th Australian Agronomy Conference*, Australian Society of Agronomy Inc., pp. 1–4, (2017).

66 Walsh MJ, Broster JC, Aves C, and Powles SB, Influence of crop competition and harvest weed seed control on rigid ryegrass (*Lolium rigidum*) seed retention height in wheat crop canopies. *Weed Sci* **66**:627–633 (2018).

67 Brunton DJ, Boutsalis P, Gill G, and Preston C, Resistance to very-long-chain fatty-acid (VLCFA)-inhibiting herbicides in multiple field-selected rigid ryegrass (*Lolium rigidum*) populations. *Weed Sci* **67**:267–272 (2019).

8

Metabolic Herbicide Resistance in *Echinochloa phyllopogon*

Nina Gracel Dimaano[1] *and Satoshi Iwakami*[2]

[1]*College of Agriculture and Food Science, University of the Philippines Los Banos, Los Banos, Philippines*
[2]*Graduate School of Agriculture, Tokyo University of Agriculture and Technology, Fuchu, Tokyo, Japan*

8.1 Introduction

Echinochloa phyllopogon (syn. *E. oryzicola*), commonly known as late watergrass, is a serious weed problem in water-seeded rice system.[1] *E. phyllopogon* is one of the weed species that has exemplified a remarkable resistance evolution to multiple herbicides due to enhanced herbicide metabolism. As is often the case with enhanced herbicide metabolism that confers cross-resistance to herbicides from distinct chemical classes, the resistant *E. phyllopogon* showed different levels of resistance to at least 15 herbicides from 6 distinct modes of action. The metabolism-based cross-resistance was even extended to relatively new herbicides, even without the application history of such herbicides in the field.[2] The rapid evolution of metabolism-based cross-resistance in *E. phyllopogon* poses a huge threat to herbicide sustainability and rice productivity.

The multiple herbicide-resistant (MHR) population of *E. phyllopogon* was first reported in Sacramento Valley, California in the 1990s.[3] Early research to elucidate the resistance mechanism in MHR *E. phyllopogon* involved physiological and metabolic studies, which provided some important clues on the metabolism-based resistance mechanism and the possible enzyme groups involved in the resistance. Gene expression analysis, gene functional characterization, and inheritance studies paved the way to the discovery of cytochrome P450s (P450s) involved in the metabolic resistance in MHR *E. phyllopogon*. Two P450 genes, *CYP81A12* and *CYP81A21*, which are constitutively highly expressed in the MHR population, confer concomitant cross-resistance to herbicides from multiple modes of action.

Resistance in Weeds from Herbicide Metabolism, First Edition.
Edited by Vijay K. Nandula and Roland Beffa.

The case of MHR *E. phyllopogon* clearly demonstrates the severity of the resistance problem and the grave threat brought about by the evolution of a few resistance genes in a weed population, i.e., *CYP81A12* and *CYP81A21*, conferring resistance to multiple herbicides. However, up to this date, genes involved in the resistance to herbicides by which the MHR population from California rice fields have been selected (e.g., molinate, thiobencarb, and fenoxaprop-*P*-ethyl) still have not been elucidated. Is the resistance trait controlled by a single or several coexisting mechanisms? Are there other enzyme groups involved in the metabolic resistance other than P450s? In the case of *CYP81A12* and *CYP81A21*, how are these genes co-upregulated? These are some of the questions left unsolved. Moreover, the mechanism of multiple herbicide resistance in other resistant populations of *E. phyllopogon* (e.g., from South Korea and Japan) remains undisclosed.

This chapter highlights the progress of metabolic herbicide resistance research in *E. phyllopogon* and explores the lingering questions and challenges to fully understand the metabolic resistance evolution in this agronomically important weed species. The discussion here is focused mainly on the resistant population collected from California, wherein significant research breakthroughs have been achieved through many years of cumulative efforts from various institutions, based on the knowledge and findings available as of November 2021.

8.2 A Brief History of Resistance Evolution in California

Echinochloa phyllopogon is a morphological mimic of rice that could cause more than 50% yield reduction if not controlled in rice paddy fields.[1,4] *E. phyllopogon* is a predominantly self-pollinated grass under the Panicoideae subfamily and is the tetraploid progenitor ($2n=4\times=36$) of *Echinochloa crus-galli* ($2n=6\times=54$).[5] In the late 1990s, *E. phyllopogon* population with resistance to multiple herbicides (thiobencarb, molinate, fenoxaprop-*P*-ethyl, and bispyribac-sodium) was first reported in Sacramento Valley, California.[3] The second largest area planted to rice in the United States is found in California with approximately 513,000 acres[6]; thus, the evolution of metabolism-based cross-resistance is a huge concern that threatens the rice productivity in the country.

Wet-seeding and continuous flooding began in California rice fields in the 1920s as a method to suppress the severe infestations of *E. crus-galli*. In this practice, the field is flooded right after land preparation; then, the rice seeds are directly seeded into standing water under continuous flooding.[7,8] Aside from *E. crus-galli*, the other *Echinochloa* species that infested the rice fields in California were *E. phyllopogon* and *Echinochloa oryzoides*.[4] However, the latter grass weeds were only partially controlled by flooding due to their hypoxia tolerance; thus, herbicide application became a necessary complementary weed control practice.[3] The first selective herbicide to be introduced in California rice fields in the 1960s was propanil.[9] Propanil was effective due to its relatively wide window of application timing and broad–spectrum weed control. However, since propanil is foliar-active, it was ineffective when used for "into-the-water" granular application and required partial or complete field drainage for early applications. Furthermore, propanil caused injury to non-target crops leading to its restriction[10]; hence, propanil was replaced by molinate in the early 1970s[11] and then by thiobencarb in the early 1980s.[3] Both molinate and thiobencarb became significant for California rice production since these herbicides can be directly applied into flood water without the need for field drainage.

At that time, molinate was regarded as the most effective and widely used herbicide to control *Echinochloa* spp.; however, molinate was *Echinochloa*-specific, thus it was required to be combined with other herbicides for the effective control of other weed species. Thiobencarb, which has a wider spectrum of weed control, became an alternative for molinate; however, thiobencarb has a narrow window of application to control *Echinochloa* spp. To expand the narrow application window of thiobencarb, lower rates of molinate and thiobencarb were applied sequentially. Then, in the early 1990s, fenoxaprop-*P*-ethyl was introduced to control *Echinochloa* spp. that had escaped previous herbicide treatments.[3] Fenoxaprop-*P*-ethyl provides good efficacy to control *Echinochloa* spp. in broad stages of plant growth; however, it has a narrow window for selectivity between crop and weeds, potentially leading to crop injury or lack of weed control in field conditions.[10]

The continuous use of few available herbicides (i.e., molinate, thiobencarb, and fenoxaprop-*P*-ethyl) in predominantly monoculture rice fields in California is believed to have caused the evolution of the MHR population of *E. phyllopogon*.[3,12] The initial confirmation of resistance was in 1996 when farmers complained that the weed could no longer be effectively controlled by the herbicides.[8] Resistance was even reported for bispyribac-sodium (acetolactate synthase [ALS] inhibitor), which was not yet released and used in the field during that time.[3] It was suggested that this event is due to the indirect selection pressure for resistance by the previous use of another ALS inhibitor, bensulfuron-methyl, for the control of broadleaves and sedges. Bensulfuron-methyl, which has minimal activity on *E. phyllopogon*, has been widely used in California for several years.[13] Later, the MHR population of *E. phyllopogon* was also confirmed to have developed resistance to bensulfuron-methyl and several herbicides that were not yet commercialized during the initial report of resistance, such as acetyl-CoA carboxylase (ACCase) inhibitors (e.g., cyhalofop-butyl), 1-deoxy-D-xylulose 5-phosphate synthase (DXPS) inhibitor (clomazone), synthetic auxin (quinclorac), and another

ALS inhibitor (penoxsulam).[14–17] Recently, resistance to herbicides belonging to all three chemical groups of ACCase inhibitors (e.g., diclofop-methyl, tralkoxydim, and pinoxaden) was reported.[18] Furthermore, resistance to relatively novel herbicides, e.g., pyrimisulfan and propyrisulfuron (ALS inhibitors) and mesotrione (4-hydroxyphenylpyruvate dioxygenase (4-HPPD) inhibitor), was also predicted and confirmed in the MHR line.[2] So far, what is known is that the cross-resistance mechanism in the MHR line encompasses at least 15 herbicides from 10 chemical groups and 6 sites of action (Table 8.1). Notably, there may be more herbicides not yet tested to which the MHR line would exhibit resistance. Unfortunately, the resistance could also extend to newly developed or soon-to-be discovered herbicides. This scenario seriously complicates the herbicide application strategy to control *E. phyllopogon* in California rice fields.

Meanwhile, based on the similarities in morphological traits and amplified fragment length polymorphism (AFLP) fingerprints among MHR and susceptible (S) *E. phyllopogon* strains from California, it was inferred that the MHR population originated from a founder population that carries the resistance alleles.[19]

8.3 Elucidation of the Mechanism of Multiple Herbicide Resistance in *E. phyllopogon*

8.3.1 Physiological and Biochemical Studies

The elucidation of the mechanism of multiple herbicide resistance in *E. phyllopogon* has been a daunting task due to the limited understanding of herbicide resistance evolution in grass weeds. Physiological studies were initially performed to elucidate the mechanism(s) endowing multiple herbicide resistance in *E. phyllopogon*. These studies revealed that the resistant populations carry no alterations in the target sites, and the resistance to some herbicides is metabolism-based and can be reversed by P450 inhibitors. Other findings also suggested that more than one mechanism is involved in the resistance. Specifically, the results of the early physiological studies in MHR *E. phyllopogon* implicated that the mechanism of resistance for bispyribac-sodium, bensulfuron-methyl, penoxsulam, fenoxaprop-*P*-ethyl, and clomazone is due to detoxification by P450s.[13,16,23,24] Hence, this implied that the evolution of resistance to multiple herbicides, with half of them not used in the field, is caused by the phenomenon known as "metabolism-based cross-resistance." Metabolism-based cross-resistance is the expression of a mechanism that endows plants the ability to tolerate multiple herbicides mediated by a single mechanism involving herbicide metabolism. Resistance to many herbicides by metabolism-based cross-resistance was conceptually established before the early 1990s when multiple herbicide

Table 8.1 Herbicides with confirmed resistance in the multiple herbicide-resistant population of *Echinochloa phyllopogon*.

Herbicide target site	Chemical group	Herbicide active ingredient	Resistance index (R/S)	Genes/mechanism conferring resistance	Reference(s)
Acetolactate synthase	Pyrimidinyl(thio)benzoate	Bispyribac-sodium	2–12	Unknown, probably via P450-based metabolism	3, 19
		Pyrimisulfan	19	*CYP81A12*, *CYP81A21*	2
	Sulfonylurea	Bensulfuron-methyl	>1000	*CYP81A12*, *CYP81A21*	12, 13
		Propyrisulfuron	28	*CYP81A12*, *CYP81A21*	2
	Triazolopyrimidine	Penoxsulam	8–9	*CYP81A12*, *CYP81A21*	16
Acetyl-CoA carboxylase	Phenylpyrazolin	Pinoxaden	36	*CYP81A12*, *CYP81A21*	18
	Cyclohexanedione	Tralkoxydim	11	*CYP81A12*, *CYP81A21*	18
	Aryloxyphenoxypropionate	Diclofop-methyl	115	*CYP81A12*, *CYP81A21*	18
		Fenoxaprop-*P*-ethyl	10	Unknown, probably via glutathione conjugation	20
		Cyhalofop-butyl	19	Unknown	14

(Continued)

Table 8.1 (Continued)

Herbicide target site	Chemical group	Herbicide active ingredient	Resistance index (R/S)	Genes/mechanism conferring resistance	Reference(s)
Auxin receptor	Quinolinecarboxylic acid	Quinclorac	6–17	Unknown; probably via mechanism(s) leading to reduced ethylene production such as P450-based metabolism	17, 21
1-Deoxy-D-xylulose 5-phosphate synthase	Unclassified	Clomazone	2–9	*CYP81A12*, *CYP81A21*	22
Very-long-chain fatty acid elongase	Thiocarbamate	Molinate	2–4	Unknown	3
		Thiobencarb	8 to >13	Unknown	3
4-Hydroxyphenylpyruvate dioxygenase	Triketone	Mesotrione	—	*CYP81A12*, *CYP81A21*	2

Resistance index (R/S) was computed as the herbicide dose by which plant growth is reduced by 50% (GR_{50}) of the multiple herbicide-resistant line over the susceptible line. Resistance index was not computed for mesotrione.

resistance in resistant populations with enhanced ability to metabolize herbicides had been documented in grass weeds,[25] although the molecular mechanism for metabolism-based cross-resistance was unknown. While the proposed mechanism of P450 activation suggested a simple genetic mechanism resulting in the evolution of multiple herbicide resistance in *E. phyllopogon*, the physiological studies (especially studies on fenoxaprop-*P*-ethyl and quinclorac[17,20]) provided substantial data suggesting that more complicated mechanisms underline the multiple herbicide resistance. The physiological studies conducted to elucidate the mechanism of cross-resistance in MHR *E. phyllopogon* are summarized below.

8.3.1.1 No Alteration in Target Sites

The involvement of a target-site resistance mechanism, particularly a less-sensitive form and/or overproduction of the target plastid isoenzyme ACCase, was disqualified as the cause of resistance to ACCase inhibitors (e.g., cyhalofop-butyl, diclofop-methyl, fenoxaprop-*P*-ethyl, tralkoxydim, and pinoxaden). For example, the plastid ACCases of MHR and S lines were similarly susceptible to fenoxaprop acid, the active form of fenoxaprop-*P*-ethyl; thus, the resistance cannot be attributed to an alteration of the target enzyme.[20] Later, the DNA regions of the carboxyltransferase domain in the four copies of plastid ACCase genes were isolated.[26] Sequence analysis revealed that there are no nucleotide sequence variations in the carboxyltransferase domain of the four ACCase genes of MHR and S lines, supporting the result of the *in vitro* enzyme assay.[20] Meanwhile, resistance to ALS inhibitors is also non-target-site-based, as the ALS of the MHR and S lines had similar sensitivity to bensulfuron-methyl, penoxsulam, and bispyribac-sodium.[13,16,19] Later, two copies of full-length ALS genes were isolated from *E. phyllopogon*, and it was confirmed that the DNA sequences of ALS genes of both MHR and S lines are identical.[26] The DNA analysis supported the *in vitro* ALS sensitivity assay, as in the study of ACCase resistance.

The sensitivities of other herbicide target sites such as very-long-chain fatty acid elongases (VLCFAEs), 4-hydroxyphenylpyruvate dioxygenase (4-HPPDs), and auxin receptors have not yet been evaluated in MHR *E. phyllopogon*. However, target-site resistance mechanism is less likely to occur, especially for VLCFAE inhibitors. VLCFAE inhibitors inhibit a series of elongation steps (from long-chain fatty acid [C16–C18] to very-long-chain fatty acid [C20–C30]) that proceed sequentially.[27] Each step is mediated by mutually distinct enzymes that are the targets of VLCFAE inhibitors. Before a target-site resistance to VLCFAE inhibitors could be in-effect, it will require concomitant mutations in several different targets.[28,29] Thus, target-site resistance to VLCFAE inhibitors such as molinate and thiobencarb is less likely the mechanism in MHR *E. phyllopogon*.

While VLCFAE inhibitors had long been acting as the selection pressure for resistance in California fields, the herbicides targeting 4-HPPD and auxin receptors have never been used in the area prior to the report and confirmation of the resistance in 1996. Thus, it is very unlikely that the MHR line carries target-site resistance mutation(s). However, for quinclorac (synthetic auxin),[17] an insensitivity along the response pathway whereby quinclorac induces ethylene production was proposed, theoretically including an insensitive target site as the resistance mechanism, and pointed out the possibility that the habitat environment worked as selection pressure for target-site mutation. Note that this can be explained by a non-target-site mechanism, as discussed later.

8.3.1.2 Non-Target-Site Mechanism

P450-Based Metabolism Involvement of non-target-site resistance mechanisms in MHR *E. phyllopogon*, such as differential herbicide absorption, translocation, and metabolism, was investigated to elucidate the resistance mechanism to multiple herbicides. One enzyme family suggested for its involvement in the cross-resistance of MHR *E. phyllopogon* to multiple herbicides is the P450s, a group of heme-thiolate proteins that are involved in a wide array of monooxygenation/hydroxylation reactions (e.g., alkyl-hydroxylation, *N*-demethylation, *O*-demethylation, and aryl-hydroxylation).[30,31] For ALS inhibitors, penoxsulam and bensulfuron-methyl, previous studies disclosed that the resistance in MHR *E. phyllopogon* was notably reduced by P450 inhibitors, suggesting the involvement of P450s in the resistance mechanism.[13,16] In the MHR plants, penoxsulam was converted into polar metabolites that co-eluted with 5-OH-penoxsulam and BSTCA-methyl.[16] Meanwhile, enzyme activity assay revealed that there was no significant difference in ALS sensitivity toward bensulfuron-methyl between MHR and S plants.[13] A more recent study of bensulfuron-methyl metabolism in *E. phyllopogon* showed that the herbicide was more rapidly *O*-demethylated in the MHR line compared to the S line.[12]

Initial investigations revealed that the mechanism of clomazone resistance in MHR *E. phyllopogon* was not driven by differential uptake, translocation, or bioactivation of inactive clomazone to active 5-ketoclomazone.[15] Later, it was disclosed that clomazone resistance mechanism was via oxidative biotransformations of the parent herbicide.[24] The isoxazolidinone ring of clomazone was hydroxylated mostly in the MHR plants, and the rate of clomazone hydroxylation activity was greater in the MHR than in the S plants. In comparison with the S plants, the MHR plants accumulated higher amount of monohydroxylated metabolites, whereas the S plants accumulated higher amount of 5-ketoclomazone, the phytotoxic compound. As mentioned above, monohydroxylation is a typical reaction catalyzed by P450s, and thus, cross-resistance to clomazone is also attributed to P450-based metabolism.

The involvement of P450s was also suggested for bispyribac-sodium, fenoxaprop-*P*-ethyl, and thiobencarb resistance from the study of herbicide-metabolizing activity in microsomes.[32] It is to note that glutathione-mediated metabolism was proposed as the major metabolic pathway for fenoxaprop-*P*-ethyl[33–35]; thus, the role of P450s at least for this herbicide needs careful investigations (see the next section).

Glutathione Conjugation-Based Metabolism Between MHR and S lines of *E. phyllopogon*, only marginal differences were observed in absorption rates after fenoxaprop-*P*-ethyl application, and no significant difference was observed in the amount of translocated fenoxaprop-*P*-ethyl.[20] On the contrary, the metabolism pattern of fenoxaprop acid was prominently different between the MHR and S lines. There were five times less fenoxaprop acid and roughly two times more metabolites in the MHR line compared to the S line after the application of fenoxaprop-*P*-ethyl. The metabolites were identified as glutathione (GSH) and cysteine conjugates, namely, *S*-(6-chlorobenzoxazole-2-yl) glutathione, *S*-(6-chlorobenzoxazole-2-yl) glycylcysteine, and *S*-(6-chlorobenzoxazole-2-yl) cysteine.[20] Based on these results, it was proposed that GSH conjugation mediated by GSTs is the major mechanism of fenoxaprop-*P*-ethyl resistance in MHR *E. phyllopogon*.

Not Clearly Defined Mechanisms Initial physiological studies on quinclorac resistance suggested that mechanisms other than metabolism may confer multiple herbicide resistance in *E. phyllopogon*. Accordingly, the mechanisms implicated for quinclorac resistance were (1) insensitivity to quinclorac stimulation of ethylene production and (2) enhanced activity of β-cyanoalanine synthase (β-CAS), a cyanide-detoxifying enzyme.[17] The former was based on the observation that the MHR line produced less ethylene despite the application of quinclorac, an ethylene inducer. Meanwhile, the latter was based on the higher β-CAS activity in the MHR line, which had been proposed to protect plants from the damage by cyanide, the toxic compound formed upon quinclorac application. However, recent functional characterization of β-CAS genes along with genetic analysis has excluded the involvement of β-CAS-enhanced activity in quinclorac resistance, while reduced ethylene production was found to be associated with resistance in the progeny of the MHR and S lines.[21] Together, it was proposed that a mechanism leading to reduced ethylene production is the instigator of quinclorac resistance in MHR *E. phyllopogon*. Also, a mechanism mediated by P450 could be one of the possible mechanisms since quinclorac resistance was reversed by a P450 inhibitor,[17] along with a mechanism in the auxin signal transduction pathway, where ethylene production is induced by quinclorac.

8.3.2 Molecular Studies

8.3.2.1 *CYP81A12* and *CYP81A21* Identification Using Bensulfuron-Methyl as a Probe

Physiological work has clearly shown that the resistance to some herbicides, such as penoxsulam and fenoxaprop-*P*-ethyl, is metabolism-based. On the contrary, the work also suggested that two distinct enzymes, P450s and GSTs, are involved in the resistance to penoxsulam and fenoxaprop-*P*-ethyl, respectively. Accordingly, the evolution of multiple herbicide resistance in California may have complicated genetic mechanisms. Although the two enzyme systems were proposed for the resistance mechanism, the fact that most of the herbicides to which the MHR line developed resistance are subject to P450-mediated metabolism has motivated the identification of P450 genes responsible for herbicide metabolism and resistance evolution.

For the identification of molecular mechanisms involved in multiple herbicide resistance, bensulfuron-methyl was used as the "probe" herbicide. While bensulfuron-methyl is an herbicide used for rice cultivation, it shows limited efficacy to *Echinochloa* spp. In other words, *Echinochloa* spp. is not within the spectrum of bensulfuron-methyl field rate. Still, bensulfuron-methyl was prioritized for the elucidation of the apparently complicated multiple herbicide resistance phenomenon due to the following reasons. First, the resistance index to bensulfuron-methyl was the highest among all the herbicides to which the MHR line exhibited resistance. The high resistance index was expected to facilitate the research in many aspects such as phenotyping, detection of metabolites, and transgenic plant assay. Second, a P450 gene involved in bensulfuron-methyl tolerance was identified in rice.[36] Although the functions and copy number of P450 genes are not always conserved among species, it was hypothesized that the rice P450 gene can be a clue to elucidate the molecular mechanism of bensulfuron-methyl resistance in *E. phyllopogon*.

Detailed investigation of bensulfuron-methyl sensitivity revealed that the MHR line tolerated as much as a 1000-fold higher dose of bensulfuron-methyl than the S line.[12] The high resistance index helped to evaluate the resistance segregation in the F_2 populations, especially in categorizing individuals into an intermediate group. The segregation ratio fitted to 1:2:1 (resistant:intermediate:sensitive), suggesting that the bensulfuron-methyl resistance trait was under the control of a single locus. Enhanced bensulfuron-methyl metabolism into an inactive form (a demethylated bensulfuron-methyl) was also observed in the MHR line, although only a few-fold difference was observed in the reduction of bensulfuron-methyl and accumulation of demethylated bensulfuron-methyl. Demethylation was the major metabolic pathway observed for bensulfuron-methyl, a reaction which is known to be mediated by P450s.[37] The results of the genetic and metabolism studies supported the previous report that the bensulfuron-methyl resistance is P450-mediated.[13] Importantly, the data also suggested that the high-level bensulfuron-methyl resistance is caused by a simple genetic mechanism unlike the metabolism-based resistance often reported in other grass weeds such as *Lolium rigidum* and *Alopecurus myosuroides*.[38,39]

The follow-up molecular investigation was focused on *CYP81A* subfamily among the huge numbers of P450 subfamilies because *CYP81A6*, one of the four *CYP81A* genes in rice, was reported to be involved in rice tolerance to multiple herbicides including bensulfuron-methyl.[36] Based on the assumption that substrate specificity is similar among the same subfamily, *CYP81A* genes were isolated from *E. phyllopogon*. Multiple sets of degenerate primers were designed based on the *CYP81A* sequences of Poaceae, which succeeded in the partial amplification of *CYP81A* fragments of *E. phyllopogon*.[12] The following full-length isolation clarified the evolutionary phylogeny of the *CYP81A* genes of *E. phyllopogon*. The repertoire of the 12 *CYP81A* members of *E. phyllopogon* was similar to that of Panicoideae such as maize and sorghum. Furthermore, no close (or orthologous) relationship to *CYP81A6* was observed in any of the 12 *CYP81A* members. Three members were excluded from further analysis as their sequences carry indels causing frameshift that improperly encode P450s. The other genes were tested for their expression by real-time polymerase chain reaction (PCR) using gene-specific primers designed on 3′-UTR, where a relatively large number of polymorphisms exist among *CYP81A* members. The real-time PCR study revealed that the three genes, *CYP81A12*, *CYP81A21*, and *CYP81A22*, were highly accumulated in the MHR line. These genes were then transformed in *Arabidopsis thaliana*. The transgenic lines with high-level expression of *CYP81A12* and *CYP81A21* exhibited >1000-fold resistance to bensulfuron-methyl, while all the lines transformed with *CYP81A22* have unchanged bensulfuron-methyl sensitivity. The high-level resistance observed in the *CYP81A12/21* transgenic lines was in parallel with the observed high-level resistance in the MHR line. Next, *CYP81A12* and *CYP81A21* were investigated for the production of demethylated bensulfuron-methyl, the product of which was accumulated more in the MHR line.[12] The yeast expressor for plant P450s, WAT11, was used for *CYP81A12* and *CYP81A21* expression. Although the expression levels of the two P450s were not very high, it was confirmed that the transformed yeast cells produced demethylated bensulfuron-methyl using liquid chromatography-mass spectrometry/mass spectrometry (LC-MS/MS) system.[12]

The higher expression and the bensulfuron-methyl metabolizing activity of *CYP81A12* and *CYP81A21* did not conflict with the observed genetic inheritance of resistance in the MHR line. For further evidence of their involvement in bensulfuron-methyl resistance, the association of the higher transcription of the two genes and bensulfuron-methyl resistance in the recombinant inbred lines (RILs) of the MHR and S lines was investigated. In agreement with the 1:2:1 segregation in F_2 generation, higher expression of the two genes was associated with bensulfuron-methyl sensitivity in the 40 RILs. No genetic linkage was detected between *CYP81A12* and *CYP81A21*, thus it was hypothesized that the two genes are on different chromosomes. This was confirmed by the recently disclosed genome sequence of *E. phyllopogon*,[40] indicating that the two genes are located on distinctive contigs that share high synteny (data not shown). Altogether, a single genetic element is likely behind the simultaneous overexpression of bensulfuron-methyl-metabolizing *CYP81A12* and *CYP81A21*, hence causing high-level bensulfuron-methyl resistance in the MHR line.

8.3.2.2 Involvement of *CYP81A12* and *CYP81A21* in Resistance to Other Herbicides: Proof of Concept of Metabolism-Based Cross-Resistance

The use of the probe herbicide bensulfuron-methyl successfully identified the two genes involved in the metabolism-based cross-resistance. The research was followed up by the investigation of the resistance mechanism to other herbicides that are primarily recommended against *Echinochloa* spp. In the same report with bensulfuron-methyl, the involvement of *CYP81A12/21* in penoxsulam resistance was investigated.[12] The resistance index of the MHR line to penoxsulam was 6.2, which was much lower than bensulfuron-methyl. As in the case of bensulfuron-methyl, a paralleled resistance index (approximately 10) was observed in the MHR line and *CYP81A12/21* transformed *A. thaliana*. Association of transcripts and penoxsulam resistance was also observed in the RILs. The study was the first demonstration that weed P450 genes are involved in the resistance evolution to an agriculturally important herbicide. Notably, the production of putative demethylated penoxsulam by *CYP81A12/21* was later confirmed using the *E. coli* expression system (see below).[2]

In a similar approach, the involvement of *CYP81A12/21* was confirmed in the cross-resistance to clomazone, pinoxaden, diclofop-methyl, and tralkoxydim.[18,22] Furthermore, the resistance in the MHR line and the herbicide metabolizing activity of the *CYP81A12/21* were revealed in propyrisulfuron, pyrimisulfan, and mesotrione.[2] The correlation of the resistance indices of *CYP81A*-expressing *A. thaliana* (against non-transformant) and the MHR line (against susceptible line) strongly supports the involvement of *CYP81A12/21* in the resistance (Fig. 8.1). Thus, *CYP81A12/21* most likely explain the resistance to the three herbicides, although the association using the RILs was not investigated. Together, the two genes were associated with resistance to nine herbicides from four sites of action (Table 8.1). The series of studies provide a proof of concept that metabolism-based resistance could endow resistance to numerous herbicides from multiple sites of action in contrast to target-site resistance where cross-resistance is limited only to a single site of action. Meanwhile, unpublished data also support that *CYP81A12/21* explain the resistance to quinclorac, although further investigation is required (Chayapakdee et al. unpublished).

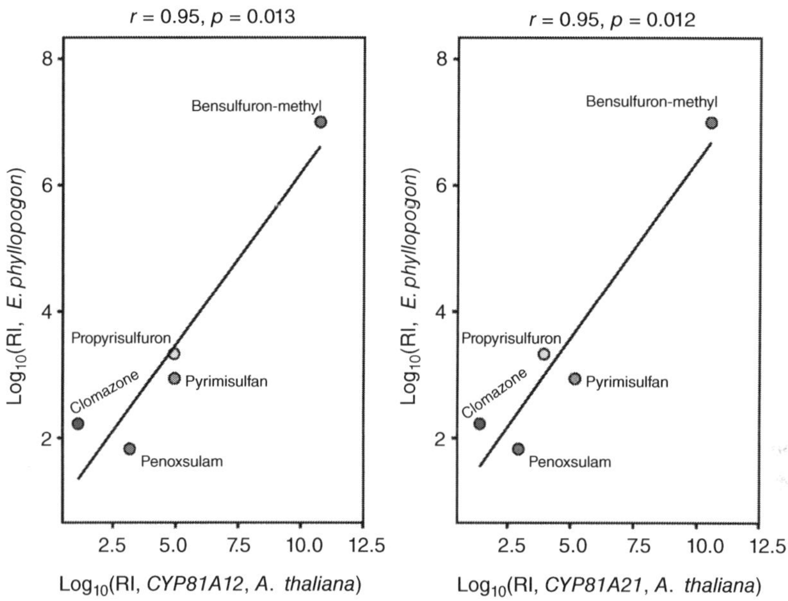

Figure 8.1 Correlation analysis of the resistance indices (RI) of *CYP81A12* and *CYP81A21* in transgenic *Arabidopsis thaliana* and in the multiple herbicide-resistant (MHR) line of *Echinochloa phyllopogon* to bensulfuron-methyl, penoxsulam, clomazone, pyrimisulfan, and propyrisulfuron. Source: Adapted from Dimaano et al.[2] (*See insert for color representation of the figure.*)

8.4 Prediction of Metabolism-Based Cross-Resistance

The unpredictability of cross-resistance evolution in weeds is a major obstacle in metabolic resistance research as well as in herbicide resistance mitigation and management. Due to limited knowledge on the enzymes involved in the metabolism of herbicides, predicting the cross-resistance pattern in weeds seems to be a very difficult task. However, improved knowledge of the substrate range of the enzymes involved in herbicide metabolism could facilitate a better prediction of cross-resistance patterns in weeds. Accordingly, *CYP81A12/21* of *E. phyllopogon* were functionally characterized for an array of possible herbicide substrates. Through ectopic expression in *A. thaliana* and *Escherichia coli*, it was disclosed that *CYP81A12/21* have a large scope of herbicide substrates encompassing at least 15 herbicide active ingredients from 10 unrelated chemical classes: all chemical groups of ALS and ACCase inhibitors, benzothiadiazinone (photosystem II inhibitor), triketone (4-HPPD inhibitor), and isoxazolidinone (DXPS inhibitor). Furthermore, the other putative functional *CYP81As* of *E. phyllopogon*, namely, *CYP81A14*, *CYP81A15*, *CYP81A18*, *CYP81A22*, *CYP81A23*, *CYP81A24*, and *CYP81A26*, were also characterized. In summary, there is differential substrate specificity among each *CYP81A* member, and the preferential substrate is biased but not limited to ALS inhibitors. Additionally, a few amino acid residues of *CYP81As* seem to play an integral role in the herbicide metabolism of multiple herbicides. The essential amino acid residues for herbicide binding and catalysis of *CYP81A* P450s, which are the key for deciphering their wide and distinct substrate specificities, are still unknown. Furthermore, it is still unclear if these *CYP81As* have evolved in *E. phyllopogon* solely for the purpose of xenobiotic detoxification.

At present (as of November 2021), the only population of *E. phyllopogon* with a reported overexpression of *CYP81A12/21* is the MHR population collected from California. The P450s are constitutively highly expressed in the MHR line compared to the S line. Based on the disclosed herbicide preference of these P450s, cross-resistance of the MHR line to relatively novel herbicides, e.g., pyrimisulfan, propyrisulfuron, and mesotrione, was predicted and confirmed. Moreover, the metabolism profiling of *CYP81As* identified various herbicides that are not substrates of *CYP81As* as potential alternative herbicides for managing MHR *E. phyllopogon*.

Notably, the accumulation of multiple mechanisms (multiple resistance) occurs even in the case of metabolic resistance. In this case, the resistance profile becomes more complex and complicated, making the prediction of resistance a more difficult undertaking. The accumulation is more prone to outcrossing species, but not limited to self-crossing species. Even in the case of the MHR *E. phyllopogon*, the promiscuous *CYP81As* do not accommodate the resistance to some herbicides such as thiobencarb, molinate, and fenoxaprop-*P*-ethyl, the herbicides known to have selected the MHR population in California rice fields.[3] Possibly, other mechanisms involved in non-target-site-based multiple resistance could explain the resistance to these herbicides, which warrant the necessity of deeper molecular understanding of metabolic resistance.

8.5 Generality and Diversity of Metabolic Resistance

The research efforts to elucidate the molecular mechanism of metabolism-based cross-resistance so far have demonstrated that *CYP81As* in *E. phyllopogon* are promiscuous enzymes, metabolizing diverse herbicides from multiple sites of action. The upregulation of one or a few *CYP81A* members potentially imparts resistance to multiple herbicides. It will be interesting to investigate the generality and diversity of the molecular mechanism of "suspected" metabolism-based resistance in *E. phyllopogon* that have been found in Asian countries, e.g., from South Korea[41] and Japan (unpublished). Furthermore, the generality and diversity in other grass weeds will be worthy of investigation. Recently, the upregulation of a member of *CYP81A*, *CYP81A10v7*, was reported in some *L. rigidum* populations exhibiting multiple herbicide resistance.[42] *CYP81A10v7* was shown to confer resistance in transgenic rice to multiple herbicides belonging to as much as five sites of action. The research implies that *CYP81A*-mediated multiple herbicide resistance is not confined to the Californian populations of *E. phyllopogon*, and that molecular convergent evolution is not limited to target-site resistance but also applicable to metabolism-based resistance. In addition to the *L. rigidum* research, the overexpression of *CYP81A* members have been reported in non-target-site resistance in *E. crus-galli*,[43,44] *E. glabrescens*,[45] *Avena fatua*,[46] *Alopecurus japonicus*,[47] and *Lolium* sp.[48,49] (Table 8.2). These observations may suggest that the upregulation of *CYP81A* explains a relatively large number of metabolism-based resistance evolution cases, although the herbicide-metabolizing functions of these *CYP81A*s need to be investigated.

Table 8.2 Overexpression of *CYP81A* genes reported in different weed species with suspected/confirmed metabolism-based resistance.

Species	Country	Herbicide target site	Overexpressed *CYP81A* gene	References
Echinochloa phyllopogon	US	(See Table 8.1)	*CYP81A12/21**	(See Table 8.1)
Echinochloa crus-galli	Brazil	ALS (imidazolinone)	*CYP81A*	43
Echinochloa crus-galli	China	ALS (penoxsulam)	EC_v6.g096099 (*CYP81A24*)	44
Echinochloa glabrescens	China	ALS (penoxsulam)	EC_v6.g096098 (*CYP81A12*)	45
Avena fatua	USA	ACCase (pinoxaden)	*CYP81A*	46
Lolium rigidum	Australia	ALS, ACCase, etc.	*CYP81A10v7*[a)]	42
Lolium sp.	France	ALS	*CYP81B1/B2* (likely *CYP81A*)[b)]	48
Lolium sp.	France	ALS	*CYP81B1/B2* (likely *CYP81A*)[b)]	49
Alopecurus japonicus	China	ACCase (fenoxaprop-*P*-ethyl)	*CYP81A*	47

ALS, acetolactate synthase; ACCase, acetyl-CoA carboxylase.

a) Herbicide-metabolizing function was confirmed.

b) Annotation in the paper was *CYP81B*. However, >70% identities to *CYP81A12/21*[49] suggest that the genes belong to CYP81A.

Further research is also needed on how the upregulation of *CYP81A* genes occurs. In the case of the MHR line of *E. phyllopogon* from California, *CYP81A12/21* upregulation is caused not by gene amplification but likely by a single trans-element.[12] The possible trans-element-mediated regulation implies that many other genes are also regulated directly or indirectly by the causal element. It would be interesting to identify the particular trans-element and reveal the components under its control. Possibly, other genes encoding the enzymes that metabolize other herbicides such as thiobencarb, molinate, fenoxaprop-*P*-ethyl, and cyhalofop-butyl are also under its control. In other words, the trans-element may simultaneously upregulate multiple herbicide-metabolizing enzymes, resulting in resistance to many and diverse herbicides. The recently disclosed draft genome of *E. phyllopogon* will help elucidate the mystery of the multiple herbicide resistance evolution in the resistant populations.[40]

8.6 Conclusion

The evolution of metabolism-based herbicide resistance is a significant threat to agriculture as the enzymes involved in the resistance could instigate the collapse of effective herbicide-based weed control in agronomically important weed species. Unfortunately, the metabolic enzymes could also extend the resistance to newly developed and yet-to-be-discovered herbicides, thus threatening the sustainability of herbicide use and the fate of novel herbicide discovery. *E. phyllopogon*, one of the weed species that have evolved metabolic herbicide resistance, has received considerable research attention in the past two decades. Through physiological and molecular approaches, two *P450* genes are revealed to be involved in the metabolism-based cross-resistance to several herbicides from distinct chemical groups. This research breakthrough pioneers gene discovery for metabolism-based herbicide resistance in grass weeds and provides a proof of concept that metabolism-based cross-resistance to several herbicides could be achieved by a single mechanism. However, some issues remain unresolved to fully comprehend the evolution of metabolism-based resistance in *E. phyllopogon*. Further studies to disclose the lingering mysteries of metabolism-based resistance will lead to better herbicide management programs for resistance mitigation. A molecular genetic mapping analysis will disclose the specific locus controlling the metabolism-based resistance to other herbicides, which will then provide rich information for the identification and functional characterization of the resistance genes. Protein structure elucidation will reveal the specific sites and features on the capacity of *CYP81A* P450s for wide herbicide binding, conformity, and catalytic reactions. Further molecular studies may also reveal the regulatory elements that dictate the overexpression of these P450s. The genes involved in herbicide metabolism may be useful for possible applications in crop improvement through biotechnological approaches. As the metabolism-based herbicide resistance in weeds is evolving at a fast pace, research efforts to elucidate its underlying mechanisms must likewise be accelerated.

References

1 Yamasue Y, Strategy of *Echinochloa oryzicola* Vasing. for survival in flooded rice. *Weed Biol Manag* **1**:28–36 (2001).

2 Dimaano NG, Yamaguchi T, Fukunishi K, Tominaga T, and Iwakami S, Functional characterization of cytochrome P450 CYP81A subfamily to disclose the pattern of cross-resistance in *Echinochloa phyllopogon*. *Plant Mol Biol* **102**:403–416 (2020).

3 Fischer AJ, Ateh CM, Bayer DE, and Hill JE, Herbicide-resistant *Echinochloa oryzoides* and *E. phyllopogon* in California *Oryza sativa* fields. *Weed Sci* **48**:225–230 (2000).

4 Hill JE, Le Strange ML, Bayer DE, and Williams JF, Integrated weed management in California, In *The Western Society of Weed Science*, Reno, pp. 100–104 (1985).

5 Yabuno T, Cytological relationship between *Echinochloa oryzicola* Vasing. and the French strain of *E. phyllopogon* Stapf subsp. *oryzicola* (Vasing.) Koss. *Cytologia* **46**:393–396 (1981).

6 Agriculture USDo, National Agricultural Statistics Service. https://quickstats.nass.usda.gov/ [accessed Dec 19, 2020].

7 Barrett SCH, and Seaman DE, The weed flora of Californian rice fields. *Aquat Bot* **9**:351–376 (1980).

8 Hill JE, Williams JF, Mutters RG, and Greer CA, The California rice cropping system: agronomic and natural resource issues for long-term sustainability. *Paddy Water Environ* **4**:13–19 (2006).

9 Valverde BE, Boddy LG, Pedroso RM, Eckert JW, and Fischer AJ, *Cyperus difformis* evolves resistance to propanil. *Crop Prot* **62**:16–22 (2014).

10 Williams J, Current and future grass herbicides in rice. In 2000 Proceedings of the California. *Weed Sci Society* **52**:123–128 (2000).

11 Soderquist CJ, Bowers JB, and Crosby DG, Dissipation of molinate in a rice field. *J Agric Food Chem* **25**:940–945 (1977).

12 Iwakami S, Endo M, Saika H, Okuno J, Nakamura N, Yokoyama M, *et al.*, Cytochrome P450 CYP81A12 and CYP81A21 are associated with resistance to two acetolactate synthase inhibitors in *Echinochloa phyllopogon*. *Plant Physiol* **165**:618–629 (2014).

13 Osuna MD, Vidotto F, Fischer AJ, Bayer DE, de Prado R, and Ferrero A, Cross-resistance to bispyribac-sodium and bensulfuron-methyl in *Echinochloa phyllopogon* and *Cyperus difformis*. *Pestic Biochem Physiol* **73**:9–17 (2002).
14 Ruiz-Santaella JP, de Prado R, Wagner J, Fischer A, and Gerhards R, Resistance mechanisms to cyhalofop-butyl in a biotype of *Echinochloa phyllopogon* (Stapf) Koss. from California. *J Plant Dis Prot* **XX**:95–100 (2006).
15 Yasuor H, TenBrook PL, Tjeerdema RS, and Fischer AJ, Responses to clomazone and 5-ketoclomazone by *Echinochloa phyllopogon* resistant to multiple herbicides in Californian rice fields. *Pest Manag Sci* **64**:1031–1039 (2008).
16 Yasuor H, Osuna MD, Ortiz A, Saldain NE, Eckert JW, and Fischer AJ, Mechanism of resistance to penoxsulam in late watergrass (*Echinochloa phyllopogon* (Stapf) Koss.). *J Agric Food Chem* **57**:3653–3660 (2009).
17 Yasuor H, Milan M, Eckert JW, and Fischer AJ, Quinclorac resistance: a concerted hormonal and enzymatic effort in *Echinochloa phyllopogon*. *Pest Manag Sci* **68**:108–115 (2012).
18 Iwakami S, Kamidate Y, Yamaguchi T, Ishizaka M, Endo M, Suda H, *et al.*, CYP81A P450s are involved in concomitant cross-resistance to acetolactate synthase and acetyl-CoA carboxylase herbicides in *Echinochloa phyllopogon*. *New Phytol* **221**:2112–2122 (2019).
19 Tsuji R, Fischer AJ, Yoshino M, Roel A, Hill JE, and Yamasue Y, Herbicide-resistant late watergrass (*Echinochloa phyllopogon*): similarity in morphological and amplified fragment length polymorphism traits. *Weed Sci* **51**:740–747 (2003).
20 Bakkali Y, Ruiz-Santaella JP, Osuna MD, Wagner J, Fischer AJ, and de Prado R, Late watergrass (*Echinochloa phyllopogon*): mechanisms involved in the resistance to fenoxaprop-*p*-ethyl. *J Agric Food Chem* **55**:4052–4058 (2007).
21 Chayapakdee P, Sunohara Y, Endo M, Yamaguchi T, Fan L, Uchino A, *et al.*, Quinclorac resistance in *Echinochloa phyllopogon* is associated with reduced ethylene synthesis rather than enhanced cyanide detoxification by β-cyanoalanine synthase. *Pest Manag Sci* **76**:1195–1204 (2020).
22 Guo F, Iwakami S, Yamaguchi T, Uchino A, Sunohara Y, and Matsumoto H, Role of CYP81A cytochrome P450s in clomazone metabolism in *Echinochloa phyllopogon*. *Plant Sci* **283**:321–328 (2019).
23 Fischer AJ, Bayer DE, Carriere MD, Ateh CM, and Yim K-O, Mechanisms of resistance to bispyribac-sodium in an *Echinochloa phyllopogon* accession. *Pestic Biochem Physiology* **68**:156–165 (2000).
24 Yasuor H, Zou W, Tolstikov VV, Tjeerdema RS, and Fischer AJ, Differential oxidative metabolism and 5-ketoclomazone accumulation are involved in *Echinochloa phyllopogon* resistance to clomazone. *Plant Physiol* **153**:319–326 (2010).

25 Hall L, Holtum J, and Powles S, Mechanisms responsible for cross resistance and multiple resistance, In Powles SB, and Holtum JAM (Eds), *Herbicide Resistance in Plants: Biology and Biochemistry*, Boca Raton, Lewis Publishers, pp. 243–261 (1994).

26 Iwakami S, Uchino A, Watanabe H, Yamasue Y, and Inamura T, Isolation and expression of genes for acetolactate synthase and acetyl-CoA carboxylase in *Echinochloa phyllopogon*, a polyploid weed species. *Pest Manag Sci* **68**:1098–1106 (2012).

27 Tanetani Y, Kaku K, Ikeda M, and Shimizu T, Action mechanism of a herbicide, thiobencarb. *J Pestic Sci* **38**:D12-047 (2012).

28 Trenkamp S, Martin W, and Tietjen K, Specific and differential inhibition of very-long-chain fatty acid elongases from *Arabidopsis thaliana* by different herbicides. *Proc Natl Acad Sci USA* **101**:11903–11908 (2004).

29 Busi R, Resistance to herbicides inhibiting the biosynthesis of very-long-chain fatty acids. *Pest Manag Sci* **70**:1378–1384 (2014).

30 Mizutani M, and Ohta D, Diversification of P450 genes during land plant evolution. *Annu Rev Plant Biol* **61**:291–315 (2010).

31 Dimaano NG, and Iwakami S, Cytochrome P450-mediated herbicide metabolism in plants: current understanding and prospects. *Pest Manag Sci* **77**:22–32 (2021).

32 Yun M-S, Yogo Y, Miura R, Yamasue Y, and Fischer AJ, Cytochrome P-450 monooxygenase activity in herbicide-resistant and -susceptible late watergrass (*Echinochloa phyllopogon*). *Pestic Biochem Physiol* **83**:107–114 (2005).

33 Romano ML, Stephenson GR, Tal A, and Hall JC, The effect of monooxygenase and glutathione S-transferase inhibitors on the metabolism of diclofop-methyl and fenoxaprop-ethyl in barley and wheat. *Pestic Biochem Physiol* **46**:181–189 (1993).

34 Tal A, Romano ML, Stephenson GR, Schwan AL, and Hall JC, Glutathione conjugation: a detoxification pathway for fenoxaprop-ethyl in barley, crabgrass, oat, and wheat. *Pestic Biochem Physiol* **46**:190–199 (1993).

35 Cummins I, Moss S, Cole DJ, and Edwards R, Glutathione transferases in herbicide-resistant and herbicide-susceptible black-grass (*Alopecurus myosuroides*). *Pestic Sci* **51**:244–250 (1997).

36 Pan G, Zhang X, Liu K, Zhang J, Wu X, Zhu J, *et al.*, Map-based cloning of a novel rice cytochrome P450 gene CYP81A6 that confers resistance to two different classes of herbicides. *Plant Mol Biol* **61**:933–943 (2006).

37 Deng F, and Hatzios KK, Characterization of cytochrome P450-mediated bensulfuron-methyl O-demethylation in rice. *Pestic Biochem Physiol* **74**:102–115 (2002).

38 Petit C, Duhieu B, Boucansaud K, and Délye C, Complex genetic control of non-target-site-based resistance to herbicides inhibiting acetyl-coenzyme A carboxylase and acetolactate-synthase in *Alopecurus myosuroides* Huds. *Plant Sci* **178**:501–509 (2010).

39 Busi R, Vila-Aiub MM, and Powles SB, Genetic control of a cytochrome P450 metabolism-based herbicide resistance mechanism in *Lolium rigidum. Heredity* **106**:817–824 (2011).

40 Ye C-Y, Wu D, Mao L, Jia L, Qiu J, Lao S, *et al.*, The genomes of the allo-hexaploid *Echinochloa crus-galli* and its progenitors provide insights into polyploidization-driven adaptation. *Mol Plant* **13**:1298–1310 (2020).

41 Song J, Lim S, Yook M, Kim J, and Kim D, Cross-resistance of *Echinochloa* species to acetolactate synthase inhibitor herbicides. *Weed Biol Manag* **17**:91–102 (2017).

42 Han H, Yu Q, Beffa R, González S, Maiwald F, Wang J, *et al.*, Cytochrome P450 CYP81A10v7 in *Lolium rigidum* confers metabolic resistance to herbicides across at least five modes of action. *Plant J* **105**:79–92 (2021).

43 Dalazen G, Markus C, and Merotto A Jr, Differential expression of genes associated with degradation enhancement of imazethapyr in barnyardgrass (*Echinochloa crus-galli*). *J Agric Sci* **10**:389 (2018).

44 Fang J, Zhang Y, Liu T, Yan B, Li J, and Dong L, Target-site and metabolic resistance mechanisms to penoxsulam in barnyardgrass (*Echinochloa crus-galli* (L.) P. Beauv). *J Agric Food Chem* **67**:8085–8095 (2019).

45 Yan B, Zhang Y, Li J, Fang J, Liu T, and Dong L, Transcriptome profiling to identify cytochrome P450 genes involved in penoxsulam resistance in *Echinochloa glabrescens. Pestic Biochem Physiol* **158**:112–120 (2019).

46 Keith BK, Burns EE, Bothner B, Carey CC, Mazurie AJ, Hilmer JK, *et al.*, Intensive herbicide use has selected for constitutively elevated levels of stress-responsive mRNAs and proteins in multiple herbicide-resistant *Avena fatua* L. *Pest Manag Sci* **73**:2267–2281 (2017).

47 Chen G, Xu H, Zhang T, Bai C, and Dong L, Fenoxaprop-*P*-ethyl resistance conferred by cytochrome P450s and target site mutation in *Alopecurus japonicus. Pest Manag Sci* **74**:1694–1703 (2018).

48 Duhoux A, Carrère S, Gouzy J, Bonin L, and Délye C, RNA-Seq analysis of rye-grass transcriptomic response to an herbicide inhibiting acetolactate-synthase identifies transcripts linked to non-target-site-based resistance. *Plant Mol Biol* **87**:473–487 (2015).

49 Duhoux A, Carrère S, Duhoux A, and Délye C, Transcriptional markers enable identification of rye-grass (*Lolium* sp.) plants with non-target-site-based resistance to herbicides inhibiting acetolactate-synthase. *Plant Sci* **257**:22–36 (2017).

9

Metabolic Resistance to Herbicides in *Echinochloa crus-galli* (and *colona*)

Alice A. Wright[1] *and Vijay K. Nandula*[2]

[1] *USDA-ARS, Houma, LA, USA*
[2] *USDA-NIFA, Kansas City, Missouri, USA*

9.1 Introduction

Echinochloa crus-galli and *Echinocloa colona*, commonly known as barnyard grass and jungle rice, respectively, are among the most problematic weeds of rice worldwide.[1,2] Through anthropomorphic selection, they have evolved to be very difficult to control; it is even thought that hand weeding in Asia inadvertently selected for seedlings that mimic rice in morphology, making them more difficult to remove even by hand.[3] *E. crus-galli* germinates anaerobically, so control measures such as flooding are not always successful.[4] Infestations by these weeds can reduce yield up to 79% in rice, depending on weed density.[1] These C_4 annual species are often difficult to distinguish from one another[1,2,5,6]; however, for the purposes of this chapter, it is meaningful to group them together.

Overuse of herbicide modes of action resulted in the evolution of resistance in populations from both species. At least 77 instances of *E. crus-galli* and *E. colona* populations with resistance to one or more modes of action are documented.[7] These reports spread across 34 countries around the world including in the areas of North, South, and Central America, Australia, Africa, the Middle East, Europe, and Asia.[5,7–9] Furthermore, resistance has been reported to several different herbicide modes of action: acetyl-CoA carboxylase (ACCase) inhibitors (HRAC Group 1/formerly A), acetolactate synthase (ALS) inhibitors (HRAC2/B), photosystem II (PS-II) inhibitors (HRAC5/C1, C2), microtubule assembly inhibitors (HRAC3/K1; HRAC13/F4), 5-enolpyruvylshikimate-3-phosphate synthase (EPSPS) inhibitors (HRAC9/G), very-long-chain fatty acid synthesis inhibitors (HRAC15/K3, N), auxin mimics (HRAC4/O), and cellulose synthesis inhibitors (HRAC29/L).[7] Most common among these is resistance to PS-II inhibitors, ACCase inhibitors, ALS

Resistance in Weeds from Herbicide Metabolism, First Edition.
Edited by Vijay K. Nandula and Roland Beffa.

inhibitors, EPSPS inhibitors, and auxin mimics (quinclorac) (Fig. 9.1). Multiple resistance in these populations is common: 17 of the 77 reports represent populations with resistance to more than one mode of action[7] (Fig. 9.1). A survey of *E. crus-galli and E. colona* populations in Arkansas found that 27% of populations examined exhibited multiple resistance.[10] Of the resistance profiles documented, resistance to propanil (HRAC5/C1, C2) was the most common, followed by quinclorac (HRAC4/O), imazethapyr (HRAC2/B), and cyhalofop (HRAC1/A),

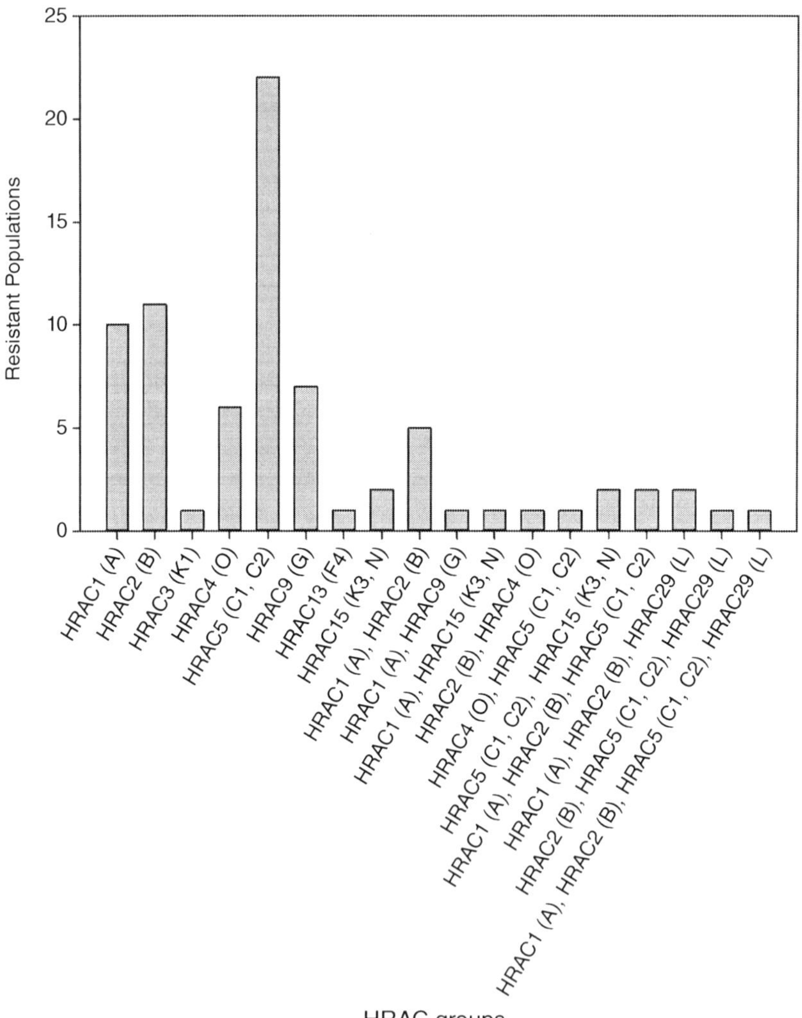

Figure 9.1 Distribution of resistant populations across HRAC groups.

in decreasing order.[10] Resistance to these and other herbicides is discussed in more detail below.

9.2 Propanil Resistance

Propanil, a PS-II inhibitor, is one of the oldest herbicides used in rice to control *E. colona* and *E. crus-galli*. Its popularity among growers stems from the fact that when applied, rice can metabolize propanil while *E. colona* and *E. crus-galli* cannot.[11,12] This metabolism is due to an aryl acylamidase protein, which has 60 times more activity in rice than *E. crus-galli* for metabolizing propanil.[11] Carbaryl, an insecticide, inhibits the breakdown of propanil into 3,4-dichloroanaline, rendering rice sensitive to propanil.[13]

Overuse of this herbicide in rice has since led to widespread resistance. In the mid-1990s, populations of *E. crus-galli* and *E. colona* were identified in North and Central America that could no longer be controlled with propanil.[5,14] Application of carbaryl two days before propanil application rendered the resistant populations sensitive to the herbicide, confirming that the same metabolic mechanism that protected rice had evolved in *E. crus-galli* and *E. colona*.[15,16] Further investigation showed that treatment with monooxygenase inhibitors, such as piperonyl butoxide, tridiphan, and prochloraz, reduced accumulation of propanil metabolites when applied with the herbicide.[16] While a specific gene has not been identified as responsible for resistance, Rouse[17] in his dissertation research identified two cytochrome P450s, CYP709B2 and CYP72A15, that may be involved based on an increase in gene expression following propanil treatment. The function of these two enzymes is unknown, and further research needs to determine if and how they are involved in propanil resistance.

9.3 Quinclorac Resistance

Resistance to quinclorac, another commonly used rice herbicide, evolved after propanil resistance. Quinclorac is an auxin mimic and, in susceptible *E. crus-galli*, is known to cause a lethal increase in cyanide.[18] Like propanil, rice is tolerant of this herbicide, so it is frequently used in rice for weed control.[18] The mechanism varies between populations, indicating that resistance has evolved more than once. High β-cyanoalanine synthase activity has been reported for some populations but not others.[17,19] β-Cyanoalanine synthase is known to prevent cyanide toxicity that is a result of ethylene production, a consequence of quinclorac treatment.[18,20] In some populations, it has been observed that treatment with a cytochrome P450 inhibitor, such as malathion, can reduce resistance to this

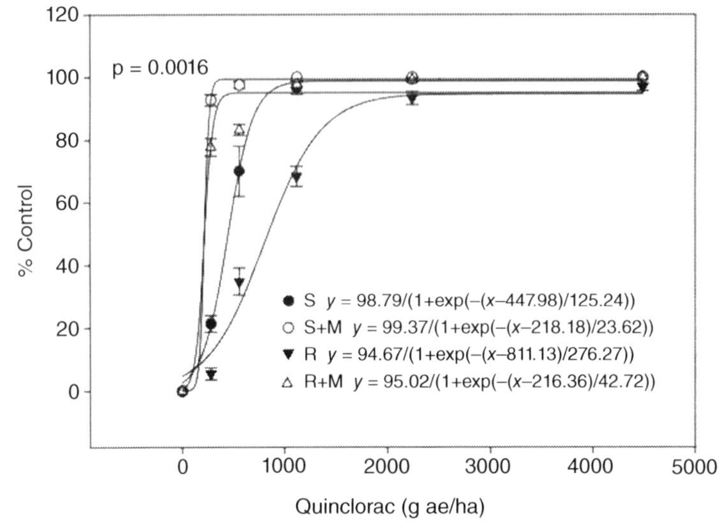

Figure 9.2 Dose–response assay for quinclorac with and without malathion (M) treatment for sensitive (S) and resistant (R) *Echinocloa colona* populations. A *t*-test value is provided indicating that the response to quinclorac treatment was significantly different between the R and S populations. Source: Adapted from Wright et al.[22]/ Cambridge University Press.

herbicide[21,22] (Fig. 9.2). Transcriptomic analysis of populations treated with quinclorac indicated increased expression of four cytochrome P450s, CYP709B2, CYP47071A1, CYP71A4 and CYP71A8, in a resistant population compared with a sensitive population[17]; however, this needs to be explored further as the function of these cytochrome P450s is unknown.

9.4 Multiple Herbicide Resistance

Most concerning is the evolution of populations with resistance to multiple herbicides, including ACCase and ALS inhibitors. Some instances of resistance, particularly for ALS inhibitor resistance, can be explained by target-site resistance (TSR) mechanisms,[21,23] while other resistant populations appear to have evolved metabolic mechanisms (which are non-target-site resistance [NTSR] mechanisms).[23–25] There are four populations that are very concerning given the broad range of herbicides to which they are resistant: a population of *E. colona* from Mississippi with resistance to propanil, quinclorac, fenoxaprop-*P*-ethyl, and imazamox[22,24,26]; a population of *E. colona* from Arkansas with resistance to propanil, quinclorac, cyhalofop-butyl, and glufosinate[17,25]; a population of *E. crus-galli* from China with resistance to pyribenoxim, imazapic, flucarbazone-sodium, penoxsulam, flucetosulfuron, cyhalofop-butyl, pinoxaden, quinclorac, and florpyrauxifen-benzyl[23]; and a population of *E. crus-galli* also from China with resistance to penoxsulam, cyhalofop-butyl, and metamifop.[27]

In Mississippi, a population of *E. colona* designated MS1 was identified as having resistance to multiple herbicides.[22,24] Resistance to the ALS inhibitor imazamox was confirmed to be a non-target site as the isolated enzyme was sensitive to the herbicide.[24] This weed was still controlled by the ACCase inhibitors, clethodim and sethoxydim, but another ACCase inhibitor, fenoxaprop-*P*-ethyl, failed to control the resistant population at the label rate; the ED_{50} was 249 g a.e./ha, nearly 11 times that of the sensitive population.[26] NTSR to the ACCase inhibitor fenoxaprop-*P*-ethyl was confirmed in a similar manner – the ACCase enzyme in isolation was just as sensitive to the herbicide as the ACCase from a population controlled by fenoxaprop-*P*-ethyl (Fig. 9.3).[26] Applying fenoxaprop-*P*-ethyl with cytochrome P450 inhibitors malathion or piperonyl butoxide or glutathione-*S*-transferase (GST) inhibitor 4-chloro-7-nitrobenzofurazan did not increase sensitivity of the resistant population to the herbicide, indicating that if it was a metabolic mechanism no enzyme inhibited by these chemicals was involved.[26] In contrast, applying malathion with imazamox or quinclorac reduced resistance to these herbicides, indicating a metabolic mechanism.[22] Resistance to imazamox was studied further using transcriptomics

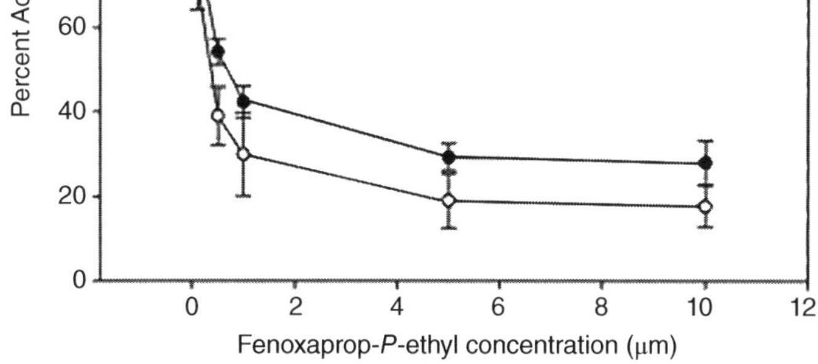

Figure 9.3 Assessment of acetyl-CoA carboxylase (ACCase) activity in the presence of increasing concentrations of fenoxaprop-*P*-ethyl for resistant (open circles) and sensitive (closed circles) *E. colona* populations. Source: Adapted from Wright et al.[26]/Cambridge University Press.

to explore gene expression response to herbicide treatment.[22] Response to herbicide treatment was very similar between the MS1 population and a sensitive population.[28] However, when examining gene expression between untreated plants, 1399 genes were differentially expressed between the two populations.[22] Specifically, a kinase and GST were constitutively upregulated in MS1 compared with three sensitive populations and an F-box protein was downregulated.[22] However, no cytochrome P450s were differentially expressed, even though malathion treatment indicated involvement of a cytochrome P450. Examination of sequence data revealed 27 cytochrome P450s that had non-synonymous single nucleotide polymorphisms (SNPs) between the resistant and sensitive populations.[22] Of these 27, CYP72A15 was also identified by Rouse[17] as being of interest in propanil resistance. Additional work will be needed to determine if any of these SNPs or differentially expressed genes contribute to herbicide resistance.

The multiple herbicide-resistant population of *E. colona* in Arkansas exhibited a different resistance profile than that of MS1 in that it was resistant to propanil, quinclorac, cyhalofop, and glufosinate.[25] Glufosinate resistance is likely due to a non-metabolic mechanism: treated resistant plants showed a similar reduction in weight to sensitive plants, except that the resistant plants would recover and produce seed.[25] Of the four herbicides, only propanil had a synergistic effect with metabolic inhibitors malathion, carbaryl, and piperonyl butoxide.[25] There was no difference in β-cyanoalanine synthase activity, indicating that reduced cyanide toxicity was not the mechanism of quinclorac resistance.[25] Transcriptomics analysis of response to each of these herbicides has been performed (see previous sections on propanil and quinclorac for cytochrome P450s of interest)[17]; however, more work needs to be done to elucidate the non-target-site mechanisms and, particularly for quinclorac and cyhalofop, determine if metabolism of the herbicide is a contributing factor to herbicide resistance.[25]

In China, an *E. crus-galli* population with resistance to several herbicides was identified.[23] While a Trp574Arg point mutation was found in the ALS gene, treatments with malathion, piperonyl butoxide, and 4-chlor-7-nitrobenzo-oxadiazole (a GST inhibitor), each increased sensitivity of the resistant population to penoxsulam, suggesting a non-target-site mechanism was also present.[23] RNA-seq analysis was used to examine gene expression response to treatments with penoxsulam, pyroxsulam, metamifop, and quinclorac. Two cytochrome P450s, CYP74B2 and CYP78A9, were upregulated across all herbicide treatments.[23] Three genes encoding GSTs were upregulated in response to penoxsulam treatment, making them good candidates for further study.[23] As with the two populations described above, additional work will be needed to determine what role, if any, these genes play in herbicide resistance.

A second population of *E. crus-galli* was identified in China with resistance to the ALS inhibitor penoxsulam and the ACCase inhibitors cyhalofop-butyl and metamifop.[27] No point mutations in either the ALS or ACCase were detected. ALS inhibitor resistance was not identified as broad spectrum: bispyribac sodium and flucetosulfuron still controlled the resistant population. Reduced resistance to penoxsulam when applied following malathion treatment suggested involvement of a cytochrome P450. The researchers identified CYP81A68 as a candidate resistance gene as it was significantly upregulated in the resistant population. Pan et al. then overexpressed the *E. crus-galli* CYP81A68 in rice, conferring resistance to penoxsulam, cyhalofop-butyl, and metamifop, and providing very strong evidence that increased expression of this gene in *E. crus-galli* was involved in the resistance mechanism. They also examined the promoter region and observed differential methylation patterns in the promoter between the sensitive and resistant plants that may contribute to increased expression of CYP81A68.[27]

These four populations highlight the complexity of multiple herbicide resistance. Recent research in *Lolium rigidum* identified a cytochrome P450, CYP81A10v7, that when overexpressed conferred resistance to herbicides representing five different mechanisms of action.[29] While it is possible that a single gene is responsible for resistance in these populations, it is unlikely the case for all of them, particularly the first three populations. Given the prevalence of resistance to propanil and quinclorac in *E. crus-galli* and *E. colona* populations,[10] it is very possible that populations resistant to one or both of these herbicides evolved resistance to other herbicides, such as ALS and ACCase inhibitors, at a later date.

9.5 Glyphosate Resistance

A population of *E. colona* in Australia has been identified as likely having both TSR and NTSR mechanisms to glyphosate.[30,31] RNA-seq analysis of this population identified an aldo-keto reductase that was overexpressed in the resistant population compared with the sensitive population.[30] Overexpression of the same gene in rice conferred glyphosate resistance, and in *Escherichia coli* it led to the metabolism of glyphosate to AMPA and glyoxylate.[30] Further investigation of this population identified an allele of EPSPS that had a point mutation known to render the enzyme resistant to glyphosate,[31] indicating that the population is probably resistant to glyphosate due to a combination of target-site and non-target-site mechanisms.

9.6 Future Work

Understanding metabolic resistance mechanisms will require an in-depth study of candidate resistance genes. *E. crus-galli* and *E. colona* present specific challenges.

Both species have hexaploid genomes,[1,2] complicating genetic studies. Only recently have genomic and transcriptomic resources become available. Two draft genomes have recently been published.[32,33] Gene number estimates range from 103,853 to 108,771.[32,33] Of these genes, 917 are predicted to be cytochrome P450s and 277 as GSTs.[32] This is larger than the number predicted for *E. colona*: 444 cytochrome P450s and 123 GSTs; however, this is likely an underestimate for *E. colona* as the counts were based on transcriptomic data.[22] The sheer number of potential candidate genes requires strategies like RNA-seq to identify genes of interest based on the differences in gene expression or SNP analysis.

Another challenge is the genetic diversity in these species. Many transcriptomic studies rely on resistant and sensitive populations that are not isogenic,[17,22,23] meaning that observed differences in gene expression may be due to genetic diversity between populations rather than herbicide resistance. Some studies that have examined morphological differences have sought to associate a fitness cost with resistance. Mahajan et al.[34] examined 10 populations of *E. colona*, some of which were resistant to glyphosate and some were not. There was a wide range in morphology, including plant height, shoot biomass, tillers per plant, and seeds per plant; however, none were associated with glyphosate resistance.[34] Fischer et al.[5] compared several morphological features among propanil-resistant and -sensitive populations. Of the many traits examined, only reduced grain weight was associated with propanil resistance.[5] To understand fitness costs associated with resistance, it will be necessary to obtain populations that are isogenic except for the alleles known to confer resistance.

Although it is technically challenging, there are strategies for determining if a specific gene or genes are involved in resistance and these studies have become more common in recent years, as was illustrated in the investigation of CYP81A68 in multiple herbicide-resistant *E. crus-galli*.[27] An excellent example in *E. colona* is the thorough investigation of the role of ABCC8 in glyphosate resistance. Pan et al.[35] identified an ABC transporter as being upregulated in glyphosate-resistant *E. colona*. Overexpression of this *E. colona* gene in rice conferred glyphosate resistance as did overexpression of corresponding gene homologs in rice, soybean, and maize. When knocked out by CRISPR/Cas9 technology, rice plants were even more susceptible to glyphosate. Analysis of glyphosate efflux in rice cells showed that efflux of the herbicide was twice as fast in rice cells overexpressing the ABC transporter. Furthermore, differential methylation of the gene between resistant and sensitive plants indicated a possible mechanism for overexpression in the resistant population.[35] It is this type of study that is needed to determine if the genes of interest identified in RNA-seq studies[17,22,23] are, in fact, involved in herbicide resistance.

9.7 Management

The increasing incidence of herbicide-resistant populations will require an integrated management approach. These weeds must be managed as yield loss in rice can be as high as 79%, depending on weed density.[1] The large number of mechanisms of action to which these weeds have evolved resistance also requires that non-chemical methods of control be implemented alongside herbicides to preserve existing chemistries – the presence of populations with resistance to multiple modes of action means that tank mixes and herbicide rotation are not enough to control these weeds. Using rice seed certified to be free of weed seed is critical to avoid introducing weeds into a field.[1] Growers can also plant cultivars that are competitive with weeds. A study by Gealy et al.[36] found that Asian cultivars were more effective at suppressing *E. crus-galli* than US cultivars. There was some variation among US cultivars in their effectiveness at suppressing weed growth. Dense planting and flooding can also suppress weed growth.[1,2,37] At harvest, one study found that the Harrington Seed Destructor effectively destroyed *E. crus-galli* seed, thereby stopping these seeds from entering the seed bank and reducing weed pressure for following seasons.[38] These strategies, combined with a rotation of different herbicide chemistries, should aid growers in reducing yield losses due to these weeds and delay the evolution of additional herbicide-resistant populations.

References

1 Bajwa AA, Jabran K, Shahid M, Ali HH, Chauhan BS, and Ehsanullah, Ecobiology and management of *Echinochloa crus-galli*. *Crop Prot* **75**:151–162 (2015).

2 Peerzada AM, Bajwa AA, Ali HH, and Chauhan BS, Biology, impact and management of *Echinochloa colona* (L.) Link. *Crop Prot* **83**:56–66 (2016).

3 SCH B, Crop mimicry in weeds. *Econ Bot* **37**:255–282 (1983).

4 Kennedy RA, Barrett SCH, Vander Zee D, and Rumpho ME, Germination and seedling growth under anaerobic conditions in *Echinochloa crus-galli* (barnyard grass). *Plant Cell Environ* **3**:243–248 (1980).

5 Fischer AJ, Granados E, and Trujillo D, Propanil resistance in populations of junglerice (*Echinochloa colona*) in Colombian rice fields. *Weed Sci* **41**:201–206 (1993).

6 Kraehmer H, Jabran K, Mennan H, and Chauhan BS, Global distribution of rice weeds – a review. *Crop Prot* **80**:73–86 (2016).

7 Heap, I International Survey of Herbicide Resistant Weeds. www.weedscience .org Accessed online July 4, 2021

8 Lopez-Martinez N, Marshall G, and de Prado R, Resistance of barnyardgrass (*Echinochloa crus-galli*) to atrazine and quinclorac. *Pestic Sci* **51**:171–175 (1997).

9 Norsworthy JK, Wilson MJ, Scott RC, and Gbur EE, Herbicidal activity on acetolactate synthase-resistant barnyardgrass (*Echinochloa crus-galli*) in Arkansas, USA. *Weed Biol Manag* **14**:50–58 (2014).

10 Rouse CE, Roma-Burgos N, Norsworthy JK, Tseng TM, Starkey CE, and Scott RC, *Echinochloa* resistance to herbicides continues to increase in Arkansas rice fields. *Weed Technol* **32**:34–44 (2018).

11 Frear DS, and Still GG, The metabolism of 3,4-dichlorpropionanilide in plants. Partial purification and properties of an aryl acylamidase from rice. *Phytochemistry* **7**:913–920 (1968).

12 Still CC, and Kurzirian O, Enzyme detoxication of 3′,4′-dichloropropionanilide in rice and barnyard grass, a factor in herbicide selectivity. *Nature* **216**:799–800 (1967).

13 Yih RY, McRae DH, and Wilson HF, Mechanism of selective action of 3′,4′-dichloropropionanilide. *Plant Physiol* **43**:1291–1296 (1968).

14 Carey VF, Hoagland RE, and Talbert RE, Verification and distribution of propanil-resistant barnyardgrass (*Echinochloa crus-galli*) in Arkansas. *Weed Technol* **9**:366–372 (1995).

15 Carey VF, Hoagland RE, and Talbert RE, Resistance mechanism of propanil-resistant barnyardgrass: II. In-vivo metabolism of the propanil molecule. *Pestic Sci* **49**:333–338 (1997).

16 Leah JM, Caseley JC, Riches CR, and Valverde BE, Effect of mono-oxygenase inhibitors on uptake, metabolism, and phytotoxicity of propanil in resistant biotypes of jungle-riceEchinochloa colona. *Pestic Sci* **49**:141–147 (1997).

17 Rouse, CE Characterization of multiple-herbicide resistant *Echinochloa colona* from Arkansas. Dissertation (2017).

18 Grossmann K, Quinclorac belongs to a new class of highly selective auxin herbicides. *Weed Sci* **46**:707–716 (1998).

19 Gao Y, Pan L, Sun Y, Zhang T, Dong L, and Li J, Resistance to quinclorac caused by the enhanced ability to detoxify cyanide and its molecular mechanism in *Echinochloa crus-galli* var. *zelayensis*. *Pestic Biochem Physiol* **143**:231–238 (2017).

20 Yip W-K, and Yang SF, Cyanide metabolism in relation to ethylene production in plant tissues. *Plant Physiol* **88**:473–476 (1988).

21 Matzenbacher FO, Bortoly ED, Kalsing A, and Merotto A, Distribution and analysis of the mechanisms of resistance of barnyardgrass (*Echinochloa* *crus-galli*) to imidazolinone and quinclorac herbicides. *J Agric Sci* DOI:10.1017/S0021859614000768 (2014).

22 Wright AA, Rodriguez-Carres M, Sasidharan R, Koski L, Peterson DG, Nandula VK, Ray JD, Bond JA, and Shaw DR, Multiple herbicide-resistant junglerice (*Echinochloa colona*): Identification of genes potentially involved in resistance through differential gene expression analysis. *Weed Sci* DOI:10.1017/wsc.2018.10 (2018).

23 Fang J, Zhang Y, Liu T, Yan B, Li J, and Dong L, Target-site and metabolic resistance mechanisms to penoxsulam in barnyardgrass (*Echinochloa crus-galli* (L.) P. Beauv). *J Agric Food Chem* **67**:8085–8095 (2019).

24 Riar DS, Norsworthy JK, Srivastava V, Nandula V, Bond JA, and Scott RC, Physiological and molecular basis of acetolactate synthase-inhibiting herbicide resistance in barnyardgrass (*Echinochloa crus-galli*). *J Agric Food Chem* **61**:278–289 (2013).

25 Rouse CE, Roma-Burgos N, and Barbosa Martins BA, Physiological assessment of non-target site resistance in multiple-resistant junglerice (*Echinochloa colona*). *Weed Sci* **67**:622–632 (2019).

26 Wright AA, Nandula VK, Grier L, Showmaker KC, Bond JA, Peterson DG, Ray JD, and Shaw DR, Characterization of fenoxaprop-*P*-ethyl-resistant junglerice (*Echinochloa colona*) from Mississippi. *Weed Sci* **64**:588–595 (2016).

27 Pan L, Guo Q, Wang J, Shi L, Yang X, Zhou Y, Yu Q, and Bai L, CYP81A68 confers metabolic resistance to ALS and ACCase-inhibiting herbicides and its epigenetic regulation in *Echinochloa crus-galli*. *J Hazard Mater* DOI:10.1016/j.jhazmat.2022.128225 (2022).

28 Wright AA, Sasidharan R, Koski L, Rodriguez-Carres M, Peterson DG, Nandula VK, Ray JD, Bond JA, and Shaw DR, Transcriptomic changes in *Echinochloa colona* in response to treatment with the herbicide imazamox. *Planta* **247**:369–379 (2018).

29 Han H, Yu Q, Beffa R, Gonzalez S, Maiwald F, Wang J, and Powles SB, Cytochrome p450 CYP81A10v7 in *Lolium rigidum* confers resistance to herbicides across at least five modes of action. *Plant J* **105**:79–92 (2021).

30 Pan L, Yu Q, Han H, Mao L, Nyporko A, Fan L, Bai L, and Powles S, Aldo-keto reductase metabolizes glyphosate and convers glyphosate resistance in *Echinochloa colona*. *Plant Physiol* **181**:1519–1534 (2019).

31 McElroy JS, and Hall ND, *Echinochloa colona* with reported resistance to glyphosate conferred by aldo-keto reductase also contains a Pro-106-Thr EPSPS target site mutation. *Plant Physiol* **183**:447–450 (2020).

32 Guo L, Qiu J, Ye C, Jin G, Mao L, Zhang H, *et al.*, *Echinochloa crus-galli* genome analysis provides insight into its adaptation and invasiveness as a weed. *Nat Commun*, DOI:10.1038/s41467-017-01067-5 (2017).

33 Ye C-Y, Wu D, Mao L, Jia L, Qui J, Lao S, *et al.*, The genomes of the allo-hexaploid *Echinochloa crus-galli* and its progenitors provide insights into polyploidization-driven adaptation. *Mol Plant* **13**:1298–1310 (2020).

34 Mahajan G, Kaur V, Thompson M, and Chauhan BS, Growth behavior and glyphosate resistance level in 10 populations of *Echinohcloa colona* in Australia. *PLoS One* DOI:10.1371/journal.pone.0221382 (2020).

35 Pan L, Yu Q, Wang J, Han H, Mao L, Nyporko A, Maguza A, Fan L, Bai L, and Powles S, An ABCC-type transporter endowing glyphosate resistance in plants. *PNAS* DOI:10.1073/pnas.2100136118 (2021).

36 Gealy DR, Wailes EJ, Estorninos LE, and RSC C, Rice cultivar differences in suppression of barnyardgrass (*Echinohcloa crus-galli*) and economics of reduced propanil rates. *Weed Sci* **51**:601–609 (2003).

37 Kent RJ, and Johnson DE, Influence of flood depth and duration of growth of lowland rice weeds Cote d'Ivoire. *Crop Protection* **20**:691–694 (2001).

38 Schwartz-Lazaro LM, Norsworthy JK, Walsh MJ, and Bagavathiannan MV, Efficacy of the integrated Harrington seed destructor on weeds of soybean and rice production systems in the Southern United States. *Crop Sci* **57**:2812–2818 (2017).

10

Metabolic Resistance to HPPD-Inhibiting Herbicides in Dioecious *Amaranthus* Species

Dean E. Riechers[1], Jeanaflor Crystal T. Concepcion[1], and Shiv S. Kaundun[2]

[1]*Department of Crop Sciences, University of Illinois, Urbana, IL, USA*
[2]*Herbicide Bioscience, Syngenta, Jealott's Hill International Research Centre, Bracknell, Berkshire, UK*

10.1 Introduction

Herbicides that inhibit the enzyme 4-hydroxyphenylpyruvate dioxygenase (HPPD) represent the most recent, novel site-of-action group of commercial herbicides discovered. Although several herbicides that inhibit HPPD had already been commercialized before the target site was known (Fig. 10.1), the commercialization of isoxaflutole in 1998 revolutionized selective weed management in maize (*Zea mays* L.).[1–3] HPPD-inhibiting herbicides provide excellent control of annual dicot weeds and some grasses, and are typically used selectively in maize, sugarcane (*Saccharum officinarum* L.), or grain sorghum (*Sorghum bicolor* (L.) Moench.).[3] Additionally, genetically modified varieties are in development that will allow certain HPPD-inhibiting herbicides to be used selectively in soybeans (*Glycine max* (L.) Merr.).[4,5] The first cases of HPPD inhibitor-resistant weeds were reported in 2009[6] in the dioecious amaranth species, waterhemp (*Amaranthus tuberculatus* (Moq.) J.D. Sauer), and Palmer amaranth (*A. palmeri* S. Watson) (Fig. 10.1), but it is likely resistant populations existed several years before 2009 based on field-use histories.[7,8] In addition to several waterhemp and Palmer amaranth populations that are resistant to HPPD-inhibiting herbicides in North America,[6] three other species (two dicots and one monocot) included recently on the list are wild radish (*Raphanus raphanistrum* L.) from Australia in 2015[6,9] and *Amaranthus retroflexus* L. from North Carolina in 2022,[6] as well as *Leptochloa chinensis* L. Nees from China in 2023.[6] The wild radish population had not been previously exposed to HPPD-inhibiting herbicides,[9] implying selection by other herbicides.

Resistance in Weeds from Herbicide Metabolism, First Edition.
Edited by Vijay K. Nandula and Roland Beffa.

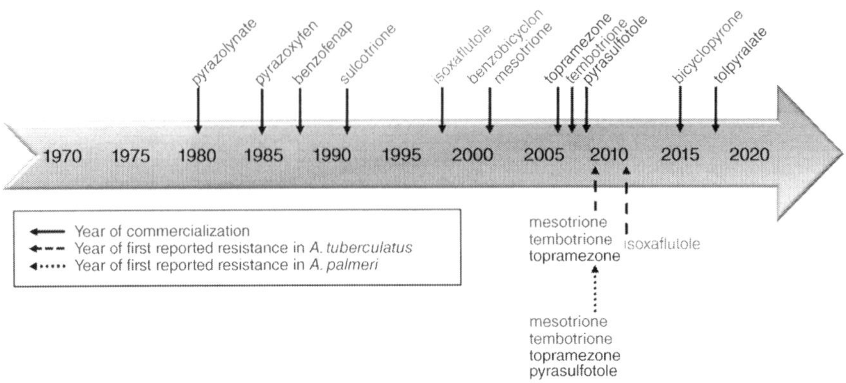

Figure 10.1 Timeline of commercialization of HPPD-inhibiting herbicides and the first year of reported resistance in waterhemp (*Amaranthus tuberculatus*) and Palmer amaranth (*A. palmeri*) populations. Year of commercialization, year of first reported resistance in waterhemp, and year of first reported resistance in Palmer amaranth are indicated by solid arrows, dashed arrows, and square-dotted arrows, respectively. (*See insert for color representation of the figure.*)

HPPD inhibitors are categorized as "very-low-risk" herbicides based on the number of resistant species reported as a percentage of all reported global cases of resistant species.[10] However, both waterhemp and Palmer amaranth represent weed species as "high risk" based on the number of resistant populations reported in these two species.[10] All reported HPPD inhibitor-resistant populations studied in detail to date exhibit enhanced rates of herbicide metabolism relative to sensitive populations (Table 10.1), including the population of wild radish recently reported from Australia.[9] HPPD-inhibiting herbicides are competitive with respect to 4-hydroxy-phenylpyruvic acid for binding to the HPPD enzyme.[1,2,11,27] This competitive binding nature may at least partially explain the lack of site-of-action-based weed resistance to date, since mutations that affect herbicide binding will likely cause a fitness cost to the plant. A recent study comparing enzyme kinetics, inhibition, and HPPD expression demonstrated mutations affecting herbicide binding also decrease the catalytic efficiency of HPPD, but this negative effect can be compensated for via increased expression of the mutant enzyme in transgenic plants.[5]

For the purpose of this chapter, we will focus on metabolism-based resistance in two dioecious amaranths, waterhemp, and Palmer amaranth. Our goals are to summarize the current knowledge regarding metabolic HPPD inhibitor resistance in waterhemp and Palmer amaranth, and to provide a framework to guide future mechanistic studies to better understand the metabolic pathways, enzymes, and genes conferring resistance to HPPD-inhibiting herbicides.

Table 10.1 Summary of crop tolerance/sensitivity and weed resistance to 4-hydroxyphenylpyruvate dioxygenase-inhibiting herbicides, as well as known herbicide metabolites and putative detoxification enzymes/genes in *Amaranthus tuberculatus*, *Amaranthus palmeri*, maize, and rice.

Structural group	Herbicide	Year of commerciali-zation	First report of *A. tuberculatus*[a] and *A. palmeri*[b] resistance or sensitivity in maize[c] or rice[d]	Phase I metabolites	Phase II metabolites	Putative genes and enzymes
Isoxazoles	Isoxaflutole ($C_{15}H_{12}F_3NO_4S$)[1,6,11]	1998	2011[a]	Diketonitrile and benzoic acid derivatives		Unknown
Pyrazoles/ benzopyrazoles	Pyrazolynate ($C_{19}H_{16}Cl_2N_2O_4S$)	1980				
	Pyrazoxyfen ($C_{20}H_{16}Cl_2N_2O_3$)	1985				
	Benzofenap ($C_{22}H_{20}Cl_2N_2O_3$)	1987				
	Topramezone ($C_{16}H_{17}N_3O_5S$)[12–14]	2006	2009[a),b)], 2010[c]	Hydroxy-topramezone 1 and 2, topramezone-hemiaminal; benzoic acid derivative; *N*-desmethyl topramezone		P450
	Pyrasulfotole ($C_{14}H_{13}F_3N_2O_4S$)	2008	2009[b]			
	Tolpyralate ($C_{21}H_{28}N_2O_9S$)[15,16]	2017	2023[c]			Unknown

(Continued)

Table 10.1 (Continued)

Structural group	Herbicide	Year of commerciali-zation	First report of *A. tuberculatus*[a] and *A. palmeri*[b] resistance or sensitivity in maize[c] or rice[d]	Phase I metabolites	Phase II metabolites	Putative genes and enzymes
Triketones	Sulcotrione ($C_{14}H_{13}ClO_5S$)[17]	1991	2019[d]	4-Hydroxy-sulcotrione		P450
	Mesotrione ($C_{14}H_{13}NO_7S$)[17–23]	2001	2009[ab], 2005[c], 2019[d]	4/5-Hydroxy-mesotrione; benzoic acid derivatives (AMBA and MNBA)	Possible sugar or sulfate conjugates	P450 (*Nsf1/CYP81A9* in maize), UGT, sulfotrans-ferase
	Tembotrione ($C_{17}H_{16}ClF_3O_6S$)[14,24]	2007	2008[c], 2009[ab]	(mono-, di-)Hydroxy-tembotrione; reduced tembotrione; benzoic acid derivative (TCMBA)	Glc-conjugate, acetyl-Glc-conjugate	P450, UGT
	Bicyclopyrone ($C_{19}H_{20}F_3NO_5$)	2015				
	Syncarpic acid-3 (SA3) ($C_{19}H_{16}F_6O_4$)[25,26]	None	2020[ac]	Hydroxy-SA3, reduced SA3, dehydrated SA3	GSH-conjugate, Cys-Gly-conjugate, Cys-conjugate	P450, GST
Other	Benzobicyclon ($C_{22}H_{19}ClO_4S_2$)[17]	2001	2019[d]			*HIS1*

a) First report of weed resistance to the corresponding HPPD-inhibiting herbicide in Column 2 occurred in waterhemp (*A. tuberculatus*).
b) First report of weed resistance to the corresponding HPPD-inhibiting herbicide in Column 2 occurred in Palmer amaranth (*A. palmeri*).
c) First report of crop hybrid/cultivar sensitivity to the corresponding HPPD-inhibiting herbicide in Column 2 occurred in maize (*Zea mays*).
d) First report of crop hybrid/cultivar sensitivity to the corresponding HPPD-inhibiting herbicide in Column 2 occurred in rice (*Oryza sativa*).

10.2 Resistance to HPPD-Inhibiting Herbicides in Waterhemp

10.2.1 Occurrence of Resistance to HPPD-Inhibiting Herbicides in Waterhemp

Waterhemp is one of the most troublesome weed species in US agronomic systems. Waterhemp is a small-seeded, summer-annual dicot that utilizes the C4 carbon fixation pathway, exhibits discontinuous germination throughout the summer months, and can produce thousands of seeds per female plant, even under suboptimal growth conditions.[28] HPPD-inhibiting herbicides are often tank-mixed with photosystem II (PS-II; HRAC Group 5)-inhibiting herbicides (most commonly atrazine) for postemergence (POST) dicot weed management in corn, since these two herbicide groups display synergistic activity in combination.[29,30] HPPD-inhibiting herbicides and atrazine are also mixed with very-long-chain fatty acid (VLCFA; HRAC Group 15) elongase inhibitors for broad spectrum, preemergence (PRE) residual weed control in corn.[3]

Resistance to HPPD-inhibiting herbicides in waterhemp was first reported in 2009 by two research groups studying two different populations: one population in Illinois[7] and the other population in Iowa (Fig. 10.1).[8] The waterhemp population in McLean County, Illinois, was named MCR. The field-use history was documented between 2003 and 2009 in continuous corn production, which included *S*-metolachlor (VLCFA inhibitor) and simazine (a triazine PSII inhibitor) applied PRE followed by various combinations of the HPPD-inhibiting herbicides mesotrione, tembotrione, or topramezone (with or without atrazine) applied POST.[7] Importantly, the MCR population did not have glyphosate treatments during this period since it was located in a continuous seed corn field, and thus remains sensitive to glyphosate. The MCR population is 10- to 35-fold resistant to POST mesotrione, depending on the sensitive population used to calculate resistance indices in the greenhouse.[7] A waterhemp population in Henry County, Iowa, evolved under a more complex field corn–soybean rotation, and the field-use history was reported from 1999 to 2009.[8] The microtubule polymerization inhibitor, pendimethalin (HRAC Group 3), was applied PRE followed by glyphosate POST in soybeans every other year from 2000 to 2008. In field corn, *S*-metolachlor was applied PRE followed by mesotrione tank-mixed with atrazine POST every other year from 2001 to 2009, but in some years the POST tank mixes included the acetolactate synthase (ALS; HRAC Group 2) inhibitor nicosulfuron plus atrazine or topramezone plus atrazine.[8]

To date, reported cases of waterhemp resistance to HPPD inhibiting herbicides also include resistance to atrazine.[31] Additionally, the MCR population is resistant to ALS inhibitors,[32] the protoporphyrinogen oxidase (PPO; HRAC Group 14) inhibitor, carfentrazone-ethyl,[33] and VLCFA-inhibiting herbicides[34] through elevated rates of herbicide metabolism. The waterhemp population from Henry County, Iowa, is resistant to HPPD inhibitors, atrazine, and the ALS inhibitor, thifensulfuron-methyl.[8] A different waterhemp population from central Illinois, named CHR, shows a similar resistance profile as MCR, except that it displays target-site-mediated PPO inhibitor resistance and resistance to the synthetic auxin, 2,4-D (HRAC Group 4).[35] Two HPPD inhibitor-resistant waterhemp populations from Nebraska have been reported and characterized in the literature: one from Columbus, Nebraska,[36–38] and the other from Tarnov, Nebraska (hereafter, the Tarnov population is referred to as NEB).[18,39] A multiple-resistant Missouri waterhemp population (MO-Ren) was indirectly determined to possess metabolic mesotrione resistance based on cytochrome P450 (CYP) monooxygenase inhibitor studies with whole plants.[40] Three additional, but uncharacterized, HPPD inhibitor-resistant waterhemp populations from North Carolina, Iowa, and Ontario, Canada, have been reported.[6] Interestingly, none of these reported HPPD inhibitor-resistant waterhemp populations exhibit resistance to the nonselective herbicide, glufosinate,[6] which may be due to minimal population exposure to glufosinate and/or a lack of weed variants able to rapidly metabolize glufosinate.

10.2.2 Mechanisms of Resistance to HPPD-Inhibiting Herbicides in Waterhemp

10.2.2.1 Mesotrione and Tembotrione

Mesotrione (Fig. 10.2A) is a triketone HPPD-inhibiting herbicide that primarily controls annual dicot weeds in corn and sugarcane.[19] In addition to foliar HPPD inhibitor resistance in MCR described above,[7] this population also displays resistance to HPPD inhibitors and atrazine applied PRE.[41,42] When conducting initial mechanistic studies, neither an altered HPPD target site (sequence, gene copy number, or gene expression) nor decreased uptake of mesotrione was detected in the MCR population.[20] Biochemical studies using excised leaves and whole plants derived from vegetative clones demonstrated that an elevated rate of oxidative metabolism (presumably via CYPs) contributes to mesotrione resistance in MCR.[20] However, due to evidence supporting the multigenic nature of mesotrione resistance in MCR[43], additional non-target-site mechanisms cannot yet be excluded.

An excised leaf assay and vegetatively cloned plants used in these studies[20,44] ensured that results obtained from herbicide metabolism assays were independent of whole-plant translocation patterns, and that identical waterhemp genotypes were analyzed within a biochemical time-course analysis.[20] Durations for 50% of absorbed mesotrione to degrade (DT_{50}) were about 12 hours for MCR (the same as

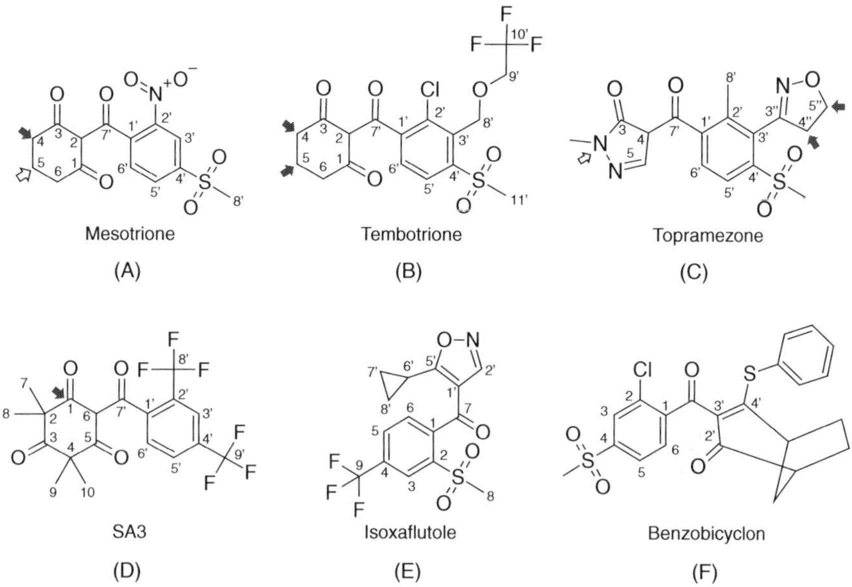

Figure 10.2 Structures of HPPD-inhibiting herbicides and reported primary sites of metabolism in weedy *Amaranthus* spp., maize, and rice. (A) Mesotrione, (B) tembotrione, (C) topramezone, (D) syncarpic acid-3 (SA3), (E) isoxaflutole, and (F) benzobicyclon. Solid arrows in (A) and (B) indicate sites for ring hydroxylation shared between resistant *Amaranthus* spp. and tolerant maize. The open arrow in (A) indicates the site of ring hydroxylation forming the 5-hydroxy-mesotrione metabolite, which occurs in greater abundance in resistant *Amaranthus palmeri* than in *Amaranthus tuberculatus*, relative to 4-hydroxy-mesotrione.[22] The open arrow in (C) indicates the site for topramezone *N*-demethylation in tolerant maize[13] whereas the solid arrows in (C) indicate two possible sites of isoxazoline ring hydroxylation of topramezone in resistant *Amaranthus tuberculatus*.[12] The arrow in (D) indicates the putative site of ketone reduction-dehydration-glutathione conjugation of SA3 in resistant *A. tuberculatus*. Source: Adapted from Concepcion et al.[25] (*See insert for color representation of the figure.*)

in tolerant corn), while DT_{50} values in two mesotrione-sensitive populations were 30 hours or >45 hours.[20] The major metabolite of mesotrione detected in MCR was 4-hydroxy-mesotrione (solid arrow in Fig. 10.2A),[20] which is the same initial metabolite formed in tolerant corn (Table 10.1). The chemical P450 inhibitors, malathion and tetcyclacis, decreased mesotrione metabolism and malathion further reduced the biomass of MCR plants when applied with mesotrione.[20]

Similar to the MCR population, rapid mesotrione metabolism via 4-hydroxylation of the cyclohexanedione ring was detected in the NEB population when compared with a sensitive population.[18] The major difference between the MCR and NEB populations is that MCR had experienced mesotrione, tembotrione, and topramezone in its field-use history, while NEB was only treated with mesotrione and tembotrione.[18] The MCR and NEB populations are resistant to topramezone, which implies that triketone HPPD inhibitors can select for resistance to a commercial HPPD inhibitor with a different (i.e., pyrazole) chemical structure, as described in Section 10.2.2.2.

Tembotrione (Fig. 10.2B) is another triketone HPPD-inhibiting herbicide used in corn production systems,[3] which provides broad-spectrum control of annual grasses and dicots. Although a mechanistic investigation of tembotrione metabolism has been reported in Palmer amaranth,[24] studies of tembotrione metabolism in waterhemp to date have been limited to indirect analyses using P450 inhibitors.[37] In this research, greenhouse studies demonstrated the chemical P450 inhibitors, malathion and piperonyl butoxide, applied in combination with tembotrione or topramezone increased HPPD inhibitor-resistant waterhemp control and decreased plant biomass greater than herbicides applied alone. However, neither of these P450 inhibitors affected the foliar activity of mesotrione on HPPD inhibitor-resistant waterhemp.[37]

10.2.2.2 Topramezone

The MCR population also displayed resistance to topramezone,[7] another HPPD-inhibiting herbicide having a distinct pyrazole structure[2] compared to the triketone structures of mesotrione and tembotrione (Fig. 10.2A–C). Initial biochemical studies with excised leaves and whole plants indicated that an elevated rate of metabolism confers topramezone resistance in MCR relative to two sensitive waterhemp populations. A comparative study on the metabolism of topramezone was then conducted in tolerant corn and multiple-resistant waterhemp populations.[12] In this study, the SIR population (a subpopulation of the MCR population, collected at the same site)[33] initially metabolized topramezone in a different manner than tolerant maize (Table 10.1).[13] At later time points (24 and 48 HAT), topramezone metabolites detected in maize were desmethyl topramezone and a benzoic acid derivative, whereas SIR predominantly formed two putative hydroxylated metabolites.[12] Mass spectral data indicated that hydroxylation of the isoxazole ring occurred in the C4″ and C5″ positions (solid arrows in Fig. 10.2C) and dealkylation occurred via *N*-demethylation of the pyrazole ring (open arrow in Fig. 10.2C).[12] Although these oxidative reactions detoxifying topramezone likely involve P450 activities, this situation is different from mesotrione resistance in that the mesotrione selectivity mechanism in maize is the same as the mesotrione resistance mechanism in waterhemp (i.e., rapid formation of 4-hydroxy-mesotrione).[5,20] The metabolic fate of topramezone has yet to be reported in Palmer amaranth.

10.2.2.3 Other HPPD-Inhibiting Herbicides

Syncarpic acid-3 (SA3; Fig. 10.2D) is a noncommercial, nonselective, synthetic compound derived from the natural product leptospermone, which was discovered in late 1970s as an allelochemical extracted from the bottlebrush plant (*Callistemon citrinus*).[27] Unique structural features of SA3 include two trifluoromethyl groups (Fig. 10.2D) that provide enhanced chemical and metabolic stability, improved lipophilicity and bioavailability, and increased protein-binding affinity in pharmaceuticals.[45] SA3 is also characterized by a cyclohexanetrione ring with alternating methyl groups (R-$(CH_3)_2$) and ketone substituents, rendering the molecule electron deficient and hence presumed as recalcitrant to plant oxidative metabolism.[27] Despite these unique structural features of SA3, waterhemp populations with multiple resistance metabolized SA3 at much faster rates compared to sensitive populations.[25]

Phases I, II, and IV metabolites of SA3 were detected and putatively identified using untargeted metabolomics following clear variations in metabolism rates of SA3 between resistant and sensitive populations. Phase I metabolites of SA3 included reduced SA3, likely formed via ketone reduction in the cyclohexanetrione ring, and a dehydrated SA3 metabolite, which possibly formed enzymatically after ketone reduction (arrow in Fig. 10.2D). A hydroxy-SA3 metabolite, possibly formed via a P450-catalyzed alkyl hydroxylation of a methyl group attached to C9 or C10 of the cyclohexanetrione ring, was also detected in this study; however, the abundance of this metabolite was not associated with resistance. A phase II metabolite, identified as a SA3–glutathione (GSH) conjugate, was more abundant in multiple resistant waterhemp compared to sensitive populations.[25] This SA3–GSH conjugate presumably formed via an unexpected glutathione *S*-transferase (GST)-catalyzed reaction. The sequential formation of reduced SA3 and dehydrated SA3 metabolites introduced a putative Michael acceptor site required for phase II GSH conjugation to occur.[25] Phase III presumably involves transport of SA3–GSH into the vacuole via ABC transporters and is therein acted upon by vacuolar peptidases forming several phase IV metabolites, representing catabolites of SA3–GSH.[25]

Resistance to the HPPD-inhibiting herbicide isoxaflutole (Fig. 10.2E) applied POST has been reported in waterhemp[6,46] but mechanistic studies involving this proherbicide, or its herbicidally active diketonitrile metabolite, have not been published. Benzobicyclon (Fig. 10.2F) is a cyclic β-triketone HPPD inhibitor discovered in 2001 (Table 10.1) and developed for selective weed control in rice (*Oryza sativa* L.) paddy fields.[47,48] Benzobicyclon displays efficacy against weeds in paddy rice that are resistant to other types of herbicides, including sulfonylureas. Benzobicyclon is a proherbicide, which upon reaction with water forms benzobicyclon hydrolysate, the active form of the herbicide.[48] Other inactive metabolites of benzobicyclon are amino- and hydroxyethylamine-substituted compounds. Resistance to benzobicyclon among weed species has not yet been reported; however, sensitivity to this herbicide has been reported in high-yielding varieties of rice.[17] Sensitivity of rice cultivars to benzobicyclon is inherited via a nonfunctional *his1* allele (from an indica variety), which was attributed to a nonsense mutation caused by a 28-base pair deletion in exon IV of the gene (*Os*02g0280700).

10.2.2.4 Molecular Genetics and Inheritance of HPPD Inhibitor Resistance in Waterhemp

Greenhouse studies have examined the genetics and inheritance of the mesotrione resistance trait in the MCR population,[43] and inheritance of mesotrione resistance was also investigated in resistant waterhemp populations from Iowa[49] and Nebraska (NEB).[38] Each of these inheritance studies concluded that mesotrione resistance in waterhemp is a multigenic trait, and Kohlhase et al.[49] demonstrated the number of functional genes governing mesotrione resistance increased as the mesotrione rate increased. The genetics and inheritance to other HPPD-inhibiting herbicides in waterhemp have not been reported; it will be of great interest to determine if the same gene(s) and alleles govern resistance to triketone, pyrazole, and syncarpic acid HPPD inhibitors (Fig. 10.2).

A transcriptome study was performed with the HPPD inhibitor-resistant NEB population,[39] which had demonstrated rapid oxidative metabolism of mesotrione in previous work.[18] Transcriptional responses in leaves from resistant and sensitive populations were compared during a time course of 3, 6, 12, and 24 hours after mesotrione treatment. Transcripts representing several classes of plant defense genes, including *GSTs* and *P450s*, as well as genes involved with reactive oxygen species-mediated signaling or detoxification were identified as differentially upregulated in the resistant population. In addition to identifying candidate genes involved in conferring HPPD inhibitor resistance in waterhemp, this study also provides a *de novo* transcriptome that can be mined in future research studies aimed at investigating herbicide resistance mechanisms or response to abiotic/biotic stresses.[39] Ultimately, identification of the underlying molecular genetic factors will greatly assist in devising new strategies to overcome HPPD inhibitor resistance, such as silencing metabolic resistance genes.[50]

10.3 Resistance to HPPD-Inhibiting Herbicides in Palmer Amaranth

10.3.1 Occurrence of Resistance to HPPD-Inhibiting Herbicides in Palmer Amaranth

Palmer amaranth is the second of three amaranth species to evolve resistance to HPPD-inhibiting herbicides. The first confirmed case of resistance was in a Kansas population from a wheat (*Triticum aestivum* L.)/sorghum field that did not have a history of HPPD-inhibiting herbicide usage.[51] The Kansas population was serendipitously discovered in 2009 following a pyrasulfotole plus bromoxynil field sorghum demonstration trial (Fig. 10.1), indicating that the highly polymorphic Palmer amaranth species may encompass populations with standing genetic variation for tolerance to some HPPD-inhibiting compounds. This scenario also implies that resistance to HPPD-inhibiting herbicides may have been selected by use of other herbicide sites of action in this population. Seven other cases of Palmer amaranth resistance to HPPD-inhibiting herbicides – Kansas (2), Nebraska (2), Wisconsin (1), North Carolina (1) and Iowa (1) – have been reported,[6] although numerous populations showing poor control with HPPD-inhibiting herbicides have also been documented. For instance, a 2016 semi-random collection and greenhouse testing of 154 Arkansas Palmer amaranth populations with a field-use rate of tembotrione showed an average mortality rate of 51% and as much as 93% of the populations displayed less than 80% control.[52] Contrary to most HPPD inhibitor-resistant *Amaranthus* spp. populations, a large number of the tembotrione-resistant populations from Arkansas were sensitive to atrazine, which indicates that cross- or multiple resistance patterns with HPPD inhibitors and triazines may not be the same between waterhemp and Palmer amaranth.[31]

Similarly, 9 out of 28 Palmer amaranth populations from a 2014 Kansas collection showed reduced sensitivity (≤59% injury) to a field-use rate of mesotrione.[53] A follow-up dose–response analysis of two Kansas populations confirmed 2.8- and 8.4-fold resistance to the HPPD-inhibiting herbicides. Testing of 110 North Carolina Palmer amaranth populations with a labeled rate of mesotrione identified plant survivors from as many as 42 populations, although most populations had survivorship of less than 5%.[54] Whole-plant mesotrione dose–response analysis of 2 of these 42 Palmer amaranth populations from North Carolina revealed low resistance indices of only 2.4- and 3.3-fold. Given the sensitivity of Palmer amaranth to herbicides when tested in controlled glasshouse conditions (S.S. Kaundun, personal observations), resistance/tolerance levels may have been underestimated in different large-scale glasshouse studies that employed a single recommended rate of HPPD-inhibiting herbicides. A more realistic picture of resistance or sensitivity shift could be obtained if the Palmer amaranth populations were analyzed with a few rates above and below a carefully established discriminating rate of HPPD-inhibiting herbicides based on several known sensitive populations.

10.3.2 Sensitivity to Mesotrione in Palmer Amaranth Plants Grown at Different Temperatures

A growth study was conducted to assess the effect of temperature on efficacy of mesotrione in Palmer amaranth.[55] A single Palmer amaranth population was grown at three different temperatures, namely, low (LT; 25/15°C, day/night temperatures), optimum (OT; 32.5/22.5°C), and high (HT; 40/30°C) temperatures.

Whole-plant dose–response analysis showed that the Palmer amaranth population was more sensitive at LT compared to OT and HT. Differences in efficacy were explored on the basis of mesotrione absorption, translocation, metabolism, and *HPPD* expression at these three temperatures.[55] Although the greatest amount of mesotrione absorption occurred at HT, a larger percentage of absorbed mesotrione translocated to meristematic tissues at LT or OT compared to HT. Plants grown at HT metabolized mesotrione faster than those at OT or LT, based on T_{50} (time required to metabolize 50% of ^{14}C radiolabeled mesotrione absorbed) values.[55] Interestingly, 48 hours after treatment (HAT), the relative ratios of *HPPD:CPS* (carbamoyl phosphate synthetase; a reference gene used for normalization) or *HPPD:β-tubulin* (the second reference gene for normalization) expression levels in mesotrione-treated plants increased more at HT compared to LT or OT (18.4-, 3.1-, and 3.5-fold relative to untreated plants, respectively).[55] Collectively, these results imply that expression of the resistant phenotype to HPPD-inhibiting herbicides in Palmer amaranth may also be temperature-dependent.[55]

10.3.3 Mechanisms of Resistance to HPPD-Inhibiting Herbicides in Palmer Amaranth

10.3.3.1 Mesotrione

The mechanism of mesotrione resistance was investigated in two Palmer amaranth populations from Kansas and Nebraska.[21] The Nebraska seed sample originated from mesotrione survivors from a seed corn production field near Shickley, Fillmore County, in 2010.[56] The Nebraska population was characterized by a four-fold resistance index based on ED_{90} values for mesotrione.[57] The population was cross-resistant to tembotrione and topramezone with respective resistance indices of 4–6-fold and 14–23-fold, depending on the sensitive population employed.[57] The Kansas population used in the mechanism of resistance study was derived from the first documented case of HPPD inhibitor resistance in a Palmer amaranth population (Fig. 10.1), which did not have a history of HPPD-inhibiting herbicide application.[51] The derived Kansas subpopulation was created by crossing male and female plants having survived a field rate of mesotrione in a glasshouse experiment. Mesotrione whole-plant dose–response tests showed the Kansas population was more than 10–18-fold resistant compared to two sensitive populations. Differential absorption and translocation were not associated with resistance in the Kansas and Nebraska populations. In contrast, within 24 HAT, metabolism of mesotrione was faster in the two Kansas and Nebraska populations compared to the sensitive populations. However, although sensitive and resistant Palmer amaranth populations completely metabolized parent ^{14}C radiolabeled mesotrione within 48–72 HAT, there was no attempt to characterize mesotrione metabolites formed in the different populations in this study.[21]

Investigation of a potential target-site-based mechanism in conferring resistance to mesotrione identified 4–12-fold higher *HPPD* transcript levels in the Kansas and Nebraska populations compared to five other sensitive populations. Therefore, it was concluded that rapid detoxification of mesotrione combined with increased *HPPD* gene expression was responsible for mesotrione resistance in the two Kansas and Nebraska Palmer amaranth populations.[21] However, given the plant-to-plant variability associated with this study,[21] the contribution of target-site gene overexpression toward conferring mesotrione resistance in *Amaranthus* spp. should be further explored with a larger number of samples within populations.

A recent study compared mesotrione metabolism profiles in several resistant and sensitive Palmer amaranth and waterhemp populations from the United States,[22] with the goal of identifying phase I and phase II mesotrione metabolites. Several phase I metabolites were identified, including 4-hydroxy-mesotrione and 5-hydroxy-mesotrione (Fig. 10.2A) and, interestingly, the ratio of 5-hydroxy-/4-hydroxy-mesotrione was higher in resistant Palmer amaranth than in resistant waterhemp.[22] Several possible, unidentified phase II metabolites were also formed in resistant Palmer amaranth,[22] but potential target-site mechanisms were not investigated in this research.

10.3.3.2 Tembotrione

Studies on the mechanism of tembotrione resistance[24] were carried out with the same Nebraska Palmer amaranth population that was previously used for the mechanism of mesotrione resistance study.[21] However, the seed collection year (2011)[24] was different than in the previous work.[21] Whole-plant dose–response analysis of the Nebraska-resistant population compared to a Nebraska-sensitive population showed a three-fold increase in resistance to tembotrione, based on both percent survival and injury data. There was no difference detected in radiolabeled mesotrione absorption or translocation between the sensitive and resistant population. In contrast, significant differences in rates of tembotrione metabolism were observed as early as 6 HAT, with 72% and 41% of parent radiolabeled mesotrione detected in the sensitive and resistant populations, respectively.[24] The resistant population was characterized by a T_{50} value for tembotrione of 7.3 hours while that of the sensitive population was 17.1 hours, amounting to 2.3-fold faster rate of tembotrione metabolism in the resistant compared to the sensitive population.[24]

Metabolite analysis identified faster 4-hydroxylation of the aryl ring of tembotrione followed by glycosylation in the resistant population (Table 10.1; Fig. 10.2B) compared to the sensitive population.[24] The corresponding time for 50% accumulation of 4-hydroxy-tembotrione was 4.9 hours for the resistant population and 11.9 hours for the sensitive population. The initial metabolic route of tembotrione in the resistant Palmer amaranth population[24] is similar to that observed for mesotrione metabolism in two multiple-resistant waterhemp populations from Illinois (MCR) and Nebraska (NEB) (Fig. 10.2A, B), whereby 4-hydroxy-mesotrione levels were the main differentiator between sensitive and resistant populations.[18,20] However, a glycosylated metabolite of mesotrione was not identified in these two multiple-resistant waterhemp populations (Table 10.1). The tembotrione study also investigated the effect of temperature (18 C or 28 C), continuous light or 16:8 hours (light:dark) conditions, and plant developmental stage (2, 4, and 8 leaves) on levels of parent tembotrione remaining 16 HAT in the resistant and sensitive populations.[24] Metabolism differences between these populations were most pronounced when plants were grown in continuous light (regardless of temperature) and treated at the four-leaf stage. As a result, this growth stage and environmental conditions were recommended for diagnosing resistance to tembotrione in Palmer amaranth populations.[24]

10.4 Conclusions and Future Research Directions

Metabolism-based herbicide resistance is frequently, but not always (as is the case with metabolic atrazine resistance in waterhemp[43]), genetically controlled as a multigenic trait,[38,42,43,50,58] which can confer unpredictable and complicated cross- or multiple resistance to herbicides with the same or different sites of action. Cross- and multiple herbicide resistance pose serious challenges for weed management in the future, especially if these resistance mechanisms do not confer a fitness cost to the plant in the absence of herbicide.[50,58] To date, significant fitness costs have not been reported for metabolic resistance to HPPD-inhibiting herbicides,[59,60] although additional research is warranted in this area. Furthermore, herbicide resistance due to rapid herbicide metabolism has the potential to confer resistance not only to existing commercial herbicides, but also to new, noncommercial, or yet-to-be discovered active ingredients.[50] Our recent work with noncommercial, syncarpic acid HPPD inhibitors, which are not selective in corn or cereal crops, clearly demonstrates this principle.[25]

The dioecious amaranths are particularly problematic from the standpoint of evolving diverse metabolic resistance mechanisms and are listed as "high-risk" weed species in a risk assessment matrix.[10] One pernicious aspect of the dioecious amaranths is their ability to transfer herbicide resistance traits through seed and pollen, owing to their obligate outcrossing nature.[28] Results from field studies demonstrated the ability of dioecious amaranths to transfer mesotrione resistance through both interspecific (from waterhemp to Palmer amaranth) and intraspecific (between waterhemp populations) hybridization.[61] A better understanding of the biochemical, molecular, and genetic mechanisms[62] conferring metabolic resistance to HPPD-inhibiting herbicides, in addition to identifying genes conferring metabolic resistance (as well as other possible non-target-site resistance mechanisms[20,43,49]), is necessary to provide new insights into rapidly evolving *Amaranthus* populations in response to herbicide selection pressures and will assist in the development of innovative and integrated resistant weed management systems. For example, unraveling metabolism-based resistance mechanisms may help identify new strategies to overcome resistance, such as silencing resistance genes, blocking metabolic enzymes with chemical inhibitors,[20,37,40,42] or developing resistance-breaking herbicide chemistries. These findings also demonstrate an urgent need for discovery of new herbicide sites of action[63] to help combat multiple-resistant weed populations.

Acknowledgments

The authors wish to thank the numerous researchers who have contributed their time, talents, resources, and efforts to investigate HPPD-inhibiting herbicide resistance mechanisms in dioecious *Amaranthus* in our laboratories. In particular, we wish to thank Dr. Anatoli V. Lygin for his dedication and outstanding accomplishments in characterizing resistance mechanisms using unlabeled topramezone in waterhemp, and for contributing insightful preliminary data regarding the metabolic fate of syncarpic acids in waterhemp.

References

1 Viviani F, Little JP, and Pallett KE, The mode of action of isoxaflutole. II. Characterization of the inhibition of carrot 4-hydroxyphenylpyruvate dioxygenase by the diketonitrile derivative of isoxaflutole. *Pestic Biochem Physiol* **62**:125–134 (1998).
2 Ndikuryayo F, Moosavi B, Yang W-C, and Yang G-F, 4-Hydroxyphenylpyruvate dioxygenase inhibitors: from chemical biology to agrochemicals. *J Agric Food Chem* **65**:8523–8537 (2017).
3 van Almsick A, New HPPD-inhibitors: a proven mode of action as a new hope to solve current weed problems. *Outlook Pest Manag* **20**:27–30 (2009).
4 Siehl DL, Tao Y, Albert H, Dong Y, Heckert M, Madrigal A, *et al.*, Broad 4-hydroxyphenylpyruvate dioxygenase inhibitor herbicide tolerance in soybean with an optimized enzyme and expression cassette. *Plant Physiol* **166**:1162–1176 (2014).
5 Hawkes TR, Langford MP, Viner R, Blain RE, Callaghan FM, Mackay EA, *et al.*, Characterization of 4-hydroxyphenylpyruvate dioxygenases, inhibition by herbicides and engineering for herbicide tolerance in crops. *Pestic Biochem Physiol* **156**:9–28 (2019).

6 Heap I, The international herbicide-resistant weed database. www.weedscience.org [Accessed May 30], 2024.

7 Hausman NE, Singh S, Tranel PJ, Riechers DE, Kaundun SS, Polge ND, *et al.*, Resistance to HPPD-inhibiting herbicides in a population of waterhemp (*Amaranthus tuberculatus*) from Illinois, United States. *Pest Manag Sci* **67**:258–261 (2011).

8 McMullan PM, and Green JM, Identification of a tall waterhemp (*Amaranthus tuberculatus*) biotype resistant to HPPD-inhibiting herbicides, atrazine, and thifensulfuron in Iowa. *Weed Technol* **25**:514–518 (2011).

9 Lu H, Yu Q, Han H, Owen MJ, and Powles SB, Evolution of resistance to HPPD-inhibiting herbicides in a wild radish population via enhanced herbicide metabolism. *Pest Manag Sci* **76**:1929–1937 (2020).

10 Moss S, Ulber L, and den Hoed I, A herbicide resistance risk matrix. *Crop Prot* **115**:13–19 (2019).

11 Pallett KE, Cramp SM, Little JP, Veerasekaran P, Crudace AJ, and Slater AE, Isoxaflutole: the background to its discovery and the basis of its herbicidal properties. *Pest Manag Sci* **57**:133–142 (2001).

12 Lygin AV, Kaundun SS, Morris JA, Mcindoe E, Hamilton AR, and Riechers DE, Metabolic pathway of topramezone in multiple-resistant waterhemp (*Amaranthus tuberculatus*) differs from naturally tolerant maize. *Front Plant Sci* **9**:1644 (2018).

13 Grossmann K, and Ehrhardt T, On the mechanism of action and selectivity of the corn herbicide topramezone: a new inhibitor of 4-hydroxyphenylpyruvate dioxygenase. *Pest Manag Sci* **63**:429–439 (2007).

14 Williams MM, and Pataky JK, Factors affecting differential sensitivity of sweet corn to HPPD-inhibiting herbicides. *Weed Sci* **58**:289–294 (2010).

15 Tsukamoto M, Kikugawa H, Nagayama S, Suganuma T, Okita T, and Miyamoto H, Discovery and structure optimization of a novel corn herbicide, tolpyralate. *J Pestic Sci* **46**:152–159 (2021).

16 Williams MM II, Hausman NE, Saballos A, Landau CA, Brooks MD, Flannery P, Tracy WF, and Thompson CJ, First report of severe tolpyralate sensitivity in corn (*Zea mays*) discovers a novel genetic factor conferring crop response to a herbicide. *Pest Manag Sci* **80**:1645–1653 (2024).

17 Maeda H, Murata K, Sakuma N, Takei S, Yamazaki A, Karim MR, *et al.*, A rice gene that confers broad-spectrum resistance to β-triketone herbicides. *Science* **365**:393–396 (2019).

18 Kaundun SS, Hutchings S-J, Dale RP, Howell A, Morris JA, Kramer VC, *et al.*, Mechanism of resistance to mesotrione in an *Amaranthus tuberculatus* population from Nebraska, USA. *PLoS One* **12**:e0180095 (2017).

19 Mitchell G, Bartlett DW, Fraser TEM, Hawkes TR, Holt DC, Townson JK, *et al.*, Mesotrione: a new selective herbicide for use in maize. *Pest Manag Sci* **57**:120–128 (2001).

20 Ma R, Kaundun SS, Tranel PJ, Riggins CW, McGinness DL, Hager AG, *et al.*, Distinct detoxification mechanisms confer resistance to mesotrione and atrazine in a population of waterhemp. *Plant Physiol* **163**:363–377 (2013).

21 Nakka S, Godar AS, Wani PS, Thompson CR, Peterson DE, Roelofs J, *et al.*, Physiological and molecular characterization of hydroxyphenylpyruvate dioxygenase (HPPD)-inhibitor resistance in Palmer amaranth (*Amaranthus palmeri* S.Wats.). *Front Plant Sci* **8**:555 (2017).

22 Concepcion JCT, Kaundun SS, Morris JA, Brandenburg AN, and Riechers DE, Metabolism of the 4-hydroxyphenylpyruvate dioxygenase inhibitor, mesotrione, in multiple-herbicide resistant Palmer amaranth (*Amaranthus palmeri*). *J Agric Food Chem* **72**:5595–5608 (2024).

23 Williams MM, Pataky JK, Nordby JN, Riechers DE, Sprague CL, and Masiunas JB, Cross-sensitivity in sweet corn to nicosulfuron and mesotrione applied postemergence. *Hort Sci* **40**:1801–1805 (2005).

24 Küpper A, Peter F, Zöllner P, Lorentz L, Tranel PJ, Beffa R, *et al.*, Tembotrione detoxification in 4-hydroxyphenylpyruvate dioxygenase (HPPD) inhibitor-resistant Palmer amaranth (*Amaranthus palmeri* S. Wats.). *Pest Manag Sci* **74**:2325–2334 (2018).

25 Concepcion JCT, Kaundun SS, Morris JA, Hutchings S-J, Strom SA, Lygin AV, and Riechers DE, Resistance to a nonselective 4-hydroxyphenylpyruvate dioxygenase-inhibiting herbicide via novel reduction-dehydration-glutathione conjugation in *Amaranthus tuberculatus*. *New Phytol* **232**:2089–2105 (2021).

26 Concepcion JCTC, Hutchings S-J, Morris J, Kaundun SS, Lygin AV, and Riechers DE, Resistance to a non-selective HPPD-inhibiting herbicide in multiple-resistant waterhemp (*Amaranthus tuberculatus*) populations. *Abstr Weed Sci Soc Am* **60**:463 (2020).

27 Beaudegnies R, Edmunds AJF, Fraser TEM, Hall RG, Hawkes TR, Mitchell G, *et al.*, Herbicidal 4-hydroxyphenylpyruvate dioxygenase inhibitors—a review of the triketone chemistry story from a Syngenta perspective. *Bioorg Med Chem* **17**:4134–4152 (2009).

28 Steckel LE, The dioecious *Amaranthus* spp.: here to stay. *Weed Technol* **21**:567–570 (2007).

29 Abendroth JA, Martin AR, and Roeth FW, Plant response to combinations of mesotrione and photosystem II inhibitors. *Weed Technol* **20**:267–274 (2006).

30 Hugie JA, Bollero GA, Tranel PJ, and Riechers DE, Defining the rate requirements for synergism between mesotrione and atrazine in redroot pigweed (*Amaranthus retroflexus*). *Weed Sci* **56**:265–270 (2008).

31 Jacobs KE, Butts-Wilmsmeyer CJ, Ma R, O'Brien SR, and Riechers DE, Association between metabolic resistances to atrazine and mesotrione in a multiple-resistant waterhemp (*Amaranthus tuberculatus*) population. *Weed Sci* **68**:358–366 (2020).

32 Guo J, Riggins CW, Hausman NE, Hager AG, Riechers DE, Davis AS, *et al.*, Nontarget-site resistance to ALS inhibitors in waterhemp (*Amaranthus tuberculatus*). *Weed Sci* **63**:399–407 (2017).

33 Obenland OA, Ma R, O'Brien SR, Lygin AV, and Riechers DE, Carfentrazone-ethyl resistance in an *Amaranthus tuberculatus* population is not mediated by amino acid alterations in the PPO2 protein. *PLoS One* **14**:e0215431 (2019).

34 Strom SA, Gonzini LC, Mitsdarfer C, Davis AS, Riechers DE, and Hager AG, Characterization of multiple herbicide–resistant waterhemp (*Amaranthus tuberculatus*) populations from Illinois to VLCFA-inhibiting herbicides. *Weed Sci* **67**:369–379 (2019).

35 Evans CM, Strom SA, Riechers DE, Davis AS, Tranel PJ, and Hager AG, Characterization of a waterhemp (*Amaranthus tuberculatus*) population from Illinois resistant to herbicides from five site-of-action groups. *Weed Technol* **33**:400–410 (2019).

36 Oliveira MC, Jhala AJ, Gaines T, Irmak S, Amundsen K, Scott JE, *et al.*, Confirmation and control of HPPD-inhibiting herbicide-resistant waterhemp (*Amaranthus tuberculatus*) in Nebraska. *Weed Technol* **31**:67–79 (2017).

37 Oliveira MC, Gaines TA, Dayan FE, Patterson EL, Jhala AJ, and Knezevic SZ, Reversing resistance to tembotrione in an *Amaranthus tuberculatus* (var. *rudis*) population from Nebraska, USA with cytochrome P450 inhibitors. *Pest Manag Sci* **74**:2296–2305 (2018).

38 Oliveira MC, Gaines TA, Jhala AJ, and Knezevic SZ, Inheritance of mesotrione resistance in an *Amaranthus tuberculatus* (var. *rudis*) population from Nebraska, USA. *Front Plant Sci* **9**:60 (2018).

39 Kohlhase DR, O'Rourke JA, Owen MDK, and Graham MA, Using RNA-seq to characterize responses to 4-hydroxyphenylpyruvate dioxygenase (HPPD) inhibitor herbicide resistance in waterhemp (*Amaranthus tuberculatus*). *BMC Plant Biol* **19**:182 (2019).

40 Shergill LS, Bish MD, Jugulam M, and Bradley KW, Molecular and physiological characterization of six-way resistance in an *Amaranthus tuberculatus* var. *rudis* biotype from Missouri. *Pest Manag Sci* **74**:2688–2698 (2018).

41 Hausman NE, Tranel PJ, Riechers DE, Mawell DJ, Gonzini LC, and Hager AG, Responses of an HPPD inhibitor-resistant waterhemp (*Amaranthus tuberculatus*) population to soil-residual herbicides. *Weed Technol* **27**:704–711 (2013).

42 Ma R, Evans AF, and Riechers DE, Differential responses to preemergence and postemergence atrazine in two atrazine-resistant waterhemp populations. *Agron J* **108**:1196–1202 (2016).

43 Huffman J, Hausman NE, Hager AG, Riechers DE, and Tranel PJ, Genetics and inheritance of nontarget-site resistances to atrazine and mesotrione in a waterhemp (*Amaranthus tuberculatus*) population from Illinois. *Weed Sci* **63**:799–809 (2015).

44 Ma R, Skelton JJ, and Riechers DE, Measuring rates of herbicide metabolism in dicot weeds with an excised leaf assay. *J Vis Exp* **103**:e53236 (2015).

45 Wang J, Sánchez-Roselló M, Aceña JL, del Pozo C, Sorochinsky AE, Fustero S, Soloshonok VA, and Liu H, Fluorine in pharmaceutical industry: fluorine-containing drugs introduced to the market in the last decade (2001–2011). *Chem Rev* **114**:2432–2506 (2014).

46 O'Brien SR, Davis AS, and Riechers DE, Quantifying resistance to isoxaflutole and mesotrione and investigating their interactions with metribuzin POST in waterhemp (*Amaranthus tuberculatus*). *Weed Sci* **66**:586–594 (2018).

47 Sekino K, Koyanagi H, Ikuta E, and Yamada Y, Herbicidal activity of a new paddy bleaching herbicide, benzobicyclon. *J Pestic Sci* **33**:364–370 (2008).

48 Komatsubara K-i, Sekino K, Yamada Y, Koyanagi H, and Nakahara S, Discovery and development of a new herbicide, benzobicyclon. *J Pestic Sci* **34**:113–114 (2009).

49 Kohlhase DR, Edwards JW, and Owen MDK, Inheritance of 4-hydroxyphenylpyruvate dioxygenase inhibitor herbicide resistance in an *Amaranthus tuberculatus* population from Iowa, USA. *Plant Sci* **274**:360–368 (2018).

50 Yu Q, and Powles S, Metabolism-based herbicide resistance and cross-resistance in crop weeds: a threat to herbicide sustainability and global crop production. *Plant Physiol* **166**:1106–1118 (2014).

51 Thompson CR, Peterson D and Lally NG. Palmer amaranth survives pyrosulfotole and bromoxynil and other HPPD herbicide treatment. In *65th Annual Meeting of the North Central Weed Science Society*, Society NCWSS. North Central Weed Science Society: Lexington, Kentucky (2010).

52 González-Torralva F, Norsworthy JK, Piveta LB, Varanasi VK, Barber T, and Brabham C, Susceptibility of Arkansas Palmer amaranth accessions to common herbicide sites of action. *Weed Technol* **34**:770–775 (2020).

53 Kumar V, Liu R, and Stahlman PW, Differential sensitivity of Kansas Palmer amaranth populations to multiple herbicides. *Agron J* **112**:2152–2163 (2020).

54 Mahoney DJ, Jordan DL, Roma-Burgos N, Jennings KM, Leon RG, Vann MC, *et al.*, Susceptibility of Palmer amaranth (*Amaranthus palmeri*) to herbicides in accessions collected from the North Carolina Coastal Plain. *Weed Sci* **68**:582–593 (2020).

55 Godar AS, Varanasi VK, Nakka S, Prasad PVV, Thompson CR, and Mithila J, Physiological and molecular mechanisms of differential sensitivity of Palmer amaranth (*Amaranthus palmeri*) to mesotrione at varying growth temperatures. *PLoS One* **10**:e0126731 (2015).

56 Sandell L, Amit J and Kruger G. Evaluation of a putative HPPD-resistant Palmer amaranth (*Amaranthus palmeri*) population in Nebraska. In *67th Annual Meeting of the North Central Weed Science Society*, Society NCWSS. North Central Weed Science Society: St Louis, Missouri, pp. 82 (2012).

57 Jhala AJ, Sandell LD, Rana N, Kruger GR, and Knezevic SZ, Confirmation and control of triazine and 4-hydroxyphenylpyruvate dioxygenase-inhibiting herbicide-resistant Palmer amaranth (*Amaranthus palmeri*) in Nebraska. *Weed Technol* **28**:28–38 (2014).

58 Délye C, Jasieniuk M, and Le Corre V, Deciphering the evolution of herbicide resistance in weeds. *Trends Genet* **29**:649–658 (2013).

59 Wu C, Davis AS, and Tranel PJ, Limited fitness costs of herbicide-resistance traits in *Amaranthus tuberculatus* facilitate resistance evolution. *Pest Manag Sci* **74**:293–301 (2018).

60 Jones EAL, Owen MDK, and Leon RG, Influence of multiple herbicide resistance on growth in *Amaranthus tuberculatus*. *Weed Res* **59**:235–244 (2019).

61 Oliveira MC, Gaines TA, Patterson EL, Jhala AJ, Irmak S, Amundsen K, *et al.*, Interspecific and intraspecific transference of metabolism-based mesotrione resistance in dioecious weedy *Amaranthus*. *Plant J* **96**:1051–1063 (2018).

62 Murphy BP, Beffa R, and Tranel PJ, Genetic architecture underlying HPPD-inhibitor resistance in a Nebraska *Amaranthus tuberculatus* population. *Pest Manag Sci* **77**:4884–4891 (2021).

63 Qu R-Y, He B, Yang J-F, Lin H-Y, Yang W-C, Wu Q-Y, *et al.*, Where are the new herbicides? *Pest Manag Sci* **77**:2620–2625 (2021).

11

Overview of Metabolism-Based Resistance to PS-II Inhibitors in Weed Species

Mithila Jugulam[1], Balaji Aravindhan Pandian[2], and P. V. Vara Prasad[3]

[1] *Texas A&M AgriLife Research and Extension Center, Beaumont*
[2] *Enko Chem Inc. Mystic, USA*
[3] *Department of Agronomy, Kansas State University, USA*

11.1 Introduction

Incorporation of chemical weed control strategies is an integral part of modern crop production across the globe. Many synthetic chemical compounds with herbicidal properties have been discovered and developed over the last eight decades. The primary mechanism of action of many herbicide chemistries involves direct or indirect inhibition of photosynthetic function. One of the most widely used and important herbicides that are known to affect the photosynthetic function is photosystem II (PS-II) inhibitors (HRAC classification Groups 5 and 6, www.hracglobal.com). The PS-II inhibitors block PS-II-dependent chemical reactions in electron transport during light reaction of photosynthesis. The PS-II inhibitors include chemical families such as ureas, amides, triazines, triazinones, pyridazinones, carbamates, and nitrophenols.[1]

Among the PS-II inhibitors, the triazine family (including atrazine) is widely used for weed management. The chemical structure of atrazine has an N-substituted ring as its base, which carries the required element for phytotoxicity.[2] The position of the N atoms determines the symmetrical or asymmetrical arrangement of the compound. Derivatives of triazines have different substituents on the ring, which impart various degrees of toxicity and persistence. Substituted ureas and uracils also have the required structural elements for toxicity but have different structures than triazines.[1] The PS-II inhibitors block photosynthetic electron transport by displacing plastoquinone (PQ) from the binding site on the D1 protein subunit of PS-II. The molecular target of these herbicides is the *psbA* gene in the chloroplast, which encodes D1

Resistance in Weeds from Herbicide Metabolism, First Edition.
Edited by Vijay K. Nandula and Roland Beffa.

protein. In the photosynthetic electron transport, D1 protein is an important component required for PQ binding. PS-II inhibitors competitively bind to the D1 protein replacing the PQ, thereby blocking the electron transport and eventually producing nicotinamide adenine dinucleotide phosphate (NADP) and adenosine triphosphate (ATP). This results in the formation of reactive oxygen species that cause membrane destruction and leakage leading to cell death. Upon application, leaves exhibit interveinal chlorosis and necrotic symptopms.[3]

11.2 Crop Use and Weed Spectrum

PS-II inhibitors are primarily used to control broadleaf weeds such as pigweeds (*Amaranthus* sp.), kochia (*Bassia scoparia* (L.) Scott), wild radish (*Raphanus raphanistrum* L.), wild buckwheat (*Polygonum convolvulus* L.), common lamb's-quarters (*Chenopodium album* L.), annual ragweed (*Ambrosia artemisiifolia* L.), and grassweeds such as Italian ryegrass (*Lolium multiflorum* Lam.), rigid ryegrass (*Lolium rigidum* Gaudin), jungle rice (*Echinochloa colona* (L.) Link), and more.[4–6] Triazines, triazinones, uracils, and phenylureas are primarily used as soil-applied or early postemergence (POST). These herbicides can be absorbed by both plant shoots and roots but are translocated only in the xylem. Benzothiadiazinone family herbicide bentazon and nitrile herbicide bromoxynil are contact herbicides used primarily as early POST.[7,8] The triazines are registered for preemergence (PRE) and POST use in corn (*Zea mays* L.), sorghum (*Sorghum bicolor* L. Moench), sugarcane (*Saccharum officinarum* L.), and cotton (*Gossypium hirsutum* L.); triazinones are labeled for PRE in soybeans (*Glycine max* L. Merr.) and POST in winter wheat (*Triticum aestivum* L.) and several horticultural and tree crops.[1] However, ureas are registered for use in cotton, alfalfa (*Medicago sativa* L.), carrot (*Daucus carota* subsp. *sativus*), rangelands, and small grains. Uracils are primarily used as PRE in sugarcane, tree crops, and non-crop areas,[9,10] while phenyl carbamates are applied as POST in sugar beet (*Beta vulgaris* L.).[11] The nitriles are used in cotton, small grain crops, corn, grain sorghum,[12,13] and benzothiadiazoles in soybeans and rice (*Oryza sativa* L.).[14] The pyridazines family herbicide pyridate are used in corn, winter wheat, and peanuts.[15]

11.3 Crop Tolerance to PS-II Inhibitors

PS-II inhibitors selectively control weeds in crops such as corn, sorghum, sugarcane, sugar beet, carrot, and more.[16] The basis for tolerance to atrazine has been well studied in crops such as corn and sorghum,[17] which can naturally metabolize atrazine by three chemical reactions, i.e., *N*-dealkylation, hydrolysis, and conjugation.[18] The conjugation reaction is more predominant, which is catalyzed by glutathione-*S*-transferase (GST) enzyme activity. Atrazine-resistant corn inbreds expressed a high level of GST activity in leaves compared to the susceptible lines.[19] The GST conjugation reaction occurs primarily in the leaves and shoots, and the nonenzymatic hydroxylation occurs in roots.[20] Apart from corn and sorghum, GST-mediated atrazine metabolism was also reported in sugarcane.[21] Metabolism of metribuzin to a dione or homo-GSH derivative mediated by GST enzymes has been reported in tolerant soybean and potato cultivars.[22,23] However, specific GST enzymes involved in PS-II inhibitor metabolism in crops have not been identified yet.

11.4 Overview of PS-II Inhibitor Resistance in Weed Species

Because of extensive and repeated use, several weed species have evolved resistance to PS-II inhibitors.[24] Specifically, to date, 79 and 29 weed species have evolved resistance to triazine and ureas/amide classes of PS-II inhibitors, respectively.[24] Many dicot weeds, including Palmer amaranth (*Amaranthus palmeri* S. Watson), common waterhemp (*Amaranthus tuberculatus* (Moq.) J. D. Sauer), redroot pigweed (*Amaranthus retroflexus* L.), common lamb's-quarters, kochia, horseweed (*Erigeron canadensis* L.), and monocot species such as barnyard grass (*Echinochloa crus-galli* (L.) P. Beauv.), goosegrass (*Eleusine indica* (L.) Gaertn.), and ryegrass species (*Lolium* sp.) were documented to have evolved resistance to these herbicides.[24] PS-II inhibitor-resistant weeds are common in many cropping systems.

Weed species can evolve resistance to herbicides via two mechanisms: first, target-site-based resistance (TSR), which is conferred by either mutation in the herbicide target site or amplification of target gene resulting in increased expression of the gene,[25] and second, non-target-site-based resistance (NTSR), which includes those mechanisms that are not related to target site, such as rapid metabolism, vacuolar sequestration, reduced absorption or translocation of herbicides, and rapid necrosis.[26] Both TSR and NTSR mechanisms have been reported in PS-II inhibitor-resistant weed species (Fig. 11.1). While TSR to PS-II inhibitors is commonly endowed because of mutations in the *psbA* gene resulting in various amino acid substitutions,[3] NTSR has been attributed to rapid metabolism of PS-II inhibitors via GST or cytochrome P450 (CYP) enzyme activity (Fig. 11.1).[27–30] The primary focus of this chapter is to review metabolism-based resistance to these herbicides in weed species, with special reference to *Amaranthus* weed species. In addition a summary of TSR to these herbicides in other weed species completes this chapter.

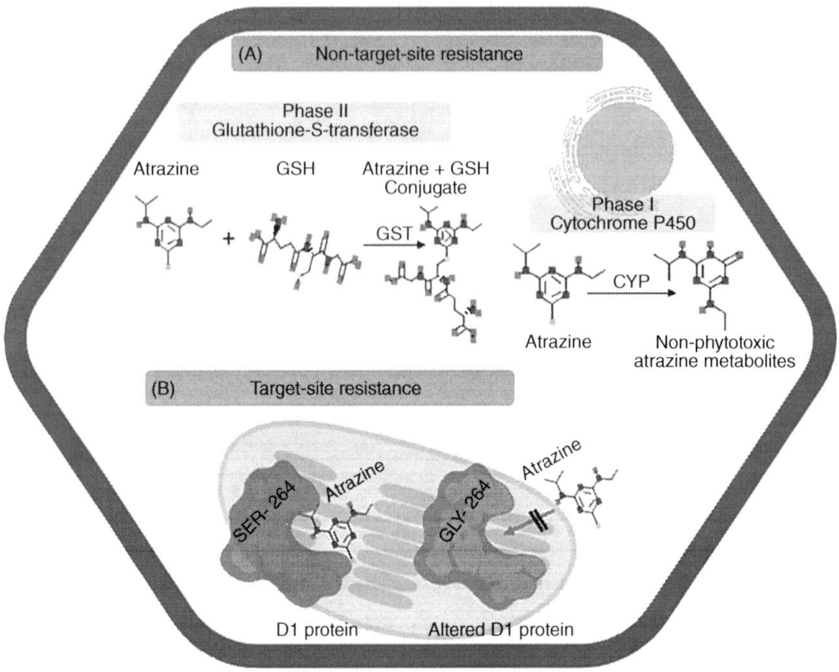

Figure 11.1 Illustration of known mechanisms conferring photosystem II (PS-II) inhibitor resistance in plants, using atrazine as a model herbicide. (A) Non-target-site resistance bestowed by the conjugation of atrazine via GSH (γ-Glu-Cys-Gly) catalyzed by the glutathione-*S*-transferase (GST) and cytochrome P450s (CYP). (B) Altered D1 protein (the target site of PS-II inhibitors), as a result of mutation in the *psbA* gene, leading to serine (SER) to glycine (GLY) substitution preventing atrazine binding to the target. (*See insert for color representation of the figure.*)

11.5 Metabolic Resistance to PS-II Inhibitors in Weed Species

Enhanced metabolism of PS-II inhibitors is the most common NTSR mechanism documented in both weeds and crops.[31,32] Apart from the metabolic resistance, reduced absorption and translocation of atrazine have also been reported in annual bluegrass (*Poa annua* L.).[33]

The GSTs are known to catalyze the metabolism of PS-II inhibitors; however, CYP and aryl acylamidase (AA)-mediated metabolism has also been documented.[34] GST enzyme superfamily is found both in plants and animals, and these enzymes are known to metabolize herbicide molecules.[29] GSTs are the basis for tolerance to PS-II inhibitors in crops such as sorghum,[35] corn,[36] and weeds such as common waterhemp,[37] Palmer amaranth,[31] and more. Plant GSTs catalyze the conjugation of tripeptide glutathione (GSH, γ-Glu-Cys-Gly) with electrophilic xenobiotic substrates forming an irreversible *S*-glutothionylated product, followed by transport to the vacuole. The conjugation of GSH, via cysteine thiol group, to the xenobiotic substrate results in water-soluble conjugates with greatly reduced toxicity.[38] The GST subunits are grouped into various classes, including theta, zeta, lambda, phi, tau, dehydroascorbate reductase (DHAR), and tetrachlorohydroquinone dehalogenase (TCHQD)-like based on sequence similarity. The phi and tau classes are commonly found in most plants and are involved in herbicide metabolism.[39,40] CYPs also constitute an enzyme family containing heme-thiolate as a cofactor that can catalyze oxidation, reduction, hydroxylation, epoxidation, dealkylation, C–C cleavage, desaturation, decarboxylation, dimerization, isomerization, and ring extension reactions.[41] CYP-mediated metabolism has been reported in PS-II inhibitor herbicides such as atrazine,[42] fluometuron, linuron, diuron, and bentazon.[43,44] Another enzyme that has been shown to induce PS-II inhibitor metabolism is AA, which hydrolyzes PS-II inhibitor propanil into 3,4-dinitroaniline, a non-phytotoxic compound. This type of metabolism was reported in several wild rice species,[45] jungle rice,[46] and barnyard grass.[47]

To date, globally, the number of weed species resistant to PS-II inhibitors ranks next only to acetolactate synthase (ALS) inhibitor-resistant weeds.[24] Many *Amaranthus* weeds (e.g., Palmer amaranth, common waterhemp, smooth pigweed (*Amaranthus hybridus* L.), Powell amaranth (*Amaranthus powelli* S. Watson), and redroot pigweed have been documented to have evolved resistance to PS-II inhibitors.[24] These are some of the most troublesome weed species in agriculture across the globe. Resistance to PS-II inhibitors in *Amaranthus* weeds has been reported as early as 1960s.[48] If sensitive, PS-II inhibitors are an excellent option to manage *Amaranthus* weeds as they are extremely susceptible to these herbicides. However, many agriculturally problematic *Amaranthus* weed species are currently resistant to these herbicides. Although TSR and NTSR to PS-II inhibitors have been reported in *Amaranthus* weeds, metabolic resistance is predominantly found in top-ranked *Amaranthus* weeds, i.e., Palmer amaranth and common waterhemp.

Palmer amaranth is ranked as the most troublesome weed across the United States and many populations of this weed are found to be resistant to PS-II inhibitors, specifically to atrazine.[24] The mechanisms of atrazine resistance in several Palmer amaranth populations were previously investigated. The results indicate that there are no alterations in *psbA* gene that can confer TSR to atrazine.[49–51] Rapid conjugation of atrazine via glutathione mediated by GST activity was found in Palmer amaranth.[49] Use of GST inhibitors (e.g., 4-chloro-7-nitrobenzofurazan [NBD-Cl]) to determine metabolic resistance to atrazine in weed species is common. The NBD-Cl derivatives have been shown to have GST inhibition in human tumor cells and are termed as suicide inhibitors of GSTs.[52] We tested the effect of NBD-Cl treatment prior to atrazine application in an atrazine-resistant Palmer amaranth population.[51] Treatment with NBD-CL significantly reduced the biomass accumulation in Palmer amaranth plants indicating the involvement of GST enzymes in degrading atrazine and reversing resistance.[51] Metabolic resistance to atrazine via such mechanism has been reported in other Palmer amaranth populations as well.[53]

Common waterhemp is another problem weed in Amaranthaceae with resistance to multiple herbicides and is distributed throughout the Midwestern United States. This weed evolved resistance to PS-II inhibitors as early as 1967.[24] Common waterhemp resistance to atrazine,[54] metribuzin,[24] and cyanazine[55] has been documented. Later, many atrazine-resistant common waterhemp populations were found throughout the Midwestern United States and Canada. However, only a few cases of resistance to metribuzin and cyanazine were reported in this species.[24] The first case of atrazine resistance in common waterhemp was documented in a population collected from Fillamore, Nebraska, in 1990.[56] Later, common waterhemp populations with resistance to atrazine were found in Kansas,[24] Illinois,[57] Iowa,[58] Missouri,[59] Texas,[60] in the United States, and in Ontario, Canada.[61] In many atrazine-resistant common water-hemp populations, predominance of metabolic resistance was reported.[30,54,59] Similar to Palmer amaranth, metabolism of atrazine by GST conjugation has been reported in common waterhemp as well.[30,62] GST-mediated metabolism of atrazine was characterized extensively using two common waterhemp populations from IL (e.g., Adams County resistant [ACR] and Mclean County resistant [MCR]). The ACR common water hemp is resistant to atrazine and MCR is resistant to atrazine and 4-hydroxyphenylpyruvate dioxygenase (HPPD) inhibitor (F2/10), mesotrione. Metabolism of atrazine by GSTs in both ACR and MCR common waterhemp and mesotrione by CYPs in MCR common waterhemp has been identified.[30] Further research revealed elevated expression of *AtuGSTF2*, a phi-class GST responsible for atrazine resistance in ACR and MCR common water hemp.[30,37] However, application of the GST inhibitor NBD-Cl reversed atrazine resistance completely in ACR but not in MCR common waterhemp, which suggests that atrazine resistance mechanisms in MCR could be distinct from ACR common waterhemp.[63] Also, genetic studies in MCR population indicate the same gene confers resistance to both atrazine and mesotrione in MCR common waterhemp.[64] This results suggest possible involvement of CYPs in atrazine metabolism along with GSTs in MCR common waterhemp.

Both GSTs and CYP are associated in multiple herbicide resistance in weed species. It was found that specific type I GST, i.e., *Am*GST2 was elevated in multiple resistant (including atrazine) black grass (*Alopecurus myosuroides*).

The *Am*GST2 had high activity as a glutathione peroxidase scavenging the hydrogen peroxides formed due to herbicide stress.[65] Similarly, enhanced metabolism of atrazine by GST conjugation has been reported in velvetleaf (*Abutilon theophrasti* Medik.).[66] In a rigid ryegrass population, metribuzin and simazine were reported to be detoxified by CYP activity[27,67]; similarly, the involvement of CYP81A10v7 in the metabolism atrazine and chlorotoluron has also been reported in a multiple herbicide-resistant rigid ryegrass population.[42] Another possibility of enhanced detoxification of atrazine, which is yet to document in weeds, may include increased expression of GST via gene amplification or through gene regulation as reported for insecticide resistance in some insects.[68]

11.6 Target-Site Resistance

High-resolution X-ray crystallographic structures of D1 protein of purple bacteria in plastoquinone and atrazine complex revealed the specific amino acids of the protein interacting with the plastoquinone and atrazine.[69] Plastoquinone is hydrogen-bonded with amino acids His-215 and Ser-264 at the plastoquinone-binding site along with other hydrophobic interactions. However, atrazine is hydrogen-bonded with Phe-265 and Ser-264 preventing the binding of plastoquinone.[70] Amino acid substitution in the *psbA* gene that alters PS-II herbicides binding to the D1 protein is the only target-site mechanism reported in weeds.[71] A total of eight amino acid substitutions in seven positions, including Ser-264-Gly,[72] Ser-264-Thr,[73] Val-219-Ile,[74] Asn-266-Thrpa,[75] Phe-255-Ile,[76] Ala-251-Val,[77] Leu-218-Val,[78] and Phe-274-Val,[79] were reported in several weeds. The most common mutation in *psbA* gene Ser-264-Gly confers resistance to the triazine herbicides but moderate to no resistance to the other PS-II inhibitor families, including the triazinones. In contrast, the Ser-264-Thr substitution confers resistance to triazines and ureas, not to the nitrile or triazinone families. Asn-266-Thr confers resistance to nitrile and uracil family of PS-II inhibitors and a low or moderate level of resistance to triazinones.[75] Val-219-Ile bestows resistance to amide, ureas, triazinones, and nitriles but negative cross-resistance to benzothiadiazole herbicide bentazon.[80] Also, Phe-255-Ile confers resistance to triazinone and hexazinones.[76] The substitution Phe-274-Val confers resistance to triazine, triazinone, and ureas.[79] TSR to PS-II inhibitors conferred by Ser-264-Gly was reported in several *Amaranthus* weeds, e.g., Powell amaranth,[81] prostrate amaranth (*Amaranthus blitoides* S. Watson),[82,83] and common waterhemp.[84]

TSR to PS-II inhibitors was also reported in other weeds such as common lamb's-quarters, annual bluegrass, and kochia.[24] Common lamb's-quarters evolved resistance to atrazine, simazine, metribuzin, terbuthylazine, terbutryn, prometon, cyanazine, lenacil, metamitron, and terbacil.[73,78,85,86] Amino acid substitutions in the *psbA* gene Ser-264-Gly, Ala-251-Val, and Leu-218-Val were reported in several populations of common lamb's-quarters.[87] Similarly, Val-219-Ile and Ser-264-Gly substitutions in the *psbA* gene was found to confer resistance to atrazine, diuron, tebuthiuron, and metribuzin in kochia from North Dakota, Minnesota, Illinois, and Kansas in the United States.[88] However, there is no case of NTSR reported to confer resistance to PS-II inhibitors in these weeds. In some weed species, e.g., annual bluegrass, both TSR and NTSR to PS-II inhibitors were reported. The Ser-264-Gly is often found to be responsible for resistance to atrazine; however, an annual bluegrass biotype from Oregon was found to confer metabolic resistance to diuron, atrazine, and amicarbazone. Co-occurrence of *psbA* mutation Ser-264-Gly and enhanced metabolism of metribuzin has been reported in an wild radish population from Western Australia.[72]

11.7 Interaction of PS-II Inhibitors with Other Herbicides for the Management of Metabolic Resistance

Improved control of PS-II inhibitor-resistant weeds was found when they are applied in combination with HPPD inhibitors. Specifically, synergism of PS-II and HPPD inhibitors was reported in weed species such as Palmer amaranth,[89] redroot pigweed,[90] common waterhemp,[91] and wild radish.[92] Mixing of HPPD inhibitors, such as mesotrione, tembotrione, or topramezone, with atrazine in POST applications resulted in 90% control of a Palmer amaranth population from Nebraska that was conferred by metabolic resistance (9–14-fold) to atrazine.[89] Similarly, control of metabolism-based PS-II and HPPD inhibitor-resistant common water hemp from Illinois was >60% when mesotrione tank-mixed with atrazine.[90] Even though the basis of synergy is not completely understood, multiple factors may be involved in such synergism. For example, the complementary site of action of these herbicides is known to be one of the primary reasons imparting synergy. As discussed earlier, the PS-II inhibitors inhibit photosynthesis by competing with PQ to bind at QB-binding site of D1,[2] whereas inhibition of HPPD enzyme leads to the depletion of α-tocopherols, PQ, and thus disrupts carotenoid biosynthesis and accumulation of reactive oxygen species followed by lipid peroxidation.[93] Inhibition of PQ biosynthesis by HPPD inhibitor leads to the unavailability of PQ, which will reduce the competition between the PS-II inhibitors and PQ for binding to D1 protein in the PS-II inhibitor-resistant weeds. Disruption of electron flow and carotenoid biosynthesis result in accumulation of reactive oxygen species leading to free radical damage of the membranes.[94]

Increased absorption of mesotrione when mixed with atrazine is also reported to be a plausible reason for atrazine and mesotrione synergism in the PS-II and HPPD inhibitor-resistant Palmer amaranth from Nebraska.[53] The relationship between the observed synergy and metabolic resistance is also not well established. Metabolism of HPPD inhibitors by CYPs is the common metabolic resistance reported[31]; however, several studies confirmed that PS-II inhibitors are metabolized by both CYPs and GSTs.[32] A recent study reported that metabolic resistance to atrazine and mesotrione is conferred by the same gene in common water hemp.[64] Interestingly, unlike in PS-II and HPPD inhibitor-resistant weeds, mixing of mesotrione, tembotrione, or topramezone with atrazine did not cause crop injury to corn, which exhibits CYP- and GST-based metabolism of PS-II and HPPD inhibitors.[38,95–97]

11.8 Conclusions

Metabolic resistance to herbicides in weed species is a challenge for the management and containment of the spread of resistance. Herbicide-detoxifying enzymes (e.g., CYP or GST) are known to have activity on a wide range of xenobiotics including insecticides. Thus, the metabolic resistance to herbicides can predispose weeds to evolve resistance to other herbicides, which are yet to be commercialized. We have previously documented such cases, where metabolic resistance to one herbicide predisposed Palmer amaranth populations to evolve resistance to other modes of action of herbicides that were not previously used.[31,49,51] Investigation of underlying biochemical and molecular regulation of metabolic resistance is difficult. However, recent advances in omics approaches will help in delineating the precise role of CYP and GST enzyme families in metabolic and multiple resistance to herbicides in both crop and weed species.

References

1 LeBaron HM, McFarland JE, and Burnside OC, The triazine herbicides: a milestone in the development of weed control technology, In LeBaron HM, McFarland JE, and Burnside OC (Eds), *The Trizane Herbicides 50 years Revolutionizing Agriculture*, Elsevier Science, San Diego, CA, pp. 1–12 (2008).

2 Shimabukuro RH, and Swanson HR, Atrazine metabolism, selectivity, and mode of action. *J Agric Food Chem* **17**:199–205 (1969).

3 Gronwald JW, Resistance to photosystem II inhibiting herbicides, In Powles SB, and JAM H (Eds), *Herbicide Resistance in Plants: Biology and Biochemistry*, CRC Press, Boca Raton, FL, pp. 27–60 (1994).

4 Richburg JT, Norsworthy JK, Barber T, Roberts TL, and Gbur EE, Tolerance of corn to PRE- and POST-applied photosystem II-inhibiting herbicides. *Weed Technol* **34**:277–283 (2020).

5 Hutchinson, Eberlein CV, and Tonks DJ, Broadleaf weed control and potato crop safety with postemergence rimsulfuron, metribuzin, and adjuvant combinations. *Weed Technol* **18**:750–756 (2004).
6 Pandey J, and Verma AK, Effect of atrazine, metribuzin, sulfosulfuron and tralkoxydim on weeds and yield of wheat (*Triticum aestivum*). *Indian Jrnl of Agron* **47**:72–76 (2002).
7 Bromilow RH, and Chamberlain K, Pathways and mechanisms of transport of herbicides in plants, In Kirkwood RC (Ed.), *Target Sites for Herbicide Action*, Springer, Boston, MA, pp. 245–284 (1991).
8 Gunsolus JL and Curran WS, Herbicide Mode of Action and Injury Symptoms. http://appliedweeds.cfans.umn.edu [Accessed January 10, 2021] (2021).
9 Tworkoski T, and Miller S, Apple and Peach Orchard establishment following multi-year use of diuron, simazine, and terbacil. *HortScience* **36**:1211–1213 (2001).
10 Matherne RJ, and Millhollon RW, Tolerance of two sugarcane cultivars to terbacil, fenac, and dalapon. *Weed Sci* **21**:139–141 (1973).
11 Abdollahi F, and Ghadiri H, Effect of separate and combined applications of herbicides on weed control and yield of sugar beet. *Weed Technol* **18**:968–976 (2004).
12 Jordan DL, Smith MC, McClelland MR, and Frans RE, Weed control with bromoxynil applied alone and with graminicides. *Weed Technol* **7**:835–839 (1993).
13 Reddy SS, Stahlman PW, Geier PW, Thompson CR, Currie RS, Schlegel AJ, *et al.*, Weed control and crop safety with premixed pyrasulfotole and bromoxynil in grain sorghum. *Weed Tech* **27**:664–670 (2013).
14 Akbar TA, and Lin H, GIS based ArcPRZM-3 model for bentazon leaching towards groundwater. *J Environ Sci* **22**:1854–1859 (2010).
15 Giménez-Espinosa R, and De-Prado R, Absorption, translocation and metabolism of pyridate in chickpea (*Cicer arietinum*). *Funct Plant Biol* **25**:105–110 (1998).
16 Shukla A, and Devine MD, Basis of crop selectivity and weed resistance to triazine herbicides, In LeBaron HM, McFarland JE, and Burnside OC (Eds), *The Trizane Herbicides 50 years Revolutionizing Agriculture*, Elsevier Science, San Diego, CA, pp. 1–12 (2008).
17 Jensen SGR, and Hunt LA, Detoxification of atrazine in three gramineae subfamilies. *Weed Sci* **25**:212–220 (1977).
18 Roeth FW, and Lavy TL, Atrazine translocation and metabolism in sudangrass, sorghum, and corn. *Weed Sci* **19**:98–101 (1971).
19 Shimabukuro RH, Frear DS, Swanson HR, Walsh WC, and Glutathione conjugation. An enzymatic basis for atrazine resistance in corn. *Plant Physiol* **47**:10–14 (1971).

20 Shimabukuro RH, Atrazine metabolism in resistant corn and sorghum. *Plant Physiol* **43**:1925–1930 (1968).

21 Lamoureux GL, Stafford LE, and Shimabukuro RH, Conjugation of 2-chloro-4,6-bis(alkylamino)-s-triazines in higher plants. *J Agric Food Chem* **20**: 1004–1010 (1972).

22 Falb LN, and Smith AE, Metribuzin metabolism in soybeans. Characterization of the intraspecific differential tolerance. *J Agric Food Chem* **32**:1425–1428 (1984).

23 Gawronski SW, Haderlie LC, Callihan RH, and Gawronska H, Mechanism of metribuzin tolerance: herbicide metabolism as a basis for tolerance in potatoes. *Weed Res* **26**:307–314 (1986).

24 Heap, I, The International Herbicide-Resistant Weed Database. http://www.weedscience.org [Accessed December 28, 2021] (2021).

25 Devine MD, and Shukla A, Altered target sites as a mechanism of herbicide resistance. *Crop Prot* **19**:881–889 (2000).

26 Yuan JS, Tranel PJ, and Stewart CN Jr, Non-target-site herbicide resistance: a family business. *Trends Plant Sci* **12**:6–13 (2007).

27 Burnet MWM, Loveys BR, Holtum JAM, and Powles SB, Increased detoxification is a mechanism of simazine resistance in *Lolium rigidum*. *Pestic Biochem Physiol* **46**:207–218 (1993).

28 Gray JA, Balke NE, and Stoltenberg DE, Increased glutathione conjugation of atrazine confers resistance in a wisconsin velvetleaf (*Abutilon theophrasti*) biotype. *Pestic Biochem Physiol* **55**:157–171 (1996).

29 Cummins I, Dixon DP, Freitag-Pohl S, Skipsey M, and Edwards R, Multiple roles for plant glutathione transferases in xenobiotic detoxification. *Drug Metab Rev* **43**:266–280 (2011).

30 Ma R, Kaundun SS, Tranel PJ, Riggins CW, McGinness DL, Hager AG, *et al.*, Distinct detoxification mechanisms confer resistance to mesotrione and atrazine in a population of waterhemp. *Plant Physiol* **163**:363–377 (2013).

31 Nakka S, Godar AS, Wani PS, Thompson CR, Peterson DE, Roelofs J, *et al.*, Physiological and molecular characterization of hydroxyphenylpyruvate dioxygenase (HPPD)-inhibitor resistance in Palmer amaranth (Amaranthus palmeri S. Wats.). *Front Plant Sci* **8**:555 (2017).

32 Jugulam M, and Shyam C, Non-target-site resistance to herbicides: recent developments. *Plants* **8** (2019).

33 Svyantek AW, Aldahir P, Chen S, Flessner ML, McCullough PE, Sidhu SS, *et al.*, Target and nontarget resistance mechanisms induce annual bluegrass (*Poa annua*) resistance to atrazine, amicarbazone, and diuron. *Weed Technol* **30**:773–782 (2016).

34 Gaines TA, Duke SO, Morran S, Rigon CAG, Tranel PJ, Küpper A, *et al.*, Mechanisms of evolved herbicide resistance. *J Biol Chem* **30** (2020).

35 Lamoureux GL, Stafford LE, Shimabukuro RH, and Zaylskie RG, Atrazine metabolism in sorghum: catabolism of the glutathione conjugate of atrazine. *J Agric Food Chem* **21**:1020–1030 (1973).

36 Hatton PJ, Dixon D, Cole DJ, and Edwards R, Glutathione transferase activities and herbicide selectivity in maize and associated weed species. *Pestic Sci* **46**:267–275 (1996).

37 Evans AF Jr, O'Brien SR, Ma R, Hager AG, Riggins CW, Lambert KN, *et al.*, Biochemical characterization of metabolism-based atrazine resistance in *Amaranthus* tuberculatus and identification of an expressed GST associated with resistance. *Plant Biotechnol J* **15**:1238–1249 (2017).

38 Timmerman KP, Molecular characterization of corn glutathione S-transferase isozymes involved in herbicide detoxication. *Physiol Plant* **77**:465–471 (1989).

39 Edwards R, and Dixon DP, Plant glutathione transferases. *Methods Enzymol* **401**:169–186 (2005).

40 Dixon DP, Skipsey M, and Edwards R, Roles for glutathione transferases in plant secondary metabolism. *Phytochemistry* **71**:338–350 (2010).

41 Pandian BA, Sathishraj R, Djanaguiraman M, Prasad PVV, and Jugulam M, Role of cytochrome P450 enzymes in plant stress response. *Antioxidants* **9**:454 (2020).

42 Han H, Yu Q, Beffa R, González S, Maiwald F, Wang J, *et al.*, Cytochrome P450 CYP81A10v7 in Lolium rigidum confers metabolic resistance to herbicides across at least five modes of action. *Plant J*:79–92 (2020).

43 Siminszky B, Corbin FT, Ward ER, Fleischmann TJ, and Dewey RE, Expression of a soybean cytochrome P450 monooxygenase cDNA in yeast and tobacco enhances the metabolism of phenylurea herbicides. *Proc Nat Acad Sci* **96**:1750–1755 (1999).

44 Didierjean L, Gondet L, Perkins R, Lau S-MC, Schaller H, O'Keefe DP, *et al.*, Engineering herbicide metabolism in tobacco and Arabidopsis with CYP76B1, a cytochrome P450 enzyme from Jerusalem artichoke. *Plant Physiol* **130**:179–189 (2002).

45 Chen JJ, and Matsunaka S, The propanil hydrolyzing enzyme aryl acylamidase in the wild rices of genus Oryza. *Pestic Biochem Physiol* **38**:26–33 (1990).

46 Leah JM, Caseley JC, Riches CR, and Valverde B, Association between elevated activity of aryl acylamidase and propanil resistance in Jungle-rice, Echinochloa colona. *Pestic Sci* **42**:281–289 (1994).

47 Hirase K, and Hoagland RE, Characterization of aryl acylamidase activity from propanil-resistant barnyardgrass (*Echinochloa crus-galli* [L.] Beauv.). *Weed Biol Manag* **6**:197–203 (2006).

48 Ritter RL, Harris TC, and Varano WJ, Influence of herbicides and tillage on the control of triazine-resistant smooth pigweed (*Amaranthus hybridus*) in corn (*Zea mays*) and soybeans (*Glycine max*). *Weed Sci* **33**:400–404 (1985).

49 Nakka S, Godar AS, and Thompson CR, Rapid detoxification via glutathione S-transferase (GST) conjugation confers a high level of atrazine resistance in Palmer amaranth (*Amaranthus palmeri*). *Pest Manag Sci* **73**:2236–2243 (2017).

50 Chaudhari S, Varanasi VK, Nakka S, Bhowmik PC, Thompson CR, Peterson DE, *et al.*, Evolution of target and non-target based multiple herbicide resistance in a single Palmer amaranth (*Amaranthus palmeri*) population from Kansas. *Weed Technol* **34**:447–453 (2020).

51 Shyam C, Borgato EA, Peterson DE, Dille JA, and Jugulam M, Predominance of metabolic resistance in a six-way-resistant Palmer amaranth (*Amaranthus palmeri*) population. *Front Plant Sci* **11**:614618 (2021).

52 Turella P, Cerella C, Filomeni G, Bullo A, De Maria F, Ghibelli L, *et al.*, Proapoptotic activity of new glutathione S-transferase inhibitors. *Cancer Res* **65**:3751–3761 (2005).

53 Chahal PS, Jugulam M, and Jhala AJ, Mechanism of atrazine resistance in atrazine- and HPPD inhibitor-resistant Palmer amaranth (Amaranthus palmeri S. Wats.) from Nebraska. *Can J Plant Sci* **99**:815–823 (2019).

54 Vennapusa AR, Faleco F, Vieira B, Samuelson S, Kruger GR, Werle R, *et al.*, Prevalence and mechanism of atrazine resistance in waterhemp (*Amaranthus tuberculatus*) from Nebraska. *Weed Sci* **66**:595–602 (2018).

55 Patzoldt WL, Dixon BS, and Tranel PJ, Triazine resistance in *Amaranthus tuberculatus* (Moq) Sauer that is not site-of-action mediated. *Pest Manag Sci* **59**:1134–1142 (2003).

56 Anderson DD, Roeth FW, and Martin AR, Occurrence and control of triazine-resistant common waterhemp (*Amaranthus rudis)* in field corn (*Zea mays*). *Weed Technol* **10**:570–575 (1996).

57 Patzoldt WL, Tranel PJ, and Hager AG, A waterhemp (*Amaranthus tuberculatus*) biotype with multiple resistance across three herbicide sites of action. *Weed Sci* **53**:30–36 (2005).

58 McMullan PM, and Green JM, Identification of a tall waterhemp (*Amaranthus tuberculatus*) biotype resistant to HPPD-inhibiting herbicides, atrazine, and thifensulfuron in Iowa. *Weed Technol* **25**:514–518 (2011).

59 Shergill LS, Bish MD, Jugulam M, and Bradley KW, Molecular and physiological characterization of six-way resistance in an *Amaranthus tuberculatus* var. rudis biotype from Missouri. *Pest Manag Sci* **74**:2688–2698 (2018).

60 Singh V, Garetson R, McGinty J, Dotray P, Morgan G, Nolte S, *et al.*, Distribution of herbicide-resistant waterhemp (*Amaranthus tuberculatus*) across row crop production systems in Texas. *Weed Technol* **34**:129–139 (2020).

61 Vyn JD, Swanton CJ, Weaver SE, and Sikkema PH, Control of herbicide-resistant common waterhemp (Amaranthus tuberculatus var. rudis) with pre- and post-emergence herbicides in soybean. *Can J Plant Sci* **87**:175–182 (2007).

62 Chahal PS, Jugulam M, and Jhala AJ, Basis of atrazine and mesotrione synergism for controlling atrazine- and HPPD inhibitor-resistant Palmer amaranth. *Agron J* **111**:3265–3273 (2019).

63 Ma R, Evans AF, and Riechers DE, Differential responses to preemergence and postemergence atrazine in two atrazine-resistant waterhemp populations. *Agron J* **108**:1196–1202 (2016).

64 Jacobs KE, Butts-Wilmsmeyer CJ, Ma R, and O'brien SR, Association between metabolic resistances to atrazine and mesotrione in a multiple-resistant waterhemp (*Amaranthus tuberculatus*) population. *Weed Sci* **68**:358–366 (2020).

65 Cummins I, Cole DJ, and Edwards R, A role for glutathione transferases functioning as glutathione peroxidases in resistance to multiple herbicides in black-grass: glutathione transferases and herbicide resistance. *Plant J* **18**:285–292 (1999).

66 Anderson MP, and Gronwald JW, Atrazine resistance in a velvetleaf (*Abutilon theophrasti*) biotype due to enhanced glutathione S-transferase activity. *Plant Physiol* **96**:104–109 (1991).

67 Ma H, Lu H, Han H, Yu Q, and Powles S, Metribuzin resistance via enhanced metabolism in a multiple herbicide resistant Lolium rigidum population. *Pest Manag Sci* **76**:3785–3791 (2020).

68 Enayati AA, Ranson H, and Hemingway J, Insect glutathione transferases and insecticide resistance. *Insect Mol Biol* **14**:3–8 (2005).

69 Lancaster CRD, and Michel H, Refined crystal structures of reaction centres from Rhodopseudomonas viridis in complexes with the herbicide atrazine and two chiral atrazine derivatives also lead to a new model of the bound carotenoid. *J Mol Biol* **286**:883–898 (1999).

70 Fuerst EP, and Norman MA, Interactions of herbicides with photosynthetic electron transport. *Weed Sci* **39**:458–464 (1991).

71 Pfister K, and Arntzen CJ, The mode of action of photosystem II-specific inhibitors in herbicide-resistant weed biotypes. *Zeitschrift für Naturforschung C* **34**:996–1009 (1979).

72 Lu H, Yu Q, Han H, Owen MJ, and Powles SB, Metribuzin resistance in a wild radish (*Raphanus raphanistrum*) population via both *psbA* gene mutation and enhanced metabolism. *J Agric Food Chem* **67**:1353–1359 (2019).

73 Mechant E, De Marez T, Hermann O, Olsson R, and Bulcke R, Target site resistance to metarnitron in *Chenopodium album* L. *J Plant Dis Prot* **21**:37–40 (2008).

74 Mengistu LW, Christoffers MJ, and Lym RG, A *psbA* mutation in *Kochia scoparia* (L) Schrad from railroad rights-of-way with resistance to diuron, tebuthiuron and metribuzin. *Pest Manag Sci* **61**:1035–1042 (2005).

75 Park KW, and Mallory-Smith CA, *psbA* mutation (Asn266 to Thr) in Senecio vulgaris L. confers resistance to several PS II-inhibiting herbicides. *Pest Manag Sci* **62**:880–885 (2006).

76 Perez-Jones A, Intanon S, and Mallory-Smith C, molecular analysis of hexazinone-resistant shepherd's-purse (*Capsella bursa-pastoris*) reveals a novel psbA mutation. *Weed Sci* **57**:574–578 (2009).

77 McMurray LS, Preston C, Vandenberg A, Mao D, Bett KE, and Paull JG, Induced novel psbA mutation (Ala251 to Thr) in higher plants confers resistance to PSII inhibitor metribuzin in *Lens culinaris*. *Pest Manag Sci* **75**:1564–1570 (2019).

78 Thiel H, and Varrelmann M, Identification of a new PSII target site psbA mutation leading to D1 amino acid Leu218Val exchange in the Chenopodium album D1 protein and comparison to cross -resistance profiles of known modifications at positions 251 and 264. *Pest Manag Sci* **70**:278–285 (2014).

79 Lu H, Yu Q, Han H, Owen MJ, and Powles SB, A novel *psbA* mutation (Phe274-Val) confers resistance to PSII herbicides in wild radish (*Raphanus raphanistrum*). *Pest Manag Sci* **75**:144–151 (2019).

80 Pedroso RM, Al-Khatib K, Alarcón-Reverte R, and Fischer AJ, A *psbA* mutation (Val219 to Ile) causes resistance to propanil and increased susceptibility to bentazon in *Cyperus difformis*. *Pest Manag Sci* **72**:1673–1680 (2016).

81 Diebold RS, McNaughton KE, Lee EA, and Tardif FJ, Multiple resistance to imazethapyr and atrazine in powell amaranth (*Amaranthus powellii*). *Weed Sci* **51**:312–318 (2003).

82 Sibony M, and Rubin B, Molecular basis for multiple resistance to acetolactate synthase-inhibiting herbicides and atrazine in *Amaranthus blitoides* (prostrate pigweed). *Planta* **216**:1022–1027 (2003).

83 Spring J, Boydston R, and Burke I, Photosystem II inhibitor resistance in the Columbia basin of Washington state. *Weed Technology* **30**(3):783–792 (2016).

84 Foes MJ, Li L, Tranel PJ, Wax LM, and Stoller EW, A biotype of common waterhemp (*Amaranthus rudis*) resistant to triazine and als herbicides. *Weed Sci* **46**:514–520 (1998).

85 Eleftherohorinos IG, Vasilakoglou IB, and Dhima KV, Metribuzin resistance in *Amaranthus retroflexus* and *Chenopodium album* in Greece. *Weed Sci* **48**:69–74 (2000).

86 McKenzie-Gopsill A, Graham G, Laforest M, Ibarra S, Hann S, and Wagg C, Occurrence and management of PSII-inhibitor-resistant *Chenopodium album* L. in atlantic canadian potato production. *Agronomy* **10**:1369 (2020).

87 Thiel H, and Varrelmann M, Identification of a new PSII target site psbA mutation leading to D1 amino acid Leu218 Val exchange in the Chenopodium album D1 protein and comparison to cross-resistance profiles of known modifications at positions 251 and 264. *Pest Manag Sci* **70**:278–285 (2014).

88 Kumar V, Jha P, Jugulam M, Yadav R, and Stahlman PW, Herbicide-resistant kochia (*Bassia scoparia*) in North America: a review. *Weed Sci* **67**:4–15 (2019).

89 Chahal PS, and Jhala AJ, Interaction of PS II- and HPPD-inhibiting herbicides for control of Palmer amaranth resistant to both herbicide sites of action. *Agron J* **110**:2496–2506 (2018).

90 Hugie JA, Bollero GA, Tranel PJ, and Riechers DE, Defining the rate requirements for synergism between mesotrione and atrazine in redroot pigweed (*Amaranthus Retroflexus*). *Weed Sci* **56**:265–270 (2008).

91 Hausman NE, Tranel PJ, Riechers DE, and Hager AG, Responses of a waterhemp (*Amaranthus tuberculatus*) population resistant to HPPD-inhibiting herbicides to foliar-applied herbicides. *Weed Technol* **30**:106–115 (2016).

92 Walsh MJ, Stratford K, Stone K, and Powles SB, Synergistic effects of atrazine and mesotrione on susceptible and resistant wild radish (*Raphanus raphanistrum*) populations and the potential for overcoming resistance to triazine herbicides. *weed Tech* **26**:341–347 (2012).

93 Mitchell G, Bartlett DW, Fraser TE, Hawkes TR, Holt DC, Townson JK, *et al.*, Mesotrione: a new selective herbicide for use in maize. *Pest Manag Sci* **57**:120–128 (2001).

94 Rutherford AW, and Krieger-Liszkay A, Herbicide-induced oxidative stress in photosystem II. *Trends Biochem Sci* **26**:648–653 (2001).

95 Chahal PS, Irmak S, Gaines T, Amundsen K, Jugulam M, Jha P, *et al.*, Control of photosystem II- and 4-Hydroxyphenylpyruvate dioxygenase inhibitor-resistant Palmer amaranth (Amaranthus palmeri) in conventional corn. *Weed Tech* **32**:326–335 (2018).

96 Chhokar RS, Sharma RK, Gill SC, and Singh GP, Tank-mix application of p-hydroxyphenylpyruvate dioxygenase (HPPD) inhibiting herbicide (mesotrione, tembotrione or topramezone) with atrazine improves weed control in maize (Zea mays L.). *Jrnl Res Weed Sci* **3**:556–581 (2020).

97 Pataky JK, Meyer MD, Bollman JD, Boerboom CM, and Williams MM, Genetic basis for varied levels of injury to sweet corn hybrids from three cytochrome P450-metabolized herbicides. *J Am Soc Hortic Sci* **133**:438–447 (2008).

12

Fitness and Ecophysiological Cost of Metabolic Herbicide Resistance

Eshagh Keshtkar[1], Roland Beffa[2], and Per Kudsk[3]

[1] *Department of Agronomy, Tarbiat Modares University, Tehran, Iran*
[2] *Senior Scientist Consultant, Liederbach, Hesse, Germany*
[3] *Department of Agroecology, Aarhus University, Slagelse, Denmark*

12.1 Introduction

The discovery of phenoxy herbicides in the 1940s was a fundamental scientific breakthrough in weed management and food production. Herbicides, alongside hybrid-seed and fertilizer technologies, fostered the Green Revolution, saving millions of lives. However, the future role of herbicides in cropping systems has been questioned by agronomists due to the widespread occurrence of herbicide-resistant weed populations. A decade after the discovery of the early herbicides, resistance to herbicides was predicted by Harper[1]; however, warnings and local reports[2–4] were ignored until 1970 when Ryan[5] reported resistance in *Senecio vulgaris* L. (common groundsel) to simazine and atrazine in a nursery in the United States. The evolution and widespread occurrence of herbicide-resistant weeds is a menace and a challenge to modern agriculture as they reduce crop yields, decrease the value of agricultural land, and lead to additional costs for alternative or supplemental weed control measures.[6] From an agronomic point of view, management of herbicide-resistant weed populations is more complicated than that of susceptible populations due to the lack of efficacy of alternative herbicides (cross-resistance and multiple resistance), which can result in a situation with no alternative herbicide options in certain cases.[7] Moreover, herbicide resistance can be a risk to biodiversity, as evolved herbicide-resistant weed species can behave similar to invasive species. Hence, herbicide resistance has been considered a wicked problem without an easy solution.[8] It has been disputed by Harker et al.[9] who concluded that herbicide overuse was the main reason for widespread herbicide resistance. Since the early reports of 2,4-dichlorophenoxyacetic acid

Resistance in Weeds from Herbicide Metabolism, First Edition.
Edited by Vijay K. Nandula and Roland Beffa.

(2,4-D)-resistant or -tolerant *Commelina diffusa* (climbing dayflower) and *Daucus carota* (wild carrot) in 1957,[2,4] more than 509 unique cases of herbicide-resistant weed populations have been documented in 71 countries,[10] and it is expected that the number of herbicide-resistant weeds will continue to increase. Currently, resistance to 21 out of the 31 known herbicide sites of action has been reported.[10]

Herbicide resistance is a result of modern agricultural production systems with a high reliance on herbicides, leading to a high selection pressure due to a continuous and extensive use of herbicides belonging to the same chemical class or targeting the same site of action. This has resulted in the selection for both target-site resistance (TSR) and non-target-site resistance (NTSR) mechanisms in herbicide-resistant weed species. The genetic characteristics of weed species, physicochemical characteristics of herbicides, and agronomic practices can influence the evolution of herbicide resistance within weed populations.[11] Of the two broadly classified herbicide resistance mechanisms, NTSR mechanisms are more complex than TSR mechanisms.[11] NTSR can be caused by one or more processes, including reduced herbicide absorption into plants, decreased herbicide translocation within plants, increased sequestration and metabolism of the herbicides (metabolic herbicide resistance) within plants,[11,12] and rapid cell death and necrosis of treated leaves, the so-called Phoenix phenomenon.[13–15] Among the NTSR mechanisms, enhanced metabolic degradation of the herbicide is the most common mechanism.[16] Metabolic herbicide resistance can involve several genes and can be a complex genetic mechanism with different combinations of homozygous and heterozygous alleles. It is worth noting that management of an NTSR weed biotype is also more complicated than that of a TSR biotype because herbicide resistance pattern of the NTSR biotypes is unpredictable.[17] An NTSR weed biotype can show resistance to multiple herbicides with completely different sites of action and can even endow resistance to new herbicides before they will be commercially available on the market. As an example, an acetyl-CoA carboxylase (ACCase) NTSR population of *Alopecurus myosuroides* Huds. (black grass) showed resistance to fluypyrsulfuron (Group 2/B) in Denmark, two years before its introduction.[18] Due to these characteristics, it has been suggested that more attention should be paid toward NTSR weeds.[19]

12.2 Definition and Types of Plant Fitness Associated with Herbicide Resistance

Plant fitness can be defined as the number of progenies produced by one individual relative to another.[20–22] Also, fitness is defined as "survival and subsequent reproductive success of one kind of individual relative to another as a consequence of natural selection."[23] Considering herbicide resistance, fitness refers to "the number of offspring produced by a herbicide-resistant biotype within a population, relative to a nonresistant biotype."[24] Fitness is defined as reproductive success of a plant in the field,[25] or the percentage of genes that an individual plant leaves in the gene pool of a population.[20,26,27] So the organisms leaving more offspring in the population are those whose genes dominate the gene pool.[26] Accordingly, fitness of a biotype can be either "positive," "negative," or "neutral" relative to another biotype within a population. A genotype with *positive fitness* will enrich future generations with its genes, i.e., it will increase its frequency within the population. In contrast, the frequency of a genotype with *negative fitness* will decrease over time due to a lower progeny production.[28] Gene frequency of plants having *neutral fitness* will remain the same within the population.

Obviously, allele(s) conferring defense mechanisms (resistance or tolerance) to stress conditions endow a positive fitness compared to wild allele(s) under stressful conditions, e.g., in the presence of xenobiotics.[29] This positive fitness or adaptation mechanisms to stressful conditions have been developed in plants through an evolutionary process. Based on the principles of the three scientific disciplines, i.e., evolutionary biology, population genetics, and physiology, adaptive mechanisms endowing resistance to xenobiotics, insect pests, and pathogens might be associated with negative fitness, the so-called "cost of resistance,"[29–31] because it is predicted that adaptation mechanisms often are not cost-free (reviewed in Reference[32]).

Accordingly, it has generally been presumed that herbicide-resistant weed biotypes may display a lack of fitness (i.e., herbicide resistance fitness cost, negative fitness, or cost of resistance) compared to susceptible individuals of the same population in the absence of herbicides. Hence, herbicide resistance fitness cost is described as the detrimental effect of herbicide resistance gene(s) on the fecundity (number of viable offsprings) of herbicide-resistant biotype leading to a reduction in frequency of the resistant individuals compared with the herbicide susceptible individuals.[28] If a gene(s) conferring resistance to a herbicide is associated with a fitness cost, theoretically back selection should occur in a weed population in the absence of the herbicide, i.e., that the wild types carrying the susceptible gene(s) will eventually dominate the population.[24] Although a significant cost of resistance was reported for triazine-resistant weeds carrying the Gly-264 *psbA* gene,[33] the hypothesis about negative fitness cannot be generalized to all herbicide-resistant weed species,[24,32] as positive fitness[34–36] and neutral fitness[17,37–39] responses have also been reported. Evidence indicates that different resistance genes/alleles may express different fitness responses[21,24,32] in different genetic backgrounds.[21] For example, three different mutant ACCase alleles showed completely different fitness responses in *A. myosuroides*.[25] Likewise, differential fitness penalty has been reported for different mutations on the same herbicide resistance gene in *Arabidopsis thaliana* (L.) Heynh. (mouse-ear

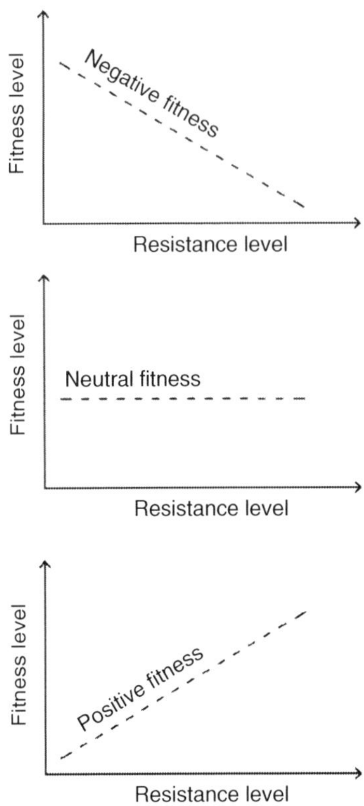

Figure 12.1 Three possible correlation scenarios between the level of herbicide resistance and the level of fitness, viz., positive, negative, and neutral fitness. Generally, it is assumed that negative fitness is the most likely scenario. Source: From Reference[24]/Cambridge University Press.

cress).[40] If a fitness cost/benefit exists in a population, it is expected that the level of fitness cost/benefit is correlated with the level of herbicide resistance (see References[41–43]). Accordingly, three different correlation scenarios can be hypothesized between level of herbicide resistance and level of fitness, viz., positive, negative, and neutral fitness, as illustrated in Fig. 12.1, as discussed in Reference.[24] In addition, fitness of a biotype is not stable because it may be altered under different environmental conditions and affected by fitness of other biotypes within the population (reviewed in Reference[24]). Nevertheless, due to very few studies performed in the fields, it is still difficult to draw indubitable conclusions.

12.3 Explanation of Herbicide Resistance Fitness Costs or Benefits

Fitness costs of herbicide resistance alleles or genes can be attributed to several causes. Here, we have classified them into five categories. Three of the five below-mentioned causes have been briefly described elsewhere (see

Reference[21]). The five causes are (1) impaired functionality of the target-site molecule, (2) resource-based allocation and evolution of gene expression theory, (3) pleiotropic effect and linkage of loci, (4) genomic position effects, and (5) genome size effects.

12.3.1 Impaired Functionality of the Target-Site Molecule

Some gene mutations endowing TSR to herbicides in plants may affect the normal function of the target-site molecule (e.g., D1 protein at photosystem II [PS-II]) or enzyme kinetics of acetolactate synthase (ALS), ACCase, and 5-enolpyruvylshikimate 3-phosphate synthase (EPSP)) and may also interfere with other plant functions and metabolism.[12,21,28,44,45] For instance, a high fitness cost in the form of a reduced photosynthetic capacity (decreased rates of CO_2 fixation and O_2 evolution) has been reported in many triazine herbicide-resistant weed species,[46] especially those carrying a mutation in the *psbA* gene, where serine is replaced by glycine at position 264 (Ser-264-Gly, at the D1 protein).[21,26,33,47–50] The mutant allele Ser-264-Gly was associated with a fitness cost of 25% in many weed species.[32] Reduced yield (up to 30%) was also observed in triazine-resistant crops such as spring canola and foxtail millet.[50] This mutation not only blocks the binding of triazine at the plastoquinone B (PQ-B, a protein-bound plastoquinone) site on the D1 protein, conferring a high level of resistance to triazine herbicides, but also impairs binding of the PQ-B to its niche on the D1 protein. A decrease of the PQ-B binding to the D1 protein decreases electron transport in PS-II and eventually decreases nicotinamide adenine dinucleotide phosphate (NADPH) and adenosine triphosphate (ATP) production and, consequently, reduces the photosynthetic activity leading to a negative fitness.[46,50] Another example is EPSPS target-site mutations, especially the EPSPS Thr-102-Ile/Pro-106-Ser (TIPS) double mutants conferring a high level of glyphosate resistance that reduces the catalytic activity of EPSPS enzyme resulting in a considerable fitness cost.[32,51,52] It has also been stated that fitness costs associated with a mutant target enzyme depend on the level of enzyme catalytic capacity (i.e., reaction velocity V_{max} and substrate affinity K_m).[32] On the contrary, mutations endowing substantial changes in the EPSPS active site not only inhibit binding of glyphosate but also binding of phosphoenol pyruvate (PEP) and, consequently, reduce the catalytic activity of EPSPS and result in plant fitness cost.[32] Theoretically, functionality of a mutant target-site molecule (e.g., the mutant D1 protein at PS-II and the mutant EPSPS enzyme) can either increase, decrease, or remain unchanged; consequently, all three scenarios illustrated in Fig. 12.1 with positive, negative, and neutral fitness may occur. It is also worth noting that the fitness cost of herbicide resistance may depend on the type of herbicide resistance mutation. For instance, in controlled conditions, it was shown that the Asn2041 ACCase mutation was not associated with any cost to seed germination in *A. myosuroides*, while the Gly2078 ACCase and Leu1781 ACCase mutations conferred positive (i.e., rapid seed germination) and negative fitness (i.e., delayed seed germination), respectively.[53]

12.3.2 Resource-Based Allocation and Evolution of Gene Expression Theory

According to resource-based allocation theory, plants can allocate resources to defense mechanisms. Fitness of a living organism is determined by its relative allocation of resources to survival (seed and plant), reproduction, and competitive ability during its life cycle.[26,27,54] Balance of costs between reproduction and defense mechanisms can be considered at two levels, namely, demographic costs (i.e., reduction of survivorship) and physiological costs (i.e., reduction of vegetative growth leading to reduction of survivorship).[55] It is generally believed that there is a negative correlation between plant defense mechanisms and fecundity of plants.[55] Resource allocation to defend against pest organisms such as herbivores and pathogens has been documented,[55] but fewer examples of allocation to herbicide resistance, e.g., fitness cost, have been reported. However, a clear-cut example was provided by Vila-Aiub et al.[30] who found that a P450-metabolism-based herbicide-resistant *Lolium rigidum* Gaud. (annual ryegrass) population (SLR 31) from Australia allocated fewer resources to growth and reproductive characteristics than susceptible plants.

Resistance to herbicides can be a plant defense mechanism.[21] Accordingly, herbicide-resistant plants may divert resources away from growth and reproduction and confer resources to (1) high production and/or increased activity of a non-target-site molecule or enzyme (e.g., detoxifying enzymes such as P450 and glutathione-*S*-transferase [GST]) and/or (2) more production of a herbicide target-site enzyme (e.g., overproduction of the EPSPS enzyme) to survive in the presence of a herbicide. Gene overexpression and gene amplification resulting in an increased production of detoxifier enzymes and high production of target-site enzymes have been reported for both NTSR and TSR weed species (reviewed in Reference[56]), e.g., TSR due to gene amplification of the *EPSPS* gene,[37,56–58] and overexpression of herbicide resistance genes such as the *EPSPS* gene,[59] *ALS* gene,[60] and *ACCase* gene.[59,61] Also, the development of NTSR weed biotypes due to gene overexpression of non-target-site genes such as *P450s*[62–65] and *GSTs*[66–69] have been confirmed in several species.

How energy is used in cells for gene expression is largely unknown.[70] However, it is clear that cells consume energy to express genes.[71] Overexpression of a gene may require a higher level of energy use,[72] i.e., imposing higher bioenergetics cost because more amino acids, ATP, RNA polymerases, ribosomes, and other cellular resources are required for protein production. This scenario can be exacerbated when an amplified gene is overexpressed because, in addition to the above-mentioned cellular resources, nucleotides, as building blocks of DNA, are needed for amplification of a gene within a genome. For instance, significant energy costs due to doubling of gene expression in the fungus *Saccharomyces cerevisiae* have been reported.[73] Limitation of cell division and proliferation is also expected due to an increased cell energy budget required for *EPSPS* gene amplification or overexpression conferring glyphosate resistance (reviewed in Reference[32]). However, contrasting results have been reported, as *Amaranthus palmeri* (Palmer amaranth) plants with an amplified EPSPS gene (more than a 100-fold increase in copy number)[57] did not show fitness cost,[37] even though it was hypothesized that if all *EPSPS* genes were expressed then fitness costs could be expected.[12] Moreover, overexpression of a gene (e.g., gene A) may indirectly lead to a lower expression of another gene (e.g., gene B) or genes (e.g., genes B and C), a secondary effect due to the limitation of cellular resources caused by overexpression of gene A.[72]

As discussed before, fitness costs of metabolic resistance (e.g., P450-based enhanced herbicide resistance in *L. rigidum*[30]) and overexpressed/amplified resistance genes (e.g., amplified *EPSPS* in *Amaranthus tuberculatus* (Moq.) Sauer var. *rudis* (water hemp)[74]) can be explained by resource-based allocation and higher biological energy demand, but it cannot be generalized to all herbicide resistance cases. For instance, metabolism-based resistance to triazine herbicides in *Abutilon theophrasti* Medik. (velvetleaf) (reviewed in Reference[46]) and *EPSPS* gene amplification in *A. palmeri*[37] was not associated with a fitness cost. In contrast, and surprisingly, overexpression of the *EPSPS* gene in glyphosate-resistant *Oryza sativa* f. *spontanea* (weedy rice)[34,35] and *A. thaliana*[75] resulted in a positive fitness. It should be noted that the results reported by Wang et al.[34] have been disputed (see Reference[76–79]).

Neutral and positive fitness of plants carrying overexpressed or amplified herbicide resistance genes can be attributed to a hypothesis assuming that the energy invested in protein production can be compensated through catabolism of the produced amino acids, e.g., the catabolism of tyrosine produced by amplification of the EPSPS gene.[32] Another reason is that herbicide resistance costs may decrease over generations due to natural selection (reviewed in Reference[21]). In other words, whenever resistance is associated with a fitness cost, beneficial mutations may drive adaptation.[80]

12.3.3 Pleiotropic Effect and Linkage of Loci

Theoretically, it can be assumed that if an allele or gene endowing herbicide resistance has either a pleiotropic effect and/or genetic linkage with other loci affecting growth, reproduction, and ecological traits of plants, a positive or negative fitness might occur. For example, a plant may become more susceptible to insect pests and plant pathogens or less attractive to pollinators due to pleiotropic effects of the resistance gene(s) and, consequently, experience a fitness cost.[21] However, examples of fitness cost of herbicide-resistant plants arising from ecological interaction have rarely been reported.[32] A classic example was reported by Gassmann[81] and Gassmann and Futuyma[82] who showed that a triazine-resistant *Amaranthus hybridus* (smooth pigweed) population carrying the *psb*A Ser-264-Gly mutation was more susceptible to the two herbivorous insects *Disonycha glabrata* (Coleoptera: Chrysomelidae) and *Trichoplusia ni* (Lepidoptera: Noctuidae) than triazine-susceptible plants. Plant materials were produced by crossing triazine-resistant and -susceptible individuals from the same population. Another example is triazine-resistant *S. vulgaris* biotype with the *psb*A Ser-264-Gly mutation, which was more sensitive to the plant pathogen *Puccinia lagenophorae*, resulting in a more rapid decline in the frequency of triazine-resistant plants than triazine-susceptible plants under pathogen pressure.[83]

Genes endowing resistance to herbicides may also be very closely linked to genes regulating plant growth rate[21,36,84] or endowing resistance to plant pathogens.[46] Accordingly, herbicide-resistant mechanisms may either be associated with positive or negative fitness. Also, lack of fitness cost of resistance might be caused by compensation, i.e., positive fitness associated with genes linked to the resistance gene.[85] For example, the positive fitness of the mutant ACCase 1781 allele endowing resistance to sethoxydim in hybrid green foxtail millet [*Setaria viridis* × *Setaria italica*] was assumed to be due to the linkage to an unknown beneficial gene.[37] It has also been proposed that the association between resistance to atrazine and resistance to powdery mildews (*Sphaerotheca epilobii* and *Erysiphe cichoracearum*) might be due to close linkage between gene(s) endowing herbicide resistance and host resistance to fungi in two weed species, namely, *Epilobium ciliatum* (willow herb) and *S. vulgaris* (reviewed in Reference[46]).

12.3.4 Genomic Position Effects

Expression of a gene can be affected by changes in its location on a chromosome. This phenomenon is called "position effects" or "chromosomal insertion site effects."[86] It implies that when a gene is amplified, inserted, or transposed within the whole genome, not only its expression and function might be influenced by nearby genes (reviewed in Reference[86]), but also it may potentially influence/disrupt expression of other genes close to the herbicide resistance gene[32] due to gene interactions. Gene expression changes due to positional effects are attributed to different mechanisms, including the activity of upstream or downstream enhancers (*cis*-acting DNA sequences, or a short sequence of DNA bound by special proteins and control gene transcription)[87] and influence of upstream genes suppressing downstream genes (reviewed in Reference[86]). Therefore, the expression of a resistance conferring gene may depend on the genomic location, i.e., position of a gene on a chromosome or among chromosomes. It has also been stated that fitness costs of transgenic plants depend on the position of the inserted transgene in a genome (reviewed in Reference[21]). Therefore, it is assumed that some of the observed fitness costs endowed by gene amplification might be attributed to positional effects. It was also concluded that positive fitness of transgenic plants overexpressing the *EPSPS* gene might be due to a random positional effect of the transgene.[32] Even though positional effects of a gene (either amplified, transposed, or inserted) cannot be attributed as a direct effect of herbicide resistance, it can change the fitness of plants[24] and alter the frequency of resistance genes within a population.

12.3.5 Genomic Size Effects

One of the unique characteristics of organisms is the amount of nuclear DNA, i.e., genome size. Genome size, which is increased by different methods such as ploidy or smaller duplication, can influence phenotype and fitness of organisms.[88] Genome size can be considered an adaptive character,[89] and it is assumed that plant adaptation can be influenced by genome size through the number and position of mutations.[90] The effect of genome size on phenotypic traits of plants, especially at cellular level, has been evidenced.[91–93] For instance, a positive relationship between genome size and guard cell length, and a negative correlation between genome size and stomatal density have been documented.[91] It is also possible that increased demand for resources in plants having large genome size may decrease plant metabolic rates.[93] Even though gene amplification can theoretically increase size of a plant genome, it must be noticed that an amplified gene within a genome may not increase the size of a plant genome substantially. However, if the above-mentioned explanations for herbicide fitness cost (i.e., impaired functionality of the target site, resource-based allocation theory, high biological energy demand of gene overexpression and amplification, pleiotropic effect, linkage of genes, and positional effect of resistance gene) cannot explain an observed fitness cost associated with herbicide resistance, the hypothesis of genome size effect can be considered a potential cause but remains to be tested.

As it was mentioned for genomic positional effect, fitness change caused by genome size cannot be considered as a direct effect of an herbicide resistance gene. However, theoretically, positional effect and genome size effect caused by gene amplification or transposition may change fitness of an organism. It is assumed that duplication of the *EPSPS* gene may affect the fitness of *A. palmeri* in several ways, including genome instability and disruption of *EPSPS* gene insertion.[56] However, to the best of our knowledge, fitness costs of herbicide resistance genes associated with positional effects or genome size effects have not been reported so far.

12.4 Methods of Measuring Fitness

As it has been highlighted by many researchers, determining fitness cost/benefit is not a straightforward task,[22,28,55] and it is often not easy to accurately determine potential fitness cost/benefit of herbicide resistance alleles.[94] Therefore, most studies failed to determine fitness of plants precisely.[21,22,95,96] Lack of control of the genetic background of plants, lack of multigenerational fitness studies, and lack of estimating plant fitness under competition are the three main shortcomings of fitness studies.[22] We strongly suggest that researchers consider these prerequisites before initiating fitness studies.[21,22,24,29]

Fitness of herbicide-resistant weeds is measured using direct or indirect methods (reviewed in Reference[24]). Before measuring fitness, it is recommended to ensure homogeneity in the genetic background of both the susceptible and resistant plants. In addition, knowledge, if available, about molecular and biochemical mechanisms of herbicide resistance is useful. For indirect methods, also called the "allele frequency approach," variations in the frequency of herbicide-resistant and -susceptible plants are determined over several generations with no herbicide applications.[21,94,97] The validity of the indirect methods has been questioned for four reasons: (1) genetic pollution of ecosystems owing to herbicide-resistant gene flow and escape, (2) fitness costs may vary at different plant life stages, (3) indirect methods are time-consuming, and (4) complexity of identifying herbicide-resistant and -susceptible plants within a population.[24] Accordingly, indirect methods may not be appropriate, especially if the aim of fitness studies is to develop management strategies for the control of herbicide-resistant weed populations.[24] With the direct methods, plant characteristics influencing fitness, such as seed germination, seed dormancy, phenological development, seedling establishment, plant growth rate, pollination, size of seed, fecundity, and biomass production, i.e., all aspects of the plant life cycle, are compared to herbicide-resistant and -susceptible plants.[24]

It should be noted that no general agreement exists among researchers, including geneticists, evolutionary ecologists, plant breeders, and weed scientists, on the method of studying fitness. For example, plant breeders most often study the effect of an inserted gene (e.g., a plant pathogen resistance gene) on crop yield in the absence and presence of the pathogen, while evolutionary ecologists conduct multigenerational studies determining the frequency of resistance and susceptible alleles within a weed population. In contrast, weed scientists, focusing on weed management, explore fitness components of herbicide-resistant and -susceptible plants/populations with a view to develop herbicide resistance management strategies.[24] Both the fitness costs and benefits of herbicide-resistant populations can be exploited in management programs for herbicide-resistant weed species.

12.5 Fitness of Metabolic Herbicide-Resistant Weeds

Metabolic herbicide resistance is one of the main NTSR mechanisms. Metabolism-based herbicide-resistant weeds prevent a lethal herbicide dose to reach the target site through an increased activity of metabolizing enzymes such as cytochrome P450s, GSTs, glucosyl transferases (GTs), aryl acylamidase, aldo-keto reductase, and β-cyanoalanine synthase.[14,98] These enzymes are encoded by key genes that are either overexpressed or mutated.[14] Among them, P450s and GSTs are the most important gene families endowing NTSR.[14] As the metabolism-based herbicide resistance mechanism is common in the grass species *L. rigidum* and *A. myosuroides*,[98] and fitness of metabolism-based herbicide resistance mechanism has been most widely studied in these weed species, we will review the studies on fitness costs/benefits in these troublesome grassweed species.

12.5.1 Fitness of P450-Metabolism-Based Resistant *Lolium rigidum*

Lolium rigidum is an annual noxious grassweed species widely spread around the world, which can reduce crop yields by more than 40% (reviewed in Reference[99]). *L. rigidum* has developed resistance (NTSR and TSR) to several herbicides across 14 different herbicide sites of action.[10] Fitness of both NTSR and TSR *L. rigidum* populations have been studied.[30,44,84,99–105]

To the best of our knowledge, only one example exists of well-designed fitness experiments with a metabolism-resistant *L. rigidum* population, which is the experiment done with the SLR31 population. The SLR31 population is a well-known multiple-resistant *L. rigidum* population originating from South Australia (140°46′ E, 36°18′ S).[84,100] The population is resistant to several herbicides including Group 1/A (diclofop-methyl, fluazifop-butyl, chlorazifop propargyl [Ref: CGA 82725], and sethoxydim) and Group 2/B (chlorsulfuron, metsulfuron-methyl [Ref: DPX T6376], and imidazolinone).[106–111] Both TSR (ACCase Ile-1781-Leu) and NTSR (P450 metabolism-based) mechanisms have been found within the SLR31 population (reviewed in Reference[100]). Vila-Aiub et al.[100] isolated a subpopulation carrying a P450-metabolism-based resistance mechanism within the SLR31 population using a plant-cloning technique called "single population approach." The single population approach is a method for selecting susceptible and resistant subpopulations within a population to minimize variation in the genetic background of plants for fitness studies.[24,29] Several fitness components of this population including seed germination, seedling emergence,[100] and physiological[84] and ecological characteristics of plants[30] were investigated, as summarized in Table 12.1 and discussed below. Recently, it was discovered that a single *P450* gene (*CYP81A10v7*) endows resistance in this *L. rigidum* population.[16]

12.5.1.1 Germination and Seedling Emergence Fitness

In a series of experiments, effects of light regimes, burial depths, and temperature regimes on germination and seedling emergence of P450-metabolism-based resistant (R) and herbicide-susceptible (S) *L. rigidum* plants were studied.[100] No difference was observed in final seedling emergence between R and S plants under a temperature regime of 25/15 °C with a photoperiod of 12 hours, where seeds were buried at a depth of up to 6 cm, while at 8 cm burial depth seedling emergence of P450-metabolism-based resistant plants (20%) was significantly lower than that of the S plants (36.6%). The time (days) to reach 50% of final seedling emergence was also similar for R and S plants at different burial depths. It was also reported that fatal germination for R plants (54%) tended to be higher than that for S plants (39%) at 8 cm burial depth. The effect of light on germinability of R and S plants was similar. Also, no differences were observed in final seedling emergence between R and S plants at 20, 15, and 12 °C, while at lower temperatures, i.e., 8 and 5 °C, germination of R plants was significantly lower than that of the S phenotype. In addition, the base temperature for P450-metabolism-based resistant plants (4 ± 0.4 °C) was significantly higher than that of the S plants (2.9 ± 0.1 °C).

12.5.1.2 Physiological Fitness in Noncompetitive Condition

Reproductive allocation, vegetative allocation, photosynthesis rate, and growth parameters of P450-metabolism-based resistant and susceptible *L. rigidum* plants have also been studied in a series of noncompetitive interference experiments in greenhouse and field conditions.[84] During the vegetative growth stage, above-ground biomass production of the R biotype was lower than that of the S biotype.[84] This trend was also observed for physiological traits, including relative growth rate (RGR), net assimilation rate (NAR), and photosynthetic rate (leaf CO_2 assimilation rates based on μmol/m^2/s) at the early growth stage of the R plants. A direct fitness cost between 13% and 25% was found for P450-metabolism-based resistance in the SLR31 population. In contrast to vegetative and physiological traits, reproductive fitness (seed number and seed mass) of P450-metabolism-based resistant and S plants was similar, suggesting

Table 12.1 Examples of studies on fitness, fitness components (i.e., life history traits), and biochemical or physiological traits of P450-metabolism-based resistant *Lolium rigidum*.

Fitness, fitness components, and biochemical or physiological traits	Environment	References
Reduced seedling emergence at deep burial depth and low temperature (8 and 5 °C)	Controlled condition	100
Equal seedling emergence at shallow burial depth and optimum temperature	Controlled condition	100
Higher fatal germination at deep burial	Controlled condition	100
Higher base temperature	Controlled condition	100
Reduced vegetative growth under noncompetitive condition	Controlled greenhouse condition	84
Reduced physiological traits, i.e., relative growth rate (RGR), net assimilation rate (NAR), and photosynthetic rates in noncompetitive condition	Controlled greenhouse condition	84
Equal reproductive fitness in noncompetitive condition	Semi-field (outdoor pot) condition	84
Reduced vegetative growth (30%) in competitive condition	Semi-field (outdoor pot) condition	30
Reduced fecundity fitness (i.e., seed number [23%], seed mass [29%], and reproductive tillers [22%]) in competitive condition	Semi-field (outdoor pot) condition	30
Reduced harvest index (13%) in competitive condition	Semi-field (outdoor pot) condition	30

The genetic background of herbicide-resistant and -susceptible plants was controlled in all studies that vegetative and physiological fitness costs were compensated during the reproductive growth stage.[84]

12.5.1.3 Vegetative and Reproductive Fitness in Competition with Wheat

The S plants produced significantly more (30%) above-ground biomass (Fig. 12.2) and greater reproductive attributes (i.e., seed number (23%), seed mass (29%), and reproductive tillers (22%)) than those of P450-metabolism-based resistant plants when grown together with an increasing density of wheat plants.[30] This means that S plants were stronger competitors than P450-metabolism-based resistant plants in competition with an increasing density of wheat, considering both vegetative and reproductive fitness. In another word, in competition with wheat, the P450-metabolism-based resistant plants allocated fewer resources to vegetative and reproductive organs than the S plants. Vila-Aiub et al.[30] provided evidence showing that a resource-based cost was associated with the P450-metabolism-based resistance mechanism in SLR31 population. Based on the observed vegetative fitness cost (i.e., 30% lower above ground biomass production) and reproductive fitness cost (i.e., 23% lower seed production), it should be possible to exploit agronomic practices such as competitive crops and varieties

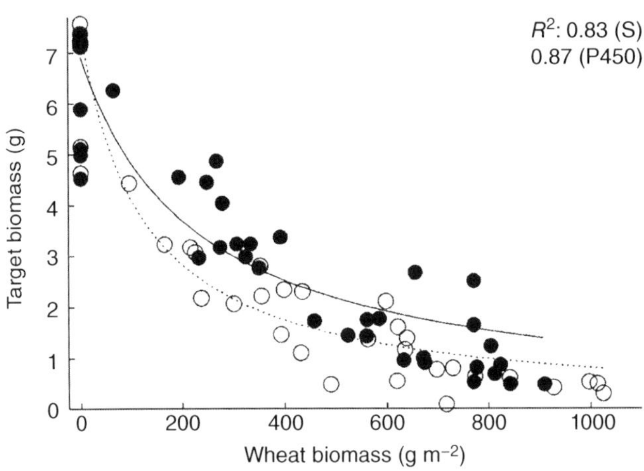

Figure 12.2 Response (vegetative biomass, g/plant) of P450-metabolism-based herbicide-resistant (R, ○) and herbicide-susceptible (S, •) *Lolium rigidum* biotypes to increasing vegetative biomass of wheat (g/m^2). The R and S plants were selected within a P450-metabolism-based resistant *L. rigidum* population using a plant cloning technique, homogenizing the genetic background of the studied plants. There was a significant difference (p-value = 0.012) in the slope of the two curves estimated by hyperbolic nonlinear regression ($y = a/[1 + bx]$). The estimated slope of equations for R and S plants were 0.06 ± 0.006 and 0.039 ± 0.004, respectively. The steeper slope reflects a lower response to an increasing biomass of wheat with the R plants compared to the S plants. Source: From Reference[30]/John Wiley & Sons.

in crop rotation programs to suppress P450-metabolism-based resistant plants and, eventually, suppress the resistant biotype in the population. It should be noted that under noncompetitive conditions, reproductive traits such as number of reproductive tillers, seed mass, and seed number were similar for R and S plants, while in the presence of wheat competition the S plants performed better, showing that herbicide resistance costs were only expressed under conditions of resource limitations. Recently, the fitness cost of insecticide resistance in insects with P450-metabolism-based resistance such as *Anopheles funestus*, a major mosquito vector of malaria in Africa, has also been reported.[112]

Even though physiological and vegetative fitness costs associated with the P450-metabolism-based resistance mechanism are in accordance with the resource allocation theory, the precise mechanisms behind the observed costs associated with this type of herbicide resistance mechanism are unknown. It is assumed that if resistance in a *L. rigidum* population is conferred by overexpression of the P450 enzyme, it can be concluded that energy expenditure in P450-metabolism-based resistant plants may be higher than in S plants, i.e., plants reduce the allocation of energy for growth, reproduction, etc. to compensate for the higher energy demand required for overexpression of the gene. For instance, differences in germinability of subpopulations at low temperatures were attributed to the variability in their base temperature requirements. This means that P450-metabolism-based resistant plants required more thermal time than S plants.[100] According to this explanation of fitness cost, it can also be concluded that the lower seedling emergence and higher fatal germination of P450-metabolism-based resistant plants at 8 cm burial depth in comparison with S plants might be due to differential nutrition/resource accumulation and partitioning within the seeds, even though the mean seed weight of R and S plants was similar. It means, despite similar seed size, that the ratio of seed compartments, i.e., seed coat (testa), embryo, and endosperm might be different due to environmental and genetic effects. This hypothesis merits to be tested in the future studies. Also, the other potential mechanisms explaining herbicide resistance costs, especially linkage and pleiotropy, cannot be completely ruled out, as a single *P450* gene (*CYP81A10v7*) endows resistance in P450-metabolism-based resistant *L. rigidum* plants.[16]

12.5.2 Fitness of Metabolic Herbicide-Resistant *Alopecurus myosuroides*

Alopecurus myosuroides is one of the most important grassweeds in Northwestern Europe and is the most important herbicide-resistant weed species in European arable cropping systems.[113,114] It can reduce winter wheat yields by up to 45%.[115] *A. myosuroides* ranks among the 15 most important weed species worldwide and has developed resistance to several herbicides covering seven different sites of action.[10] As with *L. rigidum*, both TSR and NTSR have been detected within populations of *A. myosuroides*. Considering the number of reports on resistance of *A. myosuroides*, the number of fitness cost/benefit studies is low. However, studies on the fitness of NTSR and TSR *A. myosuroides* populations from Denmark,[7,17,116–118] France,[25,53,119,120] and the United Kingdom[121,122] have been published. Most of these fitness studies have been conducted on NTSR *A. myosuroides* populations, and several fitness components have been studied as described below and summarized in Table 12.2.

12.5.2.1 Seed Dormancy, Germination, and Seedling Emergence Fitness

In a series of experiments, seed dormancy, seed germination, seedling emergence, antioxidant enzyme activity of seeds, and seed protein content of an NTSR and an S subpopulation selected within a Danish *A. myosuroides* (ID-914 population) were studied. Seeds were tested under different conditions including burial depths, temperature regimes, salinity concentrations, drought stress, and accelerated seed-aging conditions (i.e., exposing seeds to high relative humidity [100%] at 45 °C for different time intervals). It is worth noting that the NTSR and S subpopulations shared a random homogeneous genetic background as they were selected within the ID-914 population through the "single population approach."[24,29] The parent ID-914 population was resistant to several herbicides, including Group 1/A (fenoxaprop-*P*-ethyl), 2/B (flupyrsulfuron-methyl-sodium), 3/K1 (pendimethalin), and 15/K3 (prosulfocarb).[18] The precise NTSR mechanism endowing resistance in the ID-914 population remains to be studied. However, none of the 12 and 7 known point mutations endowing TSR to ACCase and ALS-inhibiting herbicides, respectively, were detected within the resistant plants.[18,118] In addition, a literature review revealed that NTSR mechanisms such as reduced herbicide penetration and translocation have not been reported in *A. myosuroides* populations.[123,124] Accordingly, it was assumed that enhanced metabolism was the main mechanism endowing NTSR in the ID-914 population.[17]

Germination of freshly harvested seeds of the NTSR subpopulation (89%) was higher than the S subpopulation (78%), suggesting that the primary seed dormancy of the NTSR subpopulation was lower than the S subpopulation.[116] From an agronomic point of view, higher germination shortly after seed shedding in the summer is a fitness cost for the NTSR subpopulation, because seedbed preparation or herbicide spraying before sowing of an autumn-sown crop will eliminate a higher percentage of the NTSR seedlings compared to the S subpopulation. In contrast, in another study with nondormant seeds (i.e., without primary dormancy), the NTSR subpopulation had lower seedling emergence than the corresponding S subpopulation, especially at low temperature and deep burial.[118] For instance, at 6 cm burial depth at 10/5 °C (day/night), seedling emergence of the NTSR and S subpopulations was 8% and 37%, respectively. It means that seedling emergence of

Table 12.2 Examples of studies on fitness, fitness components (i.e., life history traits), and biochemical and physiological traits of metabolism-based herbicide-resistant *Alopecurus myosuroides* populations.

Fitness, fitness components, and biochemical and physiological traits	Environment	References
Increased germination of freshly harvested seeds, i.e., reduced primary dormancy	Controlled condition	116
Reduced emergence of nondormant seeds at low temperature and deep burial	Controlled condition	118
Higher sensitivity of seeds to burial depth	Controlled condition	118
Equal seedling emergence at shallow burial depth and optimum temperature	Controlled condition	118
Reduced germination under salinity, drought, and accelerated aging treatments	Controlled condition	117
Equal final seed germination and germination timing at optimum condition	Controlled condition	117, 118
Reduced activity of catalase and superoxide dismutase in artificially aged seeds	Controlled condition	117
Equal activity of peroxidase and similar protein content in both non-aged and artificially aged seeds	Controlled condition	117
Longer tiller length and greater intraspecific competitive effect at vegetative growth stage	Semi-field (outdoor pot) condition	122
Equal vegetative growth in both competitive and noncompetitive conditions	Field and cold greenhouse condition	116
Lower biomass allocated to reproductive tissues at flowering stage	Semi-field (outdoor pot) condition	122
Reduced reproductive fitness (number of seed head: 27%, total seed-head length: 23%) under nitrogen deprivation condition	Semi-field (outdoor pot) condition	122
Equal reproductive fitness in both competitive and noncompetitive conditions	Field, semi-field (outdoor pot) and cold greenhouse condition	17, 116, 122
Equal herbicide resistance level of populations after three successive generations of ceasing to use herbicides, i.e., equal reproductive fitness	Semi-field (outdoor pot) condition	121
Lower harvest index		116
Delayed booting (BBCH 45) and flowering (BBCH 65) stages, i.e., required more energy and thermal time to reach the reproductive phenological stages		116

The genetic background of herbicide-resistant and -susceptible plants was controlled in all studies the NTSR subpopulation was five times lower than that of the S subpopulation.[118] Accordingly, it was concluded that deep soil cultivation and delayed planting of autumn-sown crops may provide suboptimal conditions for the NTSR subpopulation and reduce its frequency in the field more than that of the corresponding S subpopulation.[116,118] However, no substantial differences were found in the time of seedling emergence between the subpopulations. Burial depth preventing seedling emergence by 50% for the NTSR subpopulation was 4.2 cm, while it was 5.8 cm for the S subpopulation at 10/5 °C (day/night).[118] The results very clearly revealed a fitness cost in the NTSR subpopulation in terms of a higher sensitivity to soil burial depth.

Germination of the NTSR subpopulation was more suppressed by salinity, drought, and accelerated aging treatments than the S subpopulation. For instance, the time of accelerated seed-aging treatment required to reduce final germination by 50% was 90 and 67 hours for the NTSR and S subpopulations, respectively.[117] The same trend was observed with the activity of antioxidant enzymes, catalase and superoxide dismutase, that were more suppressed in the NTSR subpopulation than in the S subpopulation under accelerated seed-aging conditions.[117] However, the activity of peroxidase enzyme and protein content were similar for the two subpopulations in both stress-free and accelerated aging conditions.[117] Based on these results, it can be expected that the persistence of NTSR seeds will be lower than that of the S plants; however, this remains to be validated in field experiments.[117] In summary, it was clearly demonstrated that NTSR in the ID-914 *A. myosuroides* population was associated with biochemical and germination fitness costs.

12.5.2.2 Vegetative, Reproductive, and Phenological Fitness Under Competitive and Noncompetitive Conditions

The vegetative and reproductive growth and phenological development of NTSR *A. myosuroides* populations from Denmark and the United Kingdom have been studied.[17,116,121,122] The Danish NTSR and S subpopulations showed similar vegetative and reproductive fitness under both competitive and noncompetitive conditions (Fig. 12.3A and B).[17,116] A UK NTSR population with enhanced metabolic resistance did not show a consistent reproductive cost of resistance when compared with its S counterparts.[122] Herbicide resistance levels of two UK NTSR populations (PELD03 and PELDSS05) did not change following three successive generations, where they were grown in a field under competitive conditions with winter wheat in the absence of an herbicide selection pressure.[121] This shows that no fecundity fitness cost was associated with the UK populations. Accordingly, Comont et al.[122] suggested that the NTSR mechanism was not associated with a severe reproductive cost. However, under different environmental conditions, the NTSR mechanism may be associated with a fitness cost.

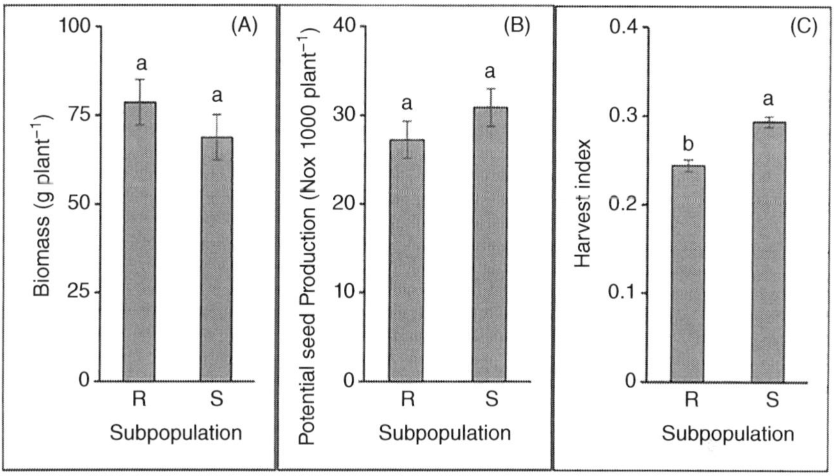

Figure 12.3 Above ground biomass (A), potential seed production (B), and harvest index (C, i.e., the ratio of reproductive production of plants to the total plant vegetative biomass + reproductive production) of herbicide-susceptible (S) and -resistant (R) *Alopecurus myosuroides* plants. The R and S plants were selected within a non-target-site-resistant population using a plant-cloning technique, minimizing the variation in the genetic background of the plant material. Significant differences ($P < 0.05$) between the R and S plants are shown by different letters on each column. Source: From Reference[116]/with permission of Elsevier.

For example, under conditions of nitrogen limitations, the UK NTSR population studied by Comont et al.[122] produced a significantly lower number of seed heads (27%) and total seed-head length (23%) in comparison with the S plants. Also, inconsistent fitness was observed in different life history traits of the UK NTSR plants. For instance, at the vegetative growth stage, the NTSR plants were taller and revealed a higher competitive effect than the S plants, while at the flowering stage, the S plants allocated proportionally more resources to reproductive tissues in comparison with the NTSR plants. The Danish NTSR subpopulation also had a lower harvest index (i.e., reproductive production of plants [cumulative weight of heads] divided by the total plant biomass [above-ground biomass + cumulative head weight]) than the corresponding S subpopulation (Fig. 12.3C), meaning that the NTSR plants allocated proportionately lower resources to reproductive output.[116] Also, phenological development of the Danish NTSR subpopulation tended to be slower than that of the S plants at late growth stages, i.e., the NTSR plants required more energy and thermal time to reach the reproductive pheno logical stages including booting (BBCH 45) and flowering (BBCH 65) stages.[116] Differential flowering time of the NTSR and S plants may prevent or hamper cross-pollination between the NTSR and S subpopulations and may reduce the spread and frequency of resistance genes.

12.6 Is NTSR Reversible?

Usually, it is expected that resistance to xenobiotics, such as herbicides, will be associated with a cost. If expression and level of fitness cost are remarkable, it is expected that herbicide metabolic resistance in the absence of an herbicide may reverse. As mentioned, fitness penalty of metabolic resistance has been reported for some metabolism-resistant weed species, e.g., an Australian P450-metabolism-based *L. rigidum*,[30,84] while reversion of metabolic resistance has not been reported so far. However, optimistic reports about reversion of resistance exist for other herbicide resistance mechanisms. For example, reversion of glyphosate resistance in four NTSR *L. rigidum* populations with reduced translocation mechanism has been reported when herbicide selection pressure was stopped for 3–4 years.[102–104]

It has also been concluded that fitness cost of herbicide resistance is not a universal phenomenon, because it depends on the mechanism of resistance (see, e.g., Reference[84]), genetic background (see, e.g., Reference[125]), type of resistance allele, gene, mutation (see, e.g., Reference[40]), fitness components (see, e.g., Reference[17,118]), etc. Therefore, reversal of resistance may not be possible in all cases. For instance, as reviewed above, no reproductive fitness cost was associated with NTSR in *A. myosuroides* populations collected from Denmark and the United Kingdom.[17,122] It should be emphasized that lack of fitness cost in these populations does not mean that there is zero fitness cost, but it merely tells us that a measurable reproductive fitness cost has not been detected.[126] Hence, reversion of resistance may not occur as shown by Moss and Hull[121] (see above). However, general conclusions should not be drawn from this result, because fitness cost of herbicide resistance depends on many factors as mentioned above. Accordingly, any small difference in fitness, be it positive or negative, in the plant life cycle between herbicide-resistant and -susceptible populations can be magnified through agronomic practices.[24] Hence, it can be assumed that exploiting agronomic practices can magnify fitness cost of resistant populations, which may lead to reversion of herbicide resistance. For instance, it was concluded that exploiting agronomic practices such as the cultivation of competitive crops and varieties in crop rotation programs will suppress P450-metabolism-based resistant *L. rigidum* plants more than the S plants, and reversion of resistance will occur as the P450-metabolism-based resistant plants showed a 30% vegetative fitness cost and a 23% reproductive fitness cost compared to the S plants in competition with wheat (see Section 5.1).

Furthermore, it has been suggested that moldboard plowing and a false seedbed prior to sowing winter wheat can reduce the density of a Danish NTSR *A. myosuroides* by approximately 10% and 30%, respectively, compared to the S subpopulation. In this case, selection for herbicide resistance is not a one-way process, even though previous studies on *A. myosuroides* concluded so.[121] However, the merit of these assumptions should be tested in future studies, e.g., by measuring frequency of herbicide-resistant and -susceptible plants within a population over several generations (see Reference[24]).

12.7 Concluding Remarks

- Before the onset of studying plant fitness, several prerequisites must be fulfilled. Among them, the control of the genetic background plays a pivotal role and without complete control of the genetic background, it is difficult to associate fitness cost with herbicide resistance.
- Herbicide resistance fitness costs cannot be generalized to all herbicide resistance cases, as it depends on the mechanism of resistance, type of allele, genes, and mutations endowing resistance.
- The main mechanisms influencing herbicide resistance fitness costs can be classified into five groups: (1) impaired functionality of the target site molecule, (2) resource-based allocation and evolution of gene expression theory, (3) pleiotropic effect and linkage of loci, (4) genomic position effects, and (5) genome size effects.
- Manifestation of potential fitness costs of herbicide resistance may depend on the plant life stage, i.e., fitness components. Even though seed production is the main endpoint when considering the fitness of plants, other plant characteristics, including seed germination, seed dormancy, phenological development, seedling establishment, plant growth rate, pollination, size of seed, fecundity, and biomass production, are of great agronomic importance.
- Manifestation of potential costs of herbicide resistance may be magnified under stressful environmental conditions such as extreme temperatures, burial depth, light, and intense resource competition. Accordingly, it is recommended to investigate fitness components of plants under different environmental conditions.
- The fitness of metabolic herbicide-resistant weed species can either be negative, positive, or neutral, but negative fitness is the most prevalent. Accordingly, reversion of herbicide resistance may or may not take place in the absence of selection pressure.
- Vegetative and reproductive fitness costs of the metabolic herbicide resistance mechanism have been reported in a P450-metabolism-based resistant *L. rigidum*

population under competitive conditions, where the estimated costs can be explained by resource-based allocation theory. Exploiting agronomic practices, such as competitive crops/varieties in crop rotation, may provide unfavorable conditions for the P450-metabolism-based *L. rigidum* subpopulation, resulting in a reduction of the frequency of the P450-metabolism-based *L. rigidum* subpopulation. In contrast, neither vegetative fitness nor reproductive fitness costs were associated with metabolic herbicide resistance in different NTSR *A. myosuroides* populations.

- Differences in seed germination, seedling emergence, seed dormancy, physiological traits, etc. between herbicide-susceptible and metabolism-based resistant plants have been documented for different populations of grass species *L. rigidum* and *A. myosuroides,* where agronomic practices may provide suboptimal conditions for the metabolism-based resistant plants. For instance, it was expected that agronomic practices, including moldboard plowing and a false seedbed may reduce the frequency of a Danish NTSR *A. myosuroides* seedlings up to 40% compared to the S subpopulation – an assumption that needs to be validated in future studies.
- To optimize Integrated Weed Management, more studies have to be performed in field conditions to find what agronomic measures can impact fitness costs and the selection of susceptible populations. This, combined with genomics to discover genes involved in fitness, might offer new perspectives to optimize weed control.

References

1 Harper JL, The evolution of weeds in relation to resistance to herbicides, In *Proceedings of the 3rd British Weed Control Conference.* British Weed Control Council: Farnham, UK, pp. 179–188 (1956).

2 Switzer CM, The existence of 2,4-D resistant strains of wild carrot, In *Proceedings of the North Eastern Weed Control Conference*, pp. 315–318 (1957).

3 Whitehead CW, and Switzer CM, The differential response of strains of wild carrot to 2,4-D and related herbicides. *Can J Plant Sci* **43**:255–262 (1963).

4 Hilton HW, Herbicide tolerant strains of weeds. Hawaiian Sugar Planters' Association. Annual Report, ed. by Report A, pp. 69 (1957).

5 Ryan GF, Resistance of common groundsel to simazine and atrazine. *Weed Sci* **18**:614–616 (1970).

6 Norsworthy JK, Ward SM, Shaw DR, Llewellyn RS, Nichols RL, Webster TM, *et al.*, Reducing the risks of herbicide resistance: best management practices and recommendations. *Weed Sci* **60**:31–62 (2012).

7 Keshtkar E, *Ecological Fitness, Molecular Basis and Selection of Resistant Black-Grass (Alopecurus myosuroides) Biotypes*, Aarhus University, Aarhus (2015).

8 Shaw DR, The "wicked" nature of the herbicide resistance problem. *Weed Sci* **64**:552–558 (2016).

9 Harker KN, Mallory-Smith C, Maxwell BD, Mortensen DA, and Smith RG, Another view. *Weed Sci* **65**:203–205 (2017).

10 Heap I, The International Survey of Herbicide Resistant Weeds. Online. Internet. 11:1. Available: www.weedscience.org [Accessed April 2021] (2021).

11 Jugulam M, and Shyam C, Non-target-site resistance to herbicides: recent developments. *Plants (Basel)* **8** (2019).

12 Powles SB, and Yu Q, Evolution in action: plants resistant to herbicides. *Annu Rev Plant Biol* **61**:317–347 (2010).

13 Duke SO, Enhanced metabolic degradation: the last evolved glyphosate resistance mechanism of weeds? *Plant Physiol* **181**:1401–1403 (2019).

14 Gaines TA, Duke SO, Morran S, Rigon CAG, Tranel PJ, Küpper A, *et al.*, Mechanisms of evolved herbicide resistance. *J Biol Chem* **295**:10307–10330 (2020).

15 Van Horn CR, Moretti ML, Robertson RR, Segobye K, Weller SC, Young BG, *et al.*, Glyphosate resistance in *Ambrosia trifida*: part 1. Novel rapid cell death response to glyphosate. *Pest Manag Sci* **74**:1071–1078 (2018).

16 Han H, Yu Q, Beffa R, González S, Maiwald F, Wang J, *et al.*, Cytochrome P450 CYP81A10v7 in *Lolium rigidum* confers metabolic resistance to herbicides across at least five modes of action. *Plant J* **105**:79–92 (2021).

17 Keshtkar E, Mathiassen SK, and Kudsk P, No vegetative and fecundity fitness cost associated with acetyl-coenzyme a carboxylase non-target-site resistance in a Black-grass (*Alopecurus myosuroides* Huds) population. *Front Plant Sci* **8** (2017).

18 Keshtkar E, Mathiassen SK, Moss SR, and Kudsk P, Resistance profile of herbicide-resistant *Alopecurus myosuroides* (black-grass) populations in Denmark. *Crop Prot* **69**:83–89 (2015).

19 Délye C, Gardin JAC, Boucansaud K, Chauvel B, and Petit C, Non-target-site-based resistance should be the Centre of attention for herbicide resistance research: *Alopecurus myosuroides* as an illustration. *Weed Res* **51**:433–437 (2011).

20 Primack RB, and Hyesoon K, Measuring fitness and natural selection in wild plant populations. *Annu Rev Ecol Syst* **20**:367–396 (1989).

21 Vila-Aiub MM, Neve P, and Powles SB, Fitness costs associated with evolved herbicide resistance alleles in plants. *New Phytol* **184**:751–767 (2009).

22 Vila-Aiub MM, Gundel PE, and Preston C, Experimental methods for estimation of plant fitness costs associated with herbicide-resistance genes. *Weed Sci* **63**:203–216 (2015).

23 Maxwell BD, and Mortimer AM, Selection for herbicide resistance, In Powles SB, and Holtum JAM (Eds), *Herbicide Resistance in Plants: Biology and Biochemistry*, Lewis Publisher, Boca Raton, FL, pp. 1–25 (1994).

24 Keshtkar E, Abdolshahi R, Sasanfar H, Zand E, Beffa R, Dayan FE, *et al.*, Assessing fitness costs from a herbicide-resistance management perspective: a review and insight. *Weed Sci* **67**:137–148 (2019).

25 Menchari Y, Chauvel B, Darmency H, and Délye C, Fitness costs associated with three mutant acetyl-coenzyme a carboxylase alleles endowing herbicide resistance in black-grass *Alopecurus myosuroides*. *J Appl Ecol* **45**:939–947 (2008).

26 Holt JS, Fitness and ecological adaptability of herbicide-resistant biotypes, In *Managing Resistance to Agrochemicals*, American Chemical Society, pp. 419–429 (1990).

27 Park KW, Mallory-Smith CA, Ball DA, and Mueller-Warrant GW, Ecological fitness of acetolactate synthase inhibitor–resistant and –susceptible downy brome (*Bromus tectorum*) biotypes. *Weed Sci* **52**:768–773 (2004).

28 Cousens RD, and Fournier-Level A, Herbicide resistance costs: what are we actually measuring and why? *Pest Manag Sci* **74**:1539–1546 (2018).

29 Vila-Aiub MM, Neve P, and Roux F, A unified approach to the estimation and interpretation of resistance costs in plants. *Heredity* **107**:386–394 (2011).

30 Vila-Aiub MM, Neve P, and Powles SB, Evidence for an ecological cost of enhanced herbicide metabolism in *Lolium rigidum*. *J Ecol* **97**:772–780 (2009).

31 Coustau C, Chevillon C, and ffrench-Constant R, Resistance to xenobiotics and parasites: can we count the cost? *Trends Ecol Evol* **15**:378–383 (2000).

32 Vila-Aiub MM, Yu Q, and Powles SB, Do plants pay a fitness cost to be resistant to glyphosate? *New Phytol* **0** (2019).

33 Gronwald JW, Resistance to photosystem II inhibiting herbicides, In Powles SB, and Holtum JAM (Eds), *Herbicide Resistance in Plants Biology and Biochemistry*, CRC Press, USA, pp. 27–60 (1994).

34 Wang W, Xia H, Yang X, Xu T, Si HJ, Cai XX, *et al.*, A novel 5-enolpyruvoylshikimate-3-phosphate (EPSP) synthase transgene for glyphosate resistance stimulates growth and fecundity in weedy rice (*Oryza sativa*) without herbicide. *New Phytol* **202**:679–688 (2014).

35 Yang X, Li L, Jiang X, Wang W, Cai X, Su J, *et al.*, Genetically engineered rice endogenous 5-enolpyruvoylshikimate-3-phosphate synthase (epsps) transgene alters phenology and fitness of crop-wild hybrid offspring. *Sci Rep* **7**:6834 (2017).

36 Wang T, Picard JC, Tian X, and Darmency H, A herbicide-resistant ACCase 1781 *Setaria* mutant shows higher fitness than wild type. *Heredity* **105**:394–400 (2010).

37 Vila-Aiub MM, Goh SS, Gaines TA, Han H, Busi R, Yu Q, *et al.*, No fitness cost of glyphosate resistance endowed by massive EPSPS gene amplification in *Amaranthus palmeri*. *Planta* **239**:793–801 (2014).

38 Goggin DE, Beckie HJ, Sayer C, and Powles SB, No auxinic herbicide–resistance cost in wild radish (*Raphanus raphanistrum*). *Weed Sci* **67**:539–545 (2019).

39 Giacomini D, Westra P, and Ward SM, Impact of genetic background in fitness cost studies: an example from glyphosate-resistant palmer Amaranth. *Weed Sci* **62**:29–37 (2014).

40 Roux F, Gasquez J, and Reboud X, The dominance of the herbicide resistance cost in several *Arabidopsis thaliana* mutant lines. *Genetics* **166**:449–460 (2004).

41 Van Etten ML, Kuester A, Chang S-M, and Baucom RS, Fitness costs of herbicide resistance across natural populations of the common morning glory, *Ipomoea purpurea*. *Evol: Int J Org Evol* **70**:2199–2210 (2016).

42 Cockerton HM, Kaundun SS, Nguyen L, Hutchings SJ, Dale RP, Howell A, *et al.*, Fitness cost associated with enhanced EPSPS gene copy number and glyphosate resistance in an *Amaranthus tuberculatus* population. *bioRxiv* **12**:651381 (2021).

43 Cockerton HM, *Investigating the Cost of Adaptation in Amaranthus tuberculatus Populations with Evolved Resistance to Glyphosate*, University of Warwick, United Kingdom (2013).

44 Vila-Aiub MM, Yu Q, Han H, and Powles SB, Effect of herbicide resistance endowing Ile-1781-Leu and Asp-2078-Gly ACCase gene mutations on ACCase kinetics and growth traits in *Lolium rigidum*. *J Exp Bot* **66**:4711–4718 (2015).

45 Yu Q, Han H, Vila-Aiub MM, and Powles SB, AHAS herbicide resistance endowing mutations: effect on AHAS functionality and plant growth. *J Exp Bot* **61**:3925–3934 (2010).

46 LeBaron HM, Chapter 10 - distribution and Management of Triazine-Resistant Weeds, In LeBaron HM, McFarland JE, and Burnside OC (Eds), *The Triazine Herbicides*, Elsevier, San Diego, pp. 119–132 (2008).

47 Warwick SI, and Black LD, Relative fitness of herbicide-résistant and susceptible biotypes of weeds. *Phytoprotection* **75**:37–49 (1994).

48 Holt JS, and Thill DC, Growth and productivity of resistant plants, In Powles SB, and Holtum JAM (Eds), *Herbicide Resistance in Plants Biology and Biochemistry*, CRC Press, Boca Raton, FL, pp. 299–316 (1994).

49 Holt JS, and Lebaron HM, Significance and distribution of herbicide resistance. *Weed Technol* **4**:141–149 (1990).

50 Warwick SI, Herbicide resistance in weedy plants: physiology and population biology. *Annu Rev Ecol Syst* **22**:95–114 (1991).

51 Yu Q, Jalaludin A, Han H, Chen M, Sammons RD, and Powles SB, Evolution of a double amino acid substitution in the 5-Enolpyruvylshikimate-3-phosphate synthase in *Eleusine indica* conferring high-Level glyphosate resistance. *Plant Physiol* **167**:1440–1447 (2015).

52 Han H, Vila-Aiub MM, Jalaludin A, Yu Q, and Powles SB, A double EPSPS gene mutation endowing glyphosate resistance shows a remarkably high resistance cost. *Plant Cell Environ* **40**:3031–3042 (2017).

53 Delye C, Menchari Y, Michel S, Cadet E, and Le Corre V, A new insight into arable weed adaptive evolution: mutations endowing herbicide resistance also affect germination dynamics and seedling emergence. *Ann Bot* **111**:681–691 (2013).

54 Maxwell BD, Roush ML, and Radosevich SR, Predicting the evolution and dynamics of herbicide resistance in weed populations. *Weed Technol* **4**:2–13 (1990).

55 Bazzaz FA, 1 - allocation of resources in plants: state of the science and critical questions, In Bazzaz FA, and Grace J (Eds), *Plant Resource Allocation*, Academic Press, San Diego, pp. 1–37 (1997).

56 Patterson EL, Pettinga DJ, Ravet K, Neve P, and Gaines TA, Glyphosate resistance and EPSPS gene duplication: convergent evolution in multiple plant species. *J Hered* **109**:117–125 (2018).

57 Gaines TA, Zhang W, Wang D, Bukun B, Chisholm ST, Shaner DL, *et al.*, Gene amplification confers glyphosate resistance in *Amaranthus palmeri*. *Proc Natl Acad Sci* **107**:1029–1034 (2010).

58 Gaines TA, Shaner DL, Ward SM, Leach JE, Preston C, and Westra P, Mechanism of resistance of evolved glyphosate-resistant palmer Amaranth (*Amaranthus palmeri*). *J Agric Food Chem* **59**:5886–5889 (2011).

59 Achary VMM, Sheri V, Manna M, Panditi V, Borphukan B, Ram B, *et al.*, Overexpression of improved EPSPS gene results in field level glyphosate tolerance and higher grain yield in rice. *Plant Biotechnol J* **18**:2504–2519 (2020).

60 Zhao N, Yan Y, Wang H, Bai S, Wang Q, Liu W, *et al.*, Acetolactate synthase overexpression in Mesosulfuron-methyl-resistant Shortawn foxtail (*Alopecurus aequalis* Sobol.): reference gene selection and herbicide target gene expression analysis. *J Agric Food Chem* **66**:9624–9634 (2018).

61 Laforest M, Soufiane B, Simard MJ, Obeid K, Page E, and Nurse RE, Acetyl-CoA carboxylase overexpression in herbicide-resistant large crabgrass (*Digitaria sanguinalis*). *Pest Manag Sci* **73**:2227–2235 (2017).

62 Iwakami S, Endo M, Saika H, Okuno J, Nakamura N, Yokoyama M, *et al.*, Cytochrome P450 CYP81A12 and CYP81A21 are associated with resistance to two Acetolactate synthase inhibitors in *Echinochloa phyllopogon. Plant Physiol* **165**:618–629 (2014).

63 Gaines TA, Lorentz L, Figge A, Herrmann J, Maiwald F, Ott M-C, *et al.*, RNA-Seq transcriptome analysis to identify genes involved in metabolism-based diclofop resistance in *Lolium rigidum. Plant J* **78**:865–876 (2014).

64 Duhoux A, Carrère S, Gouzy J, Bonin L, and Délye C, RNA-Seq analysis of rye-grass transcriptomic response to an herbicide inhibiting acetolactate-synthase identifies transcripts linked to non-target-site-based resistance. *Plant Mol Biol* **87**:473–487 (2015).

65 Gardin JAC, Gouzy J, Carrère S, and Délye C, ALOMYbase, a resource to investigate non-target-site-based resistance to herbicides inhibiting acetolactate-synthase (ALS) in the major grass weed *Alopecurus myosuroides* (black-grass). *BMC Genomics* **16**:590 (2015).

66 Dücker R, Parcharidou E, and Beffa R, Flufenacet activity is affected by GST inhibitors in blackgrass (*Alopecurus myosuroides*) populations with reduced flufenacet sensitivity and higher expression levels of GSTs. *Weed Sci* **68**:451–459 (2020).

67 Dücker R, Zöllner P, Parcharidou E, Ries S, Lorentz L, and Beffa R, Enhanced metabolism causes reduced flufenacet sensitivity in black-grass (*Alopecurus myosuroides* Huds.) field populations. *Pest Manag Sci* **75**:2996–3004 (2019).

68 Dücker R, Zöllner P, Lümmen P, Ries S, Collavo A, and Beffa R, Glutathione transferase plays a major role in flufenacet resistance of ryegrass (Lolium spp.) field populations. *Pest Manag Sci* **75**:3084–3092 (2019).

69 Cummins I, Wortley DJ, Sabbadin F, He Z, Coxon CR, Straker HE, *et al.*, Key role for a glutathione transferase in multiple-herbicide resistance in grass weeds. *Proc Natl Acad Sci USA* **110**:5812–5817 (2013).

70 Huang L, Yuan Z, Yu J, and Zhou T, Fundamental principles of energy consumption for gene expression. *Chaos: Interdiscip J Nonlinear Sci* **25**:123101 (2015).

71 Frumkin I, Schirman D, Rotman A, Li F, Zahavi L, Mordret E, *et al.*, Gene architectures that minimize cost of gene expression. *Mol Cell* **65**:142–153 (2017).

72 Rice AM, and McLysaght A, Dosage-sensitive genes in evolution and disease. *BMC Biol* **15**:78 (2017).

73 Wagner A, Energy constraints on the evolution of gene expression. *Mol Biol Evol* **22**:1365–1374 (2005).

74 Wu C, Davis AS, and Tranel PJ, Limited fitness costs of herbicide resistance traits in *Amaranthus tuberculatus* facilitate resistance evolution. *Pest Manag Sci* **74**:293–301 (2018).

75 Fang J, Nan P, Gu Z, Ge X, Feng Y-Q, and Lu B-R, Overexpressing exogenous 5-Enolpyruvylshikimate-3-phosphate synthase (EPSPS) genes increases fecundity and auxin content of transgenic *Arabidopsis* plants. *Front Plant Sci* **9** (2018).

76 Gressel J, Neal Stewart C, Giddings LV, Fischer AJ, Streibig JC, Burgos NR, *et al.*, Overexpression of epsps transgene in weedy rice: insufficient evidence to support speculations about biosafety. *New Phytol* **202**:360–362 (2014).

77 Lu B-R, Snow AA, Yang X, and Wang W, Scientific data published by a peer-reviewed journal should be properly interpreted: a reply to the letter by Gressel et al. (2014). *New Phytol* **202**:363–366 (2014).

78 Grunewald W, and Bury J, Comment on 'a novel 5-enolpyruvoylshikimate-3-phosphate (EPSP) synthase transgene for glyphosate resistance stimulates growth and fecundity in weedy rice (*Oryza sativa*) without herbicide' by Wang et al. (2014). *New Phytol* **202**:367–369 (2014).

79 Lu B-R, Snow AA, Yang X, and Wang W, Using a single transgenic event to infer fitness effects in crop–weed hybrids: a reply to the letter by Grunewald & Bury (2014). *New Phytol* **202**:370–372 (2014).

80 Qi Q, Toll-Riera M, Heilbron K, Preston GM, and MacLean RC, The genomic basis of adaptation to the fitness cost of rifampicin resistance in *Pseudomonas aeruginosa*. *Proc R Soc B Biol Sci* **283**:20152452 (2016).

81 Gassmann A, Resistance to herbicide and susceptibility to herbivores: environmental variation in the magnitude of an ecological trade-off. *Oecologia* **145**:575–585 (2005).

82 Gassmann AJ, and Futuyma DJ, Consequence of herbivory for the fitness cost of herbicide resistance: photosynthetic variation in the context of plant–herbivore interactions. *J Evol Biol* **18**:447–454 (2005).

83 Salzmann D, Handley RJ, and Müller-Schärer H, Functional significance of triazine-herbicide resistance in defence of *Senecio vulgaris* against a rust fungus. *Basic Appl Ecol* **9**:577–587 (2008).

84 Vila-Aiub MM, Neve P, and Powles SB, Resistance cost of a cytochrome P450 herbicide metabolism mechanism but not an ACCase target site mutation in a multiple resistant *Lolium rigidum* population. *New Phytol* **167**:787–796 (2005).

85 Bergelson J, Purrington CB, Palm CJ, and López-Gutiérrez JC, Costs of resistance: a test using transgenic *Arabidopsis thaliana*. *Proc R Soc London, Ser B: Biol Sci* **263**:1659–1663 (1996).

86 Betts SD, Basu S, Bolar J, Booth R, Chang S, Cigan AM, *et al.*, Uniform expression and relatively small position effects characterize sister Transformants in maize and soybean. *Front Plant Sci* **10**:1209–1209 (2019).

87 Pennacchio LA, Bickmore W, Dean A, Nobrega MA, and Bejerano G, Enhancers: five essential questions. *Nat Rev Genet* **14**:288–295 (2013).

88 Blommaert J, Genome size evolution: towards new model systems for old questions. *Proc R Soc B Biol Sci* **287**:20201441 (2020).

89 Zonneveld BJM, The systematic value of nuclear DNA content for all species of *narcissus* L. (Amaryllidaceae). *Plant Syst Evol* **275**:109 (2008).

90 Mei W, Stetter MG, Gates DJ, Stitzer MC, and Ross-Ibarra J, Adaptation in plant genomes: bigger is different. *Am J Bot* **105**:16–19 (2018).

91 Knight CA, and Beaulieu JM, Genome size scaling through phenotype space. *Ann Bot* **101**:759–766 (2008).

92 Knight CA, Molinari NA, and Petrov DA, The large genome constraint hypothesis: evolution, ecology and phenotype. *Ann Bot* **95**:177–190 (2005).

93 Leitch I, and Knight C, Genome size evolution in relation to leaf strategy and metabolic rates revisited. *Ann Bot* **99**:495–505 (2007).

94 Roux F, Giancola S, Durand S, and Reboud X, Building of an experimental cline with *Arabidopsis thaliana* to estimate herbicide fitness cost. *Genetics* **173**:1023–1031 (2006).

95 Bergelson J, and Purrington CB, Surveying patterns in the cost of resistance in plants. *Am Nat* **148**:536–558 (1996).

96 Neve P, Challenges for herbicide resistance evolution and management: 50 years after Harper. *Weed Res* **47**:365–369 (2007).

97 Roux F, Camilleri C, Berard A, and Reboud X, Multigenerational versus single generation studies to estimate herbicide resistance fitness cost in *Arabidopsis thaliana*. *Evol: Int J Org Evol* **59**:2264–2269 (2005).

98 Yu Q, and Powles S, Metabolism-based herbicide resistance and cross-resistance in crop weeds: a threat to herbicide sustainability and global crop production. *Plant Physiol* **166**:1106–1118 (2014).

99 Matzrafi M, Gerson O, Rubin B, and Peleg Z, Different mutations endowing resistance to acetyl-CoA carboxylase inhibitors results in changes in ecological fitness of *Lolium rigidum* populations. *Front Plant Sci* **8** (2017).

100 Vila-Aiub MM, Neve P, Steadman KJ, and Powles SB, Ecological fitness of a multiple herbicide-resistant *Lolium rigidum* population: dynamics of seed germination and seedling emergence of resistant and susceptible phenotypes. *J Appl Ecol* **42**:288–298 (2005).

101 Pedersen BP, Neve P, Andreasen C, and Powles SB, Ecological fitness of a glyphosate-resistant *Lolium rigidum* population: growth and seed production along a competition gradient. *Basic Appl Ecol* **8**:258–268 (2007).

102 Preston C, Wakelin AM, Dolman FC, Bostamam Y, and Boutsalis P, A decade of glyphosate-resistant Lolium around the world: mechanisms, genes, fitness, and agronomic management. *Weed Sci* **57**:435–441 (2009).

103 Wakelin AM and Preston C, The cost of glyphosate resistance: is there a fitness penalty associated with glyphosate resistance in annual ryegrass In

Fifteenth Australian Weeds Conference. Weed Management Society of South Australia: Adelaide, pp. 515-518 (2006).

104 Preston C, and Wakelin AM, Resistance to glyphosate from altered herbicide translocation patterns. *Pest Manag Sci* **64**:372–376 (2008).

105 Sabet Zangeneh H, Mohammaddust Chamanabad HR, Zand E, Asghari A, Alamisaeid K, Travlos IS, *et al.*, Study of fitness cost in three rigid ryegrass populations susceptible and resistant to acetyl-CoA carboxylase inhibiting herbicides. *Front Ecol Evol* **4** (2016).

106 Heap I, and Knight R, The occurrence of herbicide cross-resistance in a population of annual ryegrass, *Lolium rigidum*, resistant to diclofop-methyl. *Aust J Agric Res* **37**:149–156 (1986).

107 Tardif FJ, and Powles SB, Herbicide multiple-resistance in a *Lolium rigidum* biotype is endowed by multiple mechanisms: isolation of a subset with resistant acetyl-CoA carboxylase. *Physiol Plant* **91**:488–494 (1994).

108 Christopher JT, Powles SB, and Holtum JAM, Resistance to Acetolactate synthase-inhibiting herbicides in annual ryegrass (*Lolium rigidum*) involves at least two mechanisms. *Plant Physiol* **100**:1909–1913 (1992).

109 Holtum JA, Matthews JM, Häusler RE, Liljegren DR, and Powles SB, Cross-resistance to herbicides in annual ryegrass (*Lolium rigidum*) : III. On the mechanism of resistance to Diclofop-methyl. *Plant Physiol* **97**:1026–1034 (1991).

110 Häusler RE, Holtum JAM, and Powles SB, Cross-resistance to herbicides in annual ryegrass (*Lolium rigidum*): IV. Correlation between membrane effects and resistance to Graminicides. *Plant Physiol* **97**:1035–1043 (1991).

111 Christopher JT, Powles SB, Liljegren DR, and Holtum JA, Cross-resistance to herbicides in annual ryegrass (*Lolium rigidum*) : II. Chlorsulfuron resistance involves a wheat-like detoxification system. *Plant Physiol* **95**:1036–1043 (1991).

112 Tchouakui M, Riveron Miranda J, Mugenzi LMJ, Djonabaye D, Wondji MJ, Tchoupo M, *et al.*, Cytochrome P450 metabolic resistance (CYP6P9a) to pyrethroids imposes a fitness cost in the major African malaria vector *Anopheles funestus*. *Heredity* **124**:621–632 (2020).

113 Moss SR, Perryman SAM, and Tatnell LV, Managing herbicide-resistant blackgrass (*Alopecurus Myosuroides*): theory and practice. *Weed Technol* **21**:300–309 (2007).

114 Lutman PJW, Moss SR, Cook S, and Welham SJ, A review of the effects of crop agronomy on the management of *Alopecurus myosuroides*. *Weed Res* **53**:299–313 (2013).

115 Vizantinopoulos S, and Katranis N, Management of Blackgrass (*Alopecurus myosuroides*) in winter wheat in Greece. *Weed Technol* **12**:484–490 (1998).

116 Keshtkar E, Mathiassen SK, Agha Alikhani M, and Kudsk P, Differences in growth, development and innate seed dormancy of susceptible and fenoxaprop-P non-target site resistant black-grass sub-populations. *Crop Prot* **129**:105022 (2020).

117 Ghazali Z, Keshtkar E, Agha Alikhani M, and Kudsk P, Germinability and seed biochemical properties of susceptible and non-target-site herbicide-resistant blackgrass (*Alopecurus myosuroides*) sub-populations exposed to abiotic stresses. *Weed Sci* **1-31** (2020).

118 Keshtkar E, Mathiassen SK, Beffa R, and Kudsk P, Seed germination and seedling emergence of blackgrass (*Alopecurus myosuroides*) as affected by non–target-site herbicide resistance. *Weed Sci* **65**:732–742 (2017).

119 Darmency H, Menchari Y, Le Corre V, and Délye C, Fitness cost due to herbicide resistance may trigger genetic background evolution. *Evol: Int J Org Evol* **69**:271–278 (2015).

120 Colbach N, Chauvel B, Darmency H, Délye C, and Le Corre V, Choosing the best cropping systems to target pleiotropic effects when managing single-gene herbicide resistance in grass weeds. A blackgrass simulation study. *Pest Manag Sci* **72**:1910–1925 (2016).

121 Moss SR, and Hull R, Does herbicide resistance decline in *Alopecurus myosuroides* populations once herbicide use ceases? *Asp Appl Biol* **141**:23–32 (2018).

122 Comont D, Knight C, Crook L, Hull R, Beffa R, and Neve P, Alterations in life-history associated with non-target-site herbicide resistance in *Alopecurus myosuroides. Front Plant Sci* **10** (2019).

123 Hall LM, Moss SR, and Powles SB, Mechanism of resistance to Chlorotoluron in two biotypes of the grass weed *Alopecurus myosuroides. Pestic Biochem Physiol* **53**:180–192 (1995).

124 Menendez J, and De Prado R, Diclofop-methyl cross-resistance in a Chlorotoluron-resistant biotype of *Alopecurus myosuroides. Pestic Biochem Physiol* **56**:123–133 (1996).

125 Paris M, Roux F, Berard A, and Reboud X, The effects of the genetic background on herbicide resistance fitness cost and its associated dominance in *Arabidopsis thaliana. Heredity* **101**:499–506 (2008).

126 Cousens RD, A question of logic: experiments cannot prove lack of an herbicide-resistance fitness penalty. *Weed Sci* **68**:197–198 (2020).

13

Advancing Metabolic Resistance Characterization: Paving the Way for Future Technologies in Weed Management

Carlos Alberto Gonsiorkiewicz Rigon[1], *Roland Beffa*[2], *and Todd Gaines*[1]

[1] *Department of Agricultural Biology, Colorado State University, Fort Collins, CO, USA*
[2] *Senior Scientist Consultant, Liederbach, Hesse, Germany*

13.1 Introduction

Herbicide resistance in weeds poses a significant challenge to modern agriculture worldwide, compromising effective weed control strategies.[1] Herbicide resistance can arise through two main mechanisms: target-site resistance (TSR) and non-target-site resistance (NTSR).[2] TSR involves mutations in the DNA encoding the target-site enzyme, reducing herbicide-binding affinity, amplification, or higher transcription of the target gene.[2] NTSR, on the contrary, encompasses a diverse range of mechanisms that facilitate the detoxification, sequestration, or altered translocation of herbicides within weed plants, ultimately reducing their efficacy. Although both TSR and NTSR contribute to herbicide resistance in weeds, the metabolic mechanism stands out as the major and more complex resistance mechanism, owing to its extensive capacity to metabolize a wide range of herbicides within weed species.[3,4]

Metabolic resistance in weeds is a process involving the enzymatic detoxification and metabolism of xenobiotics, including herbicides.[4,5] This process is divided into three key phases: phase I – activation, phase II – conjugation, and phase III – sequestration (Fig. 13.1). During the activation phase, xenobiotics, such as herbicides, undergo chemical modifications to generate more reactive forms. In this phase, cytochrome P450 (P450) monooxygenases (CYPs) play a central role in metabolizing herbicides by adding a small functional group on the herbicide molecule. P450 enzymes catalyze various reactions, including hydroxylation, dealkylation, deamination, and others, which modify the chemical structure of herbicides and reduce their toxicity.[2] In phase II, other enzymes, such as glutathione-*S*-transferases (GSTs) and esterases, contribute to herbicide

Resistance in Weeds from Herbicide Metabolism, First Edition.
Edited by Vijay K. Nandula and Roland Beffa.

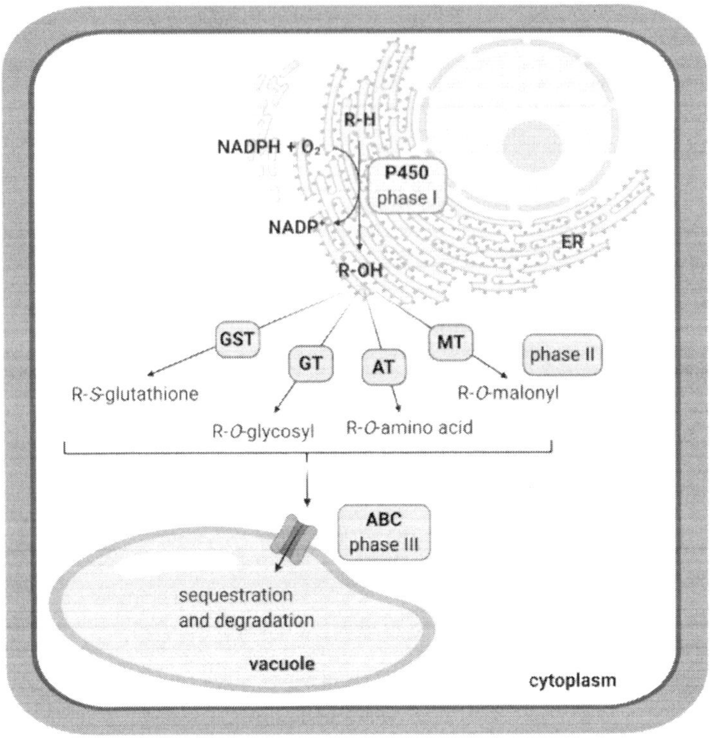

Figure 13.1 Phases of herbicide metabolic process, which herbicides can undergo when inside the plant cells. Source: Rigon et al.[3]/Ingenta. (*See insert for color representation of the figure.*)

detoxification through conjugation and hydrolysis reactions, respectively.[6] In phase III, modified herbicides or their metabolites are sequestered and compartmentalized within specific cellular compartments, mainly by ATP Binding Cassette (ABC) transporters, preventing their interaction with essential plant targets.[2,3]

In the past decade, there has been a growing recognition of the increasing importance of metabolic resistance mechanisms in weeds. The emergence and spread of herbicide-resistant weed populations have posed significant challenges to sustainable agriculture and weed management practices. While TSR has long been studied and documented, the significance of metabolic resistance mechanisms has gained prominence due to their ability to confer broad-spectrum resistance to multiple herbicides. The complexity and diversity of metabolic pathways involved in herbicide metabolism highlight the adaptability and evolutionary potential of weeds. Understanding and addressing the metabolic resistance mechanisms in weeds have become crucial for developing effective strategies to combat herbicide resistance and enable global crop production.

Our current understanding of metabolic resistance in weeds has significantly advanced in recent decades, thanks to various scientific breakthroughs discussed in the next section.

13.2 Milestones in Metabolic Resistance in Weeds

In 1957, the first herbicide-resistant weed was documented in wild carrot (*Daucus carota*) with resistance to 2,4-Dichlorophenoxyacetic acid (2,4-D)[7] (Fig. 13.2). Subsequently, additional cases of resistance emerged; however, the underlying

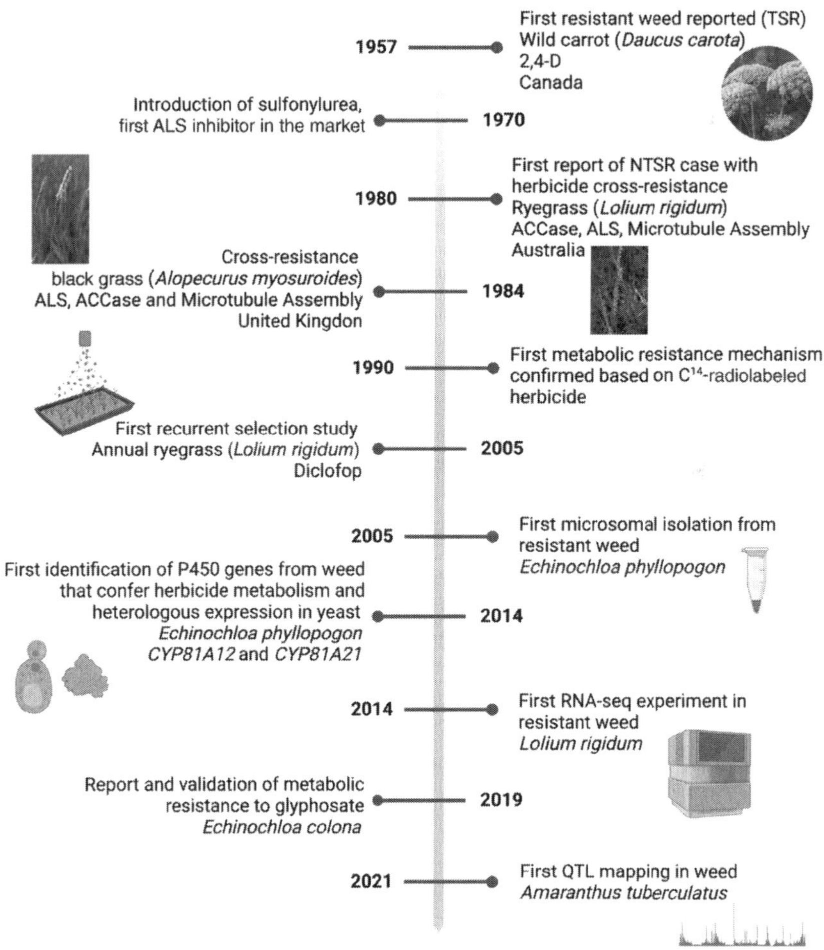

Figure 13.2 Timeline of key milestones in understanding metabolic resistance mechanisms in weeds over the past decades. Source: Patrik Stedrak/Adobe Stock Photos. (*See insert for color representation of the figure.*)

resistance mechanisms remained elusive. The introduction and widespread use of acetyl-CoA carboxylase (ACCase) and acetolactate synthase (ALS) inhibitors in the early 1970s and 1980s, respectively, resulted in a substantial increase in reported cases of herbicide resistance in weeds.[8] Besides the growing number of reported cases, many of them were not extensively investigated.

The continued use of selective herbicides in early 1980s, especially ALS inhibitors in cereals, selected weeds resistant to herbicides. The mechanisms of selectivity in tolerant crops were well known to be enhanced metabolism.[9] However, cases of weeds resistant to different herbicides started to arise. The term "cross-resistance" was not utilized in weeds but well-documented and understood in insects.[10] At that time, the first cases of weeds exhibiting cross-resistance started to evolve. Scientists concerned that resistant biotypes were not only cross-resistant to different chemical families, but in some cases, cross-resistant to herbicides that had never been used before. More than 20 years after the first resistance report, one of the first papers published assuming metabolic resistance was in a ryegrass (*Lolium rigidum*) biotype from Australia collected in 1980 (Fig. 13.2). This biotype was resistant to diclofop, fluazifop, haloxyfop, quizalofop, sethoxydim, and tralkoxydim (ACCase inhibitors) and cross-resistant to chlorsulfuron (ALS inhibitor) and trifluralin (microtubule inhibitor).[11,12] A similar resistance mechanism was identified in a black grass (*Alopecurus myosuroides*) biotype at Peldon, Essex, United Kingdom, in 1984. This biotype was resistant to chlorotoluron, and interestingly it was cross-resistant to pendimethalin and diclofop-methyl.[13]

At that time, knowledge about metabolic resistance was limited. P450 inhibitors like 1-aminobenzotriazole (ABT), piperonylbutoxide (PBO), and malathion were used to start the metabolic resistance studies. In both *Lolium* and *Alopecurus* species, the herbicides had synergistic effects with P450 inhibitors, reducing the level of resistance, indicating the involvement of metabolism resistance mechanism.[13,14] These were the first cases showing the involvement of metabolic resistance in weeds. However, during the middle 1980s and early 1990s, the use of ^{14}C herbicides and identification of the metabolites produced by resistant weeds helped to confirm the metabolic resistance mechanism and to infer which metabolic enzymes were involved based on the metabolites, such as P450s or GSTs. Resistance to chlorotoluron and isoproturon in black-grass population from Peldon was attributed to metabolism due to a faster degradation of C^{14} herbicides in the leaves to different metabolites.[13] *L. rigidum* populations were identified to be resistant to simazine (photosystem II [PS-II] inhibitor) by a faster rate of metabolism of simazine to *N*-de-ethyl derivatives,[15] probably catalyzed by P450 enzymes. Populations of *L. rigidum* and *A. myosuroides* resistant to chlorsulfuron and chlortoluron, respectively, had the same detoxification pathway as wheat, by hydroxylation of the phenyl ring followed by conjugation to glucose in *L. rigidum*,[16] or by *N*-demethylation in *A. myosuroides*[17] inferring the participation of P450 and GSTs enzymes. Populations of *Abutilon theophrasti* resistant to atrazine were found to conjugate atrazine in a faster rate than susceptible plants, showing the involvement of GST enzymes in the metabolic resistance mechanism.[18]

With the passage of time and repeated herbicide use, new cases of metabolic resistance have been identified in different species. Metabolic resistance in *Amaranthus hybridus* was reported in 1999. Resistant plants metabolized ^{14}C-chlorimuron faster than susceptible biotypes.[19] Enhanced metabolism in *Lolium multiflorum* was first reported in 2001. Resistant plants from the United Kingdom rapidly detoxified diclofop by glycosyl ester formation and were cross-resistant to tralkoxydim and cycloxydim.[20] In early 2000, a population of *Avena* spp. evolved resistance to fenoxaprop and cross-resistance to tralkoxydim.[21] *Bromus tectorum* was reported resistance to propoxycarbazone (ALS inhibitor) by a rapid rate of metabolism compared to sensitive plants and probably P450 enzymes were involved in the mechanism.[22]

Questions about the evolution of metabolic resistance in weeds were arising. It was suggested that high pesticide rates would select major genes, as TSR had been selected. However, applications of low herbicide rates were suggested to evolve polygenic herbicide resistance.[23] One notable experiment on metabolic resistance was a study using recurrent selection with low herbicide rates in *L. rigidum* (Fig. 13.2). This experiment was performed in the early 2000s with susceptible plants. After only three cycles of plant selection using low doses of diclofop-methyl, the progenies had high level of resistance by enhanced metabolism. In addition, the progenies evolved cross-resistance to fluazifop-*P*-butyl, haloxyfop-*R*-methyl, cletodim, and imazethapyr.[24] These results were a milestone on this subject and showed the rapid evolution of this type of mechanism of resistance and the consequence of using the same herbicide mode of action for consecutive years. It proved that low rates could favor the selection of minor alleles and hence stack up those alleles in the progenies, thereby evolving resistance due to a polygenic trait.

Early studies on herbicide metabolism in weeds primarily relied on *in vivo* P450 inhibitors, lacking evidence of detoxification enzymes. In the 1990s, microsome isolation from crops facilitated *in vitro* investigations of P450-mediated metabolism with various herbicides.[25] It took around 15 more years before microsome isolation from herbicide-resistant weeds was performed. In 2005, isolation of microsome from resistant *Echinochloa phyllopogon* showed that microsomes had higher P450 activity when incubated with bispyribac, fenoxaprop, and thiobencarb, which may correspond with the hydroxylation of the herbicide molecules.[26] Heterologous expression in yeast was another method used to understand P450 function and substrate affinity (Fig. 13.2). Numerous P450 genes from crops were randomly isolated, expressed in yeast, and subjected to incubation with herbicides;[25] however, due to the limited availability of isolated P450 enzymes from weeds at that time, this method was not extensively employed for weed-related studies.

Around 2010, through the isolation and identification of highly induced genes in the resistant population of *E. phyllopogon*, two significant genes, *CYP81A12* and *CYP81A21*, were discovered as the first weed genes capable of conferring herbicide metabolism (Fig. 13.2). By transforming these genes into *Arabidopsis thaliana* plants, resistance to multiple herbicides was observed, including bensulfuron-methyl through *O*-demethylation and penoxsulam.[27] The same results were observed when these genes were heterologously expressed in yeast.[27] Recently, recombinant expression in *Escherichia coli* of these *CYPs* from the 81A family has demonstrated their ability to metabolize a wide range of herbicides, including those targeting ALS, ACCase, phytoene desaturase (PDS), 4-hydroxyphenylpyruvate dioxygenase (HPPD), protoporphyrinogen oxidase (PPO), and PS-II inhibitors.[28]

At the same time, in 2014, whole transcriptome analysis was performed to compare the transcript levels between resistant and susceptible weeds aiming to identify metabolic candidate genes (Fig. 13.2). The first study was performed in *L. rigidum* resistant to diclofop.[29] Subsequently, numerous studies have utilized RNA-seq analysis to identify candidate genes responsible for metabolic resistance in various weed species, including *Lolium* spp.,[30] *Polypogon fugax*,[31] *Alopecurus aequalis*,[32] *Echinochloa crus-galli*,[33] and others.

The advanced scientific tolls of biotechnologies make it easier, but not less challenging, to validate metabolic candidate genes associated with herbicide resistance found in RNA-seq analysis using heterologous expression systems. Transcriptome studies often yield numerous upregulated genes that may be linked to herbicide resistance trait.[29,30] Nevertheless, testing and validating all of them are a labor-intensive process. In recent years, rice and *Arabidopsis* are the most common plants used for transformation. As cited before, *Arabidopsis* carrying *CYP81A12* and *CYP81A21* from *E. phyllopogon* was the first successful validation of metabolic genes from weeds. One case that was important in the history of metabolic resistance in weeds was found in an *Echinochloa colona* biotype resistant to glyphosate.[34] Glyphosate is the most commonly used herbicide in the world[35] and was not known to be metabolized by plants. This *E. colona* biotype evolved resistance to glyphosate, and TSR mechanism was not detected earlier through subsequent DNA sequencing. By RNA-seq analysis, the authors identified an aldo-keto-reductase (*AKR*) gene that was upregulated in resistant plants. Rice plants were transformed with this *AKR* from *E. colona* and became resistant to glyphosate. The mechanism of resistance is based on an oxidation reaction of glyphosate to aminomethylphosphonic acid (AMPA) performed by *AKR*. This result was a milestone because detoxification of glyphosate had not previously been demonstrated as a resistance mechanism.[34] In the same population of *E. colona*, an ABC transporter gene (*EcABCC8*) was identified in the RNA-seq analysis and transformed rice, maize, and soybean overexpressing *EcABCC8* became glyphosate-resistant. The authors used CRISPR/Cas9 to knockout the OsABCC8 in rice, and the plants became resistant to glyphosate. The mechanism of resistance is based on the movement of glyphosate molecules to extracellular space of the cell (apoplast).[36]

With advancements in technology, the identification of the underlying causal genes responsible for metabolic resistance in most weed species has become increasingly feasible. The availability of genome sequences for weed species, with the expectation of more through initiatives like the International Weed Genomics Consortium (https://www.weedgenomics.org/), has opened new possibilities for research. Previously unattainable studies such as QTL mapping can now be conducted in weeds. The first quantitative trait locus (QTL) mapping study was performed in *Amaranthus tuberculatus* with metabolic resistance to HPPD inhibitors.[37] Such studies contribute to understanding the regulation of metabolic genes, representing the next step in studying metabolic mechanisms.

Since 1980, research has been focused on the identification of metabolic resistance in grasses. Recently, metabolic resistance has been reported in various important dicot weed species. Metabolic resistance was reported in *A. tuberculatus* in 2017,[38] *Amaranthus palmeri*[39] in 2018, and *Raphanus raphanistrum* in 2020,[40] all of which exhibited resistance to HPPD inhibitors. Additionally, in 2018, *A. tuberculatus* was identified as resistant to 2,4-D through metabolic resistance.[41,42] Early efforts involved the use of P450 inhibitors, ^{14}C herbicides, and the analysis of metabolites. Several works using multiresistant populations of water hemp (*A. tuberculatus*) showed that P450s and GSTs were involved in the detoxification of several HPPD inhibitors, atrazine, and *S*-metolachlor.[43–45] Even a novel reduction-dehydration-GSH conjugation-detoxification pathway was revealed in an syncarpic acid-3 (SA3) HPPD inhibitor metabolism.[46] However, we have now entered an era of gene identification, validation, and regulation.[2,3,47] Recently, researchers have successfully identified and validated numerous genes as potential contributors to metabolic resistance (detailed in the next section). To gain further insights into the underlying mechanisms, protein modeling techniques were employed, particularly for P450, to better understand substrate affinity and identify potential substrate-binding sites. Looking ahead, we anticipate that in the coming years, advancements in our understanding will allow researchers to identify specific genes and predict their broad substrate recognition solely based on their DNA sequences. However, it is important to note that our knowledge of the genetic regulation of these resistance-related genes in plants is still incomplete. Therefore, we believe that future studies should prioritize investigating the genetic regulation of these genes to enhance our understanding of metabolic resistance. Overall, research in metabolism-based resistance has made significant strides in unraveling the complexities of metabolic resistance in both grasses and dicots. The future is optimistic about the prospects of gene identification and substrate recognition and remains committed to furthering our understanding of the genetic regulation underlying metabolic resistance in plants.

13.3 Genes Involved in Metabolic Herbicide Resistance

In the field of herbicide resistance, significant progress has been made in the identification and functional validation of numerous genes associated with herbicide metabolism in both resistant weeds and crops. This section focuses on the key genes that have been discovered in weeds and crops, demonstrating their ability to metabolize different herbicides. The identification and validation of these important genes in both weeds and crops represent significant advancements in our knowledge of metabolic resistance and provide a foundation for further research in this field. By understanding the specific gene families involved in herbicide metabolism, researchers can gain insights into the mechanisms underlying resistance and potentially develop strategies to overcome it.

The P450 gene family is widely recognized as the predominant detoxification gene family in plants.[48] *Arabidopsis* and rice have 244 and 326 P450s,[48,49] respectively, which exhibit diverse functions and play essential roles in various metabolic pathways, including the metabolism of herbicides. In weeds, the number of P450 genes are overrepresented in some species. For example, *E. crus-galli* and *E. colona* have a total of 867 and 694 P450s, respectively, which is significantly higher when compared with other species.[50] The extensive number of P450 genes in weeds highlights their metabolic versatility and potential to adapt to various environmental challenges.

P450 genes from various families have been identified and validated in both crops and weeds (Fig. 13.3). In crops like rice, *OsCYP72A31* and *OsCYP76C6* were found to metabolize ALS inhibitors such as bispyribac and bensulfuron,[51] while *OsCYP81A6* exhibited a broad spectrum of activity, metabolizing several ALS inhibitors and clomazone.[52,80] In soybean, a mutation in the *GmCYP81E22* gene was identified as the cause of high sensitivity to bentazon in specific cultivars[53] and *GmCYP71A10*, when expressed in yeast and transformed into tobacco plants, demonstrated the ability to metabolize chlorotoluron.[54] Among corn genes, *ZmCYP81A2* was found to hydroxylate bentazon at multiple positions, and *ZmCYP81A9* exhibited a wide range of activity, metabolizing HPPD, ALS, and PS-II inhibitors.[55] In weeds, P450s from

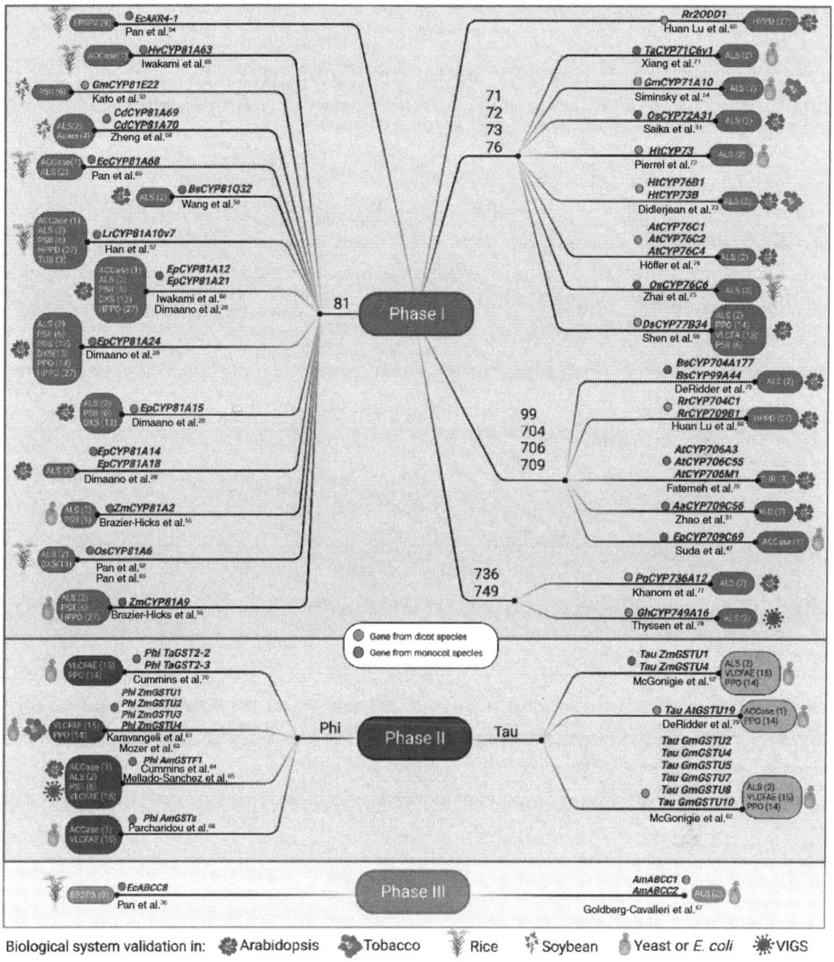

Figure 13.3 Illustration of herbicide metabolic resistance genes in weeds and their corresponding herbicide mechanisms of action (HRAC) for metabolism. Genes are separated by their function in the metabolic process as phase I (activation), phase II (conjugation), and phase III (compartmentalization and incorporation). Dark gray and light gray circles represent genes found in grass and broadleaf species, respectively. Note: VIGS – virus-induced gene silencing, Aa – *Alopecurus aequalis*, Am – *Alopecurus myosuroides*, At – *Arabidopsis thaliana*, Bs – *Beckmannia syzigachne*, Cd – *Cynodon dactylon*, Ds – *Descurainia sophia*, Ep – *Echinochloa phyllopogon*, Ec – *Echinochloa colona*, Gh – *Gossypium hirsutum*, Gm – *Glycine max*, Ht – *Helianthus tuberosus*, Hv – *Hordeum vulgare*, Lr – *Lolium rigidum*, Os – *Oryza sativa*, Pg – *Panax ginseng*, Rr – *Raphanus raphanistrum*, Ta – *Triticum aestivum*, Zm – *Zea mays*. Genes: *AKR* – aldo-keto reductase, *CYP* – cytochrome P450 monooxygenase, *GST* – glutathione-*S*-transferase, *ABC* – ATP-binding cassette. (*See insert for color representation of the figure.*)

multiple herbicide-resistant *E. phyllopogon* (e.g., *EpCYP81A12*, *EpCYP81A21*, and *EpCYP81A24*) metabolized a wide range of substrates, including ACCase, ALS, PS-II, 1-deoxy-D-xylulose-5-phosphate synthase (DXS), PDS, and HPPD inhibitors.[27,28] *EpCYP81A14* and *EpCYP81A18* shared high similarity with the previous enzymes but exhibited different substrate specificity, metabolizing only ALS inhibitors.[28] Similarly, *Cynodon dactylon* genes *CdCYP81A69* and *CdCYP81A70*, from the same family, when transformed into soybean, conferred resistance to ALS inhibitors and 2,4-D.[56] Another gene with a broad spectrum of substrates was found in multiple herbicide-resistant annual ryegrass, where transgenic rice overexpressing *Lolium LrCYP81A10v7* exhibited high resistance to ACCase and ALS, and moderate resistance to PS-II, HPPD, and tubulin inhibitors.[57] Genes from *Beckmannia syzigachne* including *BsCYP81Q32*, *BsCYP704A177*, and *BsCYP99A44* when transformed in *Arabidopsis* plants confer resistance specific to ALS inhibitor herbicides.[58,81] In dicot weed species, fewer P450 genes have been identified. For instance, *DsCYP77B34* from *Descurainia sophia*, a broadleaf weed in Asian wheat fields, rendered transformed *Arabidopsis* resistant to ALS, PPO, very-long-chain fatty acid (VLCFA), and PS-II inhibitors.[59] *Arabidopsis* expressing *R. raphanistrum RrCYP704C1* or *RrCYP709B1* demonstrated resistance to HPPD inhibitors such as mesotrione (*RrCYP704C1*), tembotrione (*RrCYP709B1*), and isoxaflutole (*RrCYP709B1*).[60] The evolution of multiple genes in grassweed species can be attributed to prolonged exposure to herbicide modes of action, particularly ACCase and ALS inhibitors, over the years. It is evident that the P450 family 81 plays a pivotal role in grasses, while other families are found in both grasses and dicots (Fig. 13.3).

Fewer genes have been identified and validated for phases II and III of herbicide metabolism, with significant research focused on corn, soybean, and wheat (Fig. 13.3). In these crops, efforts have been made to isolate and study *GSTs*. Expression of various *Phi* and *Tau ZmGSTUs*, *Phi TaGSTs*, and *Tau* class *GmGSTUs* in *Escherichia coli* demonstrated some degree of conjugation to ALS, PPO, and VLCFA inhibitors.[61–63] Among weeds, *AmGSTF1* from *A. myosuroides* has been first identified and validated. This black-grass population exhibited multiple herbicide resistance, and transgenic *Arabidopsis* overexpressing *AmGSTF1* displayed resistance to chlortoluron (ALS), alachlor (VLCFA), and atrazine (PSII), the same herbicides to which the resistant black-grass plants were resistant.[64,65] Knocking out *AmGSTF1* through virus-mediated transient expression in resistant black-grass plants restored sensitivity to fenoxaprop (ACCase), confirming the essential role of *AmGSTF1* in herbicide resistance.[61] The sequencing of black grass[82,83] offers the opportunity to further study the role of GSTs in multiple-resistant populations to pre- and postemergence herbicides.[66] Several GSTs were overexpressed in black-grass-resistant populations to the VLCFA synthesis flufenacet, and ALS and ACCase inhibitors. In addition to

AmGSTF1, four other GSTs showed activity in the detoxification of flufenacet, acetochlor, and fenoxaprop acid but have no activity on metolachlor or clodinafop acid.[66] Furthermore, one GST showed significant activity in the detoxification of pyroxasulfone, an herbicide to which this black-grass population had never been exposed.[66] These results suggest that different GSTs can act together to increase the resistance strength toward certain herbicides and/or to broaden the resistance spectrum of a given population. In addition, it showed that metabolic resistance is not related to a given herbicide mode of action but to particular chemistries. The genes associated with phase III, such as ABC transporters, which sequester herbicides into the vacuole, have received limited attention as mechanisms of resistance. In glyphosate-resistant *E. colona*, it was found that *ABCC8* was constitutively expressed at higher levels in resistant plants. Transgenic rice expressing this gene exhibited resistance to glyphosate by extruding the herbicide from the cytoplasm to the apoplast.[36] Similarly, three ABC transporters (*AmABCC1*, *AmABCC2*, and *AmABCC3*) were recently identified in multiple-resistant populations of *A. myosuroides*. Transgenic yeast carrying *AmABCC1* or *AmABBC2* had enhanced tolerance to mesosulfuron-methyl (ALS), highlighting the role of these proteins in sequestering the herbicide into the vacuole in resistant black grass.[67]

Ongoing research advancements are expected to uncover more genes, especially in dicots, benefiting from advancements in technology and genome availability. Comprehending their functions and correlation will aid in combating herbicide resistance evolution. Armed with this understanding, we can devise focused approaches to efficiently manage weeds and impede resistance development. Through the utilization of advanced tools and scientific methodologies, we are steadily advancing toward sustainable agriculture and improved weed control.

13.4 International Weed Genomics Consortium

The International Weed Genomics Consortium (IWGC) is a collaboration among academic, government, and industry researchers focused on producing genomic tools for weedy species around the world.[84] The IWGC is providing chromosome-level reference genome assemblies for at least 50 important weedy species across the globe. Each genome includes gene annotations and all are publicly available after a one-year confidentiality embargo. Species were selected for genome assembly based on member input, economic impact, and global prevalence. Future activities in the IWGC will supplement these reference genomes with additional analytical tools to increase their utility.

13.4.1 Reference Genomes and Data Analysis Tools

The first objective of the IWGC is to provide high-quality genomic resources for agriculturally important weeds. The IWGC maintains a general website with information on weed genome availability; genomic details including genome size and ploidy; links to herbicide resistance status; and information on upcoming training events. The second main resource is an online genome database called WeedPedia that hosts IWGC-generated and publicly accessible genomic data as well as various bioinformatics tools. WeedPedia is a cloud-based omics database management platform built from the "CropPedia" software, and licensed from KeyGene (Wageningen, The Netherlands). The interface allows users to access, visualize, and download genome assemblies along with structural and functional annotations. Additionally, all genome assemblies and annotations produced through the IWGC are uploaded to NCBI GenBank upon release.

13.4.2 Training and Collaboration in Weed Genomics

The second objective of the IWGC is to build capacity for scientists to utilize genomics in weed science, both through training for current weed scientists and by encouraging scientists from outside weed science to engage with the field. To accomplish these training goals, the IWGC delivers programs aimed at the full range of IWGC stakeholders and covering a breadth of relevant topics. We have taken care to ensure that our approaches are diverse as to provide training to researchers with all levels of existing experience and differing reasons for engaging with these tools. Throughout, the focus is on ensuring that our training and outreach result in impacts that benefit a wide range of stakeholders. Recordings of training sessions are available at the IWGC website (www.weedgenomics.org).

13.5 Functional Genomics

Functional genomics is playing a crucial role in agronomic weed science, particularly in understanding evolution of different resistance mechanisms. With the advancements in high-throughput sequencing technologies, scientists can now obtain complete genomes of weed species more efficiently. These new genomic resources will provide a wealth of information about the genetic variations and functional elements in weed populations, enabling researchers to identify key genes involved in metabolic resistance and develop targeted management approaches.

Scientists employ various techniques to study functional genomics in the context of weed science. One commonly used approach is transcriptomics, which involves analyzing the expression patterns of genes in different weed populations, including both resistant and susceptible biotypes. This helps identify genes that are upregulated or downregulated in resistant weeds, providing insights into the metabolic pathways and processes involved in herbicide resistance.[29,32] Another technique is proteomics, which involves studying the complete set of proteins produced by weed species.[85] Proteomics can shed light on the proteins involved in herbicide metabolism and target-site interactions, contributing to a better understanding of resistance mechanisms. Additionally, metabolomics, the study of small molecule metabolites, allows researchers to explore the metabolic changes associated with resistance.[86] By applying these techniques, scientists can gather comprehensive data on the functional elements of weed genomes and their interactions with herbicides.

Gene validation techniques such as *Arabidopsis* or tobacco transformation, virus-induced gene silencing (VIGS), and mutant databases are valuable tools for determining gene functions involved in the evolution of herbicide resistance in weeds as described in Section 13.2. Furthermore, recent advancements in gene editing technologies such as RNA interference (RNAi) and CRISPR have revolutionized our ability to understand and address herbicide resistance in weeds.[87,88] RNAi allows for targeted gene knockdown, enabling researchers to assess the impact of specific genes on resistance phenotypes.[88] CRISPR, on the contrary, enables precise gene editing, facilitating the study of gene functions and the discovery of new targets for sustainable weed management. These technologies hold great promise for both unraveling the mechanisms behind metabolic resistance evolution and developing innovative strategies for weed control.

13.6 Conclusions and Perspectives

Herbicide resistance in weeds poses a global challenge to agriculture, necessitating a deep understanding of resistance mechanisms, especially metabolic resistance, for effective weed management. The future holds promising prospects for gene identification, substrate recognition, and the prediction of broad substrate recognition based on DNA sequences. Advancements in technology and the availability of genomic databases are expected to expedite gene discovery, particularly in dicot weed species. Investigating the genetic regulation of resistance-related genes should be a priority to enhance our comprehension of metabolic resistance.

Continuous advancements in metabolic research provide valuable insights into the complexities of metabolic resistance in weeds. Combined with an improved understanding of genetic regulation, this knowledge will aid in combating the evolution of herbicide resistance and developing targeted strategies for effective weed management. By leveraging advanced tools and scientific methodologies, we are making strides toward achieving sustainable agriculture and enhanced weed control practices.

In addition, the public–private partnership model, exemplified by the IWGC, plays a crucial role in advancing weed science. Through efforts such as hosting comprehensive databases, integrating new datasets specific for metabolic resistance, providing training, and facilitating research coordination, IWGC promotes the sharing of advancements in metabolic resistance research across the scientific community. Utilizing this collaborative approach will further enhance our understanding of metabolic resistance and contribute to improved weed management strategies.

By embracing these perspectives and leveraging the power of interdisciplinary collaboration, we can effectively address the challenges posed by herbicide resistance in weeds, ensuring sustainable agriculture and enhanced weed control practices for the future.

References

1 Peterson, M.A., Collavo, A., Ovejero, R. et al. (2018). The challenge of herbicide resistance around the world: a current summary. *Pest Manag Sci* **10**: 2246–2259.

2 Gaines, T.A., Duke, S.O., Morran, S. et al. (2020). Mechanisms of evolved herbicide resistance. *J Biol Chem* **30**: 10307–10330.

3 Rigon, C.A.G., Gaines, T.A., Küpper, A., and Dayan, F.E. (2020). Metabolism-based herbicide resistance, the major threat among the non-target site resistance mechanisms. *Outlooks on Pest Management* **4**: 162–168.

4 Yu, Q. and Powles, S. (2014). Metabolism-based herbicide resistance and cross-resistance in crop weeds: a threat to herbicide sustainability and global crop production. *Plant Physiol* **3**: 1106–1118.

5 Yuan, J.S., Tranel, P.J., and Stewart, C.N. (2007). Non-target-site herbicide resistance: a family business. *Trends Plant Sci* **1**: 6–13.

6 Edwards, R. and Dixon, D.P. (2005). Plant glutathione transferases. *Methods Enzymol* **401**: 169–186.

7 Switzer, C.M. (1957). The existence of 2,4-D resistant strains of wild carrot. In: *Proceeding of the 11th Northeast. Weed Conference*, 315–318.

8 Heap I. The International Survey of Herbicide Resistant Weeds http://www.weedscience.com [Accessed Jan 10, 2023] (2023).

9 Sweetser, P.B., Schow, G.S., and Hutchison, J.M. (1982). Metabolism of chlorsulfuron by plants: biological basis for selectivity of a new herbicide for cereals. *Pestic Biochem Physiol* **17**: 18–23.

10 Georghiou, G. (1986). Current problems, trends, and developments in pesticide metabolism in plants. In: *Pesticide resistance: strategies and tactics for management*, 471. Washington, DC: National Academy Press.

11 Heap, I. and Knight, R. (1982). A population of ryegrass tolerant to the herbicide diclofop-methyl [in South Australia]. *J Aust Inst Agric Sci* **48**: 156–157.

12 Heap, I. and Knight, R. (1986). The occurrence of herbicide cross-resistance in a population of annual ryegrass, *Lolium rigidum*, resistant to diclofop-methyl. *Aust J Agric Res* **349**: 149–156.

13 Kemp, M.S., Moss, S.R., and Thomas, T.H. (1990). Herbicide resistance in *Alopecurus myosuroides*. In: *Managing resistance to agrochemicals: from fundamentals research to practival strategies*, ACS Symposium Series, 1ste (ed. M.B. Green, H.M. LeBaron, and W.K. Moberg), 498. Washington, DC: American Chemical Society.

14 Powles, S.B., Holtum, J.A.M., Matthews, J.M., and Liljegren, D.R. (1990). Herbicide cross-resistance in annual ryegrass *Lolium rigidum Gaud*. In: *Managing resistance to agrochemicals*, ACS Symposium Series, vol. **421** (ed. I. Denholm, A.L. Devonshire, and D.W. Hollomon), 394–406. American Chemical Society.

15 Burnet, M.W.M., Loveys, B.R., Holtum, J.A.M., and Powles, S.B. (1993). Increased detoxification is a mechanism of simazine resistance in *Lolium rigidum*. *Pestic Biochem Physiol* **46**: 207–218.

16 Christopher, J.T., Powles, S.B., Liljegren, D.R., and Holtum, J.A. (1991). Cross-resistance to herbicides in annual ryegrass (*Lolium rigidum*) : II. Chlorsulfuron resistance involves a wheat-like detoxification system. *Plant Physiol* **95**: 1036–1043.

17 Hall, L.M., Moss, S.R., and Powles, S.B. (1995). Mechanism of resistance to chlorotoluron in two biotypes of the grass weed *Alopecurus myosuroides*. *Pestic Biochem Physiol* **53**: 180–192.

18 Gronwald, J.W., Andersen, R.N., and Yee, C. (1989). Atrazine resistance in velvetleaf (*Abutilon theophrasti*) due to enhanced atrazine detoxification. *Pestic Biochem Physiol* **34**: 149–163.

19 Manley, B.S., Hatzios, K.K., and Wilson, H.P. (1999). Absorption, translocation, and metabolism of chlorimuron and nicosulfuron in imidazolinone-resistant and -susceptible smooth pigweed (*Amaranthus hybridus*). *Weed Technol* **13**: 759–764.

20 Cocker, K.M., Northcroft, D.S., Coleman, J.O., and Moss, S.R. (2001). Resistance to ACCase-inhibiting herbicides and isoproturon in UK populations of Lolium multiflorum: mechanisms of resistance and implications for control. *Pest Manag Sci* **57**: 587–597.

21 Cocker, K.M., Coleman, J.O.D., Blair, A.M. et al. (2000). Biochemical mechanisms of cross-resistance to aryloxyphenoxypropionate and cyclohexanedione herbicides in populations of Avena spp. *Weed Res* **40**: 323–334.

22 Park, K.W., Fandrich, L., and Mallory-Smith, C.A. (2004). Absorption, translocation, and metabolism of propoxycarbazone-sodium in ALS-inhibitor resistant Bromus tectorum biotypes. *Pestic Biochem Physiol* **79**: 18–24.

23 Georghiou G, Taylor C. Factors influencing the evolution of resistance. In: Metcalf RL, WH L (Eds), *Pesticide resistance*. Washington, DC: National Academy Press; 1986. p. 157–169.

24 Neve, P. and Powles, S. (2005). Recurrent selection with reduced herbicide rates results in the rapid evolution of herbicide resistance in *Lolium rigidum*. *Theor Appl Genet* **110**: 1154–1166.

25 Siminsky, B. (2006). Plant cytochrome P450-mediated herbicide metabolism. *Phytochem Rev* **2–3**: 445–448.

26 Yun, M.S., Yogo, Y., Miura, R. et al. (2005). Cytochrome P-450 monooxygenase activity in herbicide-resistant and-susceptible late watergrass (*Echinochloa phyllopogon*). *Pestic Biochem Physiol* **83**: 107–114.

27 Iwakami, S., Endo, M., Saika, H. et al. (2014). Cytochrome P450 *CYP81A12* and *CYP81A21* are associated with resistance to two acetolactate synthase inhibitors in *Echinochloa phyllopogon*. *Plant Physiol* **165**: 618–629.

28 Dimaano, N.G., Yamaguchi, T., Fukunishi, K. et al. (2020). Functional characterization of cytochrome P450 *CYP81A* subfamily to disclose the pattern of cross-resistance in *Echinochloa phyllopogon*. *Plant Mol Biol* **102**: 403–416.

29 Gaines, T.A., Lorentz, L., Figge, A. et al. (2014). RNA-Seq transcriptome analysis to identify genes involved in metabolism-based diclofop resistance in *Lolium rigidum*. *Plant J* **78**: 865–876.

30 Duhoux, A., Carrère, S., Gouzy, J. et al. (2015). RNA-Seq analysis of rye-grass transcriptomic response to an herbicide inhibiting acetolactate-synthase identifies transcripts linked to non-target-site-based resistance. *Plant Mol Biol* **87**: 473–487.

31 Zhao, N., Yang, J., Jiang, M. et al. (2022). Identification of essential genes involved in metabolism-based resistance mechanism to fenoxaprop-P-ethyl in Polypogon fugax. *Pest Manag Sci* **78**: 1164–1175.

32 Zhao, N., Li, W., Bai, S. et al. (2017). Transcriptome profiling to identify genes involved in mesosulfuron-methyl resistance in *Alopecurus aequalis*. *Front Plant Sci* **8**.

33 Fang, J., Zhang, Y., Liu, T. et al. (2019). Target-Site and metabolic resistance mechanisms to penoxsulam in barnyardgrass (*Echinochloa crus-galli* (L.) P. Beauv). *J Agric Food Chem* **67**: 8085–8095.

34 Pan, L., Yu, Q., Han, H.P. et al. (2019). Aldo-keto reductase metabolizes glyphosate and confers glyphosate resistance in *Echinochloa colona*. *Plant Physiol* **181**: 1519–1534.

35 Duke, S.O., Powles, S.B., and Sammons, R.D. (2018). Glyphosate - How it became a once in a hundred year herbicide and its future. *Outlooks on Pest Management* **29**: 247–251.

36 Pan, L., Yu, Q., Wang, J. et al. (2021). An ABCC-type transporter endowing glyphosate resistance in plants. *Proc Natl Acad Sci* **118**: e2100136118.

37 Murphy, B.P., Beffa, R., and Tranel, P.J. (2021). Genetic architecture underlying HPPD-inhibitor resistance in a Nebraska *Amaranthus tuberculatus* population. *Pest Manag Sci* **77**: 4884–4891.

38 Kaundun, S.S., Hutchings, S.J., Dale, R.P. et al. (2017). Mechanism of resistance to mesotrione in an *Amaranthus tuberculatus* population from Nebraska, USA. *PLoS One* **12**: 22.

39 Küpper, A., Peter, F., Zollner, P. et al. (2018). Tembotrione detoxification in 4-hydroxyphenylpyruvate dioxygenase (HPPD) inhibitor-resistant Palmer amaranth (*Amaranthus palmeri* S. Wats.). *Pest Manag Sci* **74**: 2325–2334.

40 Lu, H., Yu, Q., Han, H. et al. (2020). Evolution of resistance to HPPD-inhibiting herbicides in a wild radish population via enhanced herbicide metabolism. *Pest Manag Sci* **76**: 1929–1937.

41 Figueiredo, M.R., Leibhart, L.J., Reicher, Z.J. et al. (2018). Metabolism of 2,4-dichlorophenoxyacetic acid contributes to resistance in a common waterhemp (*Amaranthus tuberculatus*) population. *Pest Manag Sci* 2356–2362.

42 Figueiredo, M.R.A., Barnes, H., Boot, C.M. et al. (2022). Identification of a novel 2,4-D metabolic detoxification pathway in 2,4-D-resistant waterhemp (*Amaranthus tuberculatus*). *J Agric Food Chem* **70**: 15380–15389.

43 Ma, R., Kaundun, S.S., Tranel, P.J. et al. (2013). Distinct detoxification mechanisms confer resistance to mesotrione and atrazine in a population of waterhemp. *Plant Physiol* **163**: 363–377.

44 Evans, A.F. Jr., O'Brien, S.R., Ma, R. et al. (2017). Biochemical characterization of metabolism-based atrazine resistance in *Amaranthus tuberculatus* and identification of an expressed GST associated with resistance. *Plant Biotechnol J* **15**: 1238–1249.

45 Strom, S.A., Hager, A.G., Concepcion, J.C.T. et al. (2021). Metabolic Pathways for S-metolachlor detoxification differ between tolerant corn and multiple-resistant waterhemp. *Plant Cell Physiol* **62**: 1770–1785.

46 Concepcion, J.C.T., Kaundun, S.S., Morris, J.A. et al. (2021). Resistance to a nonselective 4-hydroxyphenylpyruvate dioxygenase-inhibiting herbicide via novel reduction-dehydration-glutathione conjugation in *Amaranthus tuberculatus*. *New Phytol* **232**: 2089–2105.

47 Suda, H., Kubo, T., Yoshimoto, Y. et al. (2023). Transcriptionally linked simultaneous overexpression of P450 genes for broad-spectrum herbicide resistance. *Plant Physiol* **192**: 3017–3029.

48 Werck-Reichhart, D., Bak, S., and Paquette, S. (2002). Cytochromes p450. *Arabidopsis Book* e0028.

49 Wei, K. and Chen, H. (2018). Global identification, structural analysis and expression characterization of cytochrome P450 monooxygenase superfamily in rice. *BMC Genomics* **19**: 35.

50 Wu, D., Shen, E., Jiang, B. et al. (2022). Genomic insights into the evolution of *Echinochloa* species as weed and orphan crop. *Nat Commun* **13**: 689.

51 Saika, H., Horita, J., Taguchi-Shiobara, F. et al. (2014). A novel Rrce cytochrome P450 gene, CYP72A31, confers tolerance to acetolactate synthase-inhibiting herbicides in rice and arabidopsis. *Plant Physiol* **166**: 1232–1240.

52 Pan, G., Zhang, X.Y., Liu, K.D. et al. (2006). Map-based cloning of a novel rice cytochrome P450 gene *CYP81A6* that confers resistance to two different classes of herbicides. *Plant Mol Biol* **61**: 933–943.

53 Kato, S., Yokota, Y., Suzuki, R. et al. (2020). Identification of a cytochrome P450 hydroxylase, *CYP81E22*, as a causative gene for the high sensitivity of soybean to herbicide bentazon. *Theor Appl Genet* **133**: 2105–2115.

54 Siminszky, B., Corbin, F.T., Ward, E.R. et al. (1999). Expression of a soybean cytochrome P450 monooxygenase cDNA in yeast and tobacco enhances the metabolism of phenylurea herbicides. *Proc Natl Acad Sci USA* **96**: 1750–1755.

55 Brazier-Hicks, M., Franco-Ortega, S., Watson, P. et al. (2022). Characterization of cytochrome P450s with key roles in determining herbicide selectivity in maize. *ACS Omega* .

56 Zheng, T., Yu, X., Sun, Y. et al. (2022). Expression of a cytochrome P450 gene from Bermuda grass *Cynodon dactylon* in soybean confers tolerance to multiple herbicides. *Plants* **11**: 949.

57 Han, H., Yu, Q., Beffa, R. et al. (2021). Cytochrome P450 *CYP81A10v7* in *Lolium rigidum* confers metabolic resistance to herbicides across at least five modes of action. *Plant J* **105**: 79–92.

58 Wang, J., Lian, L., Qi, J. et al. (2023). Metabolic resistance to acetolactate synthase inhibitors in *Beckmannia syzigachne*: identification of *CYP81Q32* and its transcription regulation. *Plant J* **115**: 317–334.

59 Shen, J., Yang, Q., Hao, L. et al. (2022). The metabolism of a novel cytochrome P450 (*CYP77B34*) in tribenuron-methyl-resistant *Descurainia sophia* L. to herbicides with different mode of actions. *Int J Mol Sci* **23**: 5812.

60 Lu, H., Liu, Y., Li, M. et al. (2023). Multiple metabolic enzymes can be involved in cross-resistance to 4-hydroxyphenylpyruvate-dioxygenase-inhibiting herbicides in wild radish. *J Agric Food Chem* **74**: 9302–9313.

61 Mozer, T.J., Tiemeier, D.C., and Jaworski, E.G. (1983). Purification and characterization of corn glutathione S-transferase. *Biochemistry* **22**: 1068–1072.

62 McGonigle, B., Keeler, S.J., Lau, S.M. et al. (2000). A genomics approach to the comprehensive analysis of the glutathione S-transferase gene family in soybean and maize. *Plant Physiol* **124**: 1105–1120.

63 Karavangeli, M., Labrou, N.E., Clonis, Y.D., and Tsaftaris, A. (2005). Development of transgenic tobacco plants overexpressing maize glutathione *S*-transferase I for chloroacetanilide herbicides phytoremediation. *Biomol Eng* **22**: 121–128.

64 Cummins, I., Wortley, D.J., Sabbadin, F. et al. (2013). Key role for a glutathione transferase in multiple-herbicide resistance in grass weeds. *Proc Natl Acad Sci* **110**: 5812–5817.

65 Mellado-Sánchez M, McDiarmid F, Cardoso V, Kanyuka K, MacGregor DR. Virus-mediated transient expression techniques enable genetic modification of *Alopecurus myosuroides*. *bioRxiv* DOI: https://doi.org/10.1101/2020.01.28.923466 2020.

66 Parcharidou, E., Dücker, R., Zöllner, P. et al. (2023). Recombinant glutathione transferases from flufenacet-resistant black-grass *(Alopecurus myosuroides* Huds.) form different flufenacet metabolites and differ in their interaction with pre- and post-emergence herbicides. *Pest Manag Sci* **79**: 3376–3386.

67 Goldberg-Cavalleri, A., Onkokesung, N., Franco-Ortega, S., and Edwards, R. (2023). ABC transporters linked to multiple herbicide resistance in blackgrass (*Alopecurus myosuroides*). Front Plant Sci **14**: 1082761.

68 Iwakami, S., Kamidate, Y., Yamaguchi, T. et al. (2019). CYP81A P450s are involved in concomitant cross-resistance to acetolactate synthase and acetyl-CoA carboxylase herbicides in Echinochloa phyllopogon. *New Phytol.* **221**: 2112–2122.

69 Pan, L., Guo, Q., Wang, J. et al. (2022). CYP81A68 confers metabolic resistance to ALS and ACCase-inhibiting herbicides and its epigenetic regulation in Echinochloa crus-galli. *J. Hazardous Mat.* **428**: 128225.

70 Cummins, I., O'Hagan, D., Jablonkai, I. et al. (2002). Cloning, characterization, and regulation of a family of phi class glutathione transferases from wheat. *Plant Mol Biol* **52**: 591–603.

71 Xiang, W., Wang, X., and Ren, T. (2006). Expression of a wheat cytochrome P450 monooxygenase cDNA in yeast catalyzes the metabolism of sulfonylurea herbicides. *Pest. Biochem. Physiol.* **85**: 1–6.

72 Pierrel, M.A., Batard, Y., Kazmaier, M. et al. (1994). Catalytic properties of the plant cytochrome P450 CYP73 expressed in yeast. Substrate specificity of a Cinnamate hydroxylase. *FEBS* **224**: 835–844.

73 Didierjean, L., Gondet, L., Perkins, R. et al. (2002). Engineering herbicide metabolism in tobacco and Arabidopsis with CYP76B1, a cytochrome P450 enzyme from Jerusalem artichoke. *Plant Physsiol.* **130**: 179–189.

74 Höfer, R., Boachon, B., Renault, R. et al. (2014). Dual function of the cytochrome P450 CYP76 family from Arabidopsis thaliana in the metabolism of monoterpenols and phenylurea herbicides. *Plant Physiol.* **166**: 1149–1161.

75 Zhai, X.Y., Chen, Z.J., Liu, J. et al. (2022). Expression of CYP76C6 facilitates isoproturon metabolism and detoxification in rice. *J. Agric. Food Chem.* **70**: 4599–4610.

76 Fatemeh, A., Alebrahim, M., Ngov, C. et al. (2020). Innate promiscuity of the CYP706 family of P450 enzymes provides a suitable context for the evolution of dinitroaniline resistance in weed. *New Phytol.* **229**: 3253–3268.

77 Khanom, S., Jang, J., and Lee, O.R. (2019). Overexpression of ginseng cytochrome P450 CYP736A12 alters plant growth and confers phenylurea herbicide tolerance in Arabidopsis. *J. Ginseng Res.* **43**: 645–653.

78 Thyssen, G.N., Naoumkina, M., McCarty, J.C. et al. (2018). The P450 gene CYP749A16 is required for tolerance to the sulfonylurea herbicide trifloxysulfuron sodium in cotton (Gossypium hirsutum L.). *BMC Plant Biol.* **18** 186.

79 DeRidder, B.P., Dixon, D.P., Beussman, D.J. et al. (2002). Induction of glutathione S-transferases in Arabidopsis by herbicide safeners get access arrow. *Plant Physiol.* **130**: 1497–1505.

80 Guo, F., Endo, M., Yamaguchi, T. et al. (2021). Investigation of clomazone-tolerance mechanism in a long-grain cultivar of rice. *Pest Manag Sci* **77**: 2454–2461.

81 Bai, S., Yin, M., Lyu, Q. et al. (2022). Cytochrome P450 *BsCYP99A44* and *BsCYP704A177* confer metabolic resistance to ALS herbicides in *Beckmannia syzigachne. Int J Mol Sci* **23**: 12175.

82 Kersten, S., Chang, J., Huber, C.D. et al. (2023). Standing genetic variation fuels rapid evolution of herbicide resistance in blackgrass. *Proc Natl Acad Sci* **120**: e2206808120.

83 Cai, L., Comont, D., MacGregor, D. et al. (2023). The blackgrass genome reveals patterns of non-parallel evolution of polygenic herbicide resistance. *New Phytol* **237**: 1891–1907.

84 Montgomery JS, Morran S, MacGregor DR, McElroy JS, Neve P, Neto C, et al. The international weed genomics consortium: community resources for weed genomics research. *bioRxiv* DOI:https://doi.org/10.1101/2023.07.19.549613 2023.

85 González-Torralva, F., Brown, A.P., and Chivasa, S. (2017). Comparative proteomic analysis of horseweed (*Conyza canadensis*) biotypes identifies candidate proteins for glyphosate resistance. Sci Rep **7**: 42565.

86 Hikosaka, M., Iwahashi, F., and Yamato, S. (2021). Metabolomic analysis of Schoenoplectus juncoides reveals common markers of acetolactate synthase inhibition among paddy weeds. Pestic Biochem Physiol **174**: 104827.

87 Dong, H., Huang, Y., and Wang, K. (2021). The development of herbicide resistance crop plants using CRISPR/Cas9-mediated gene editing. Gene (Basel) **12**: 912. https://doi.org/10.3390/genes12060912.

88 Zabala-Pardo D, Gaines T, Lamego FP, Avila LA. RNAi as a tool for weed management: challenges and opportunities. Adv Weed Sci. **40(Spec1)**: e020220096.

Index

Resistance in Weeds from Herbicide Metabolism, First Edition.
Edited by Vijay K. Nandula and Roland Beffa.

f

g

h

i

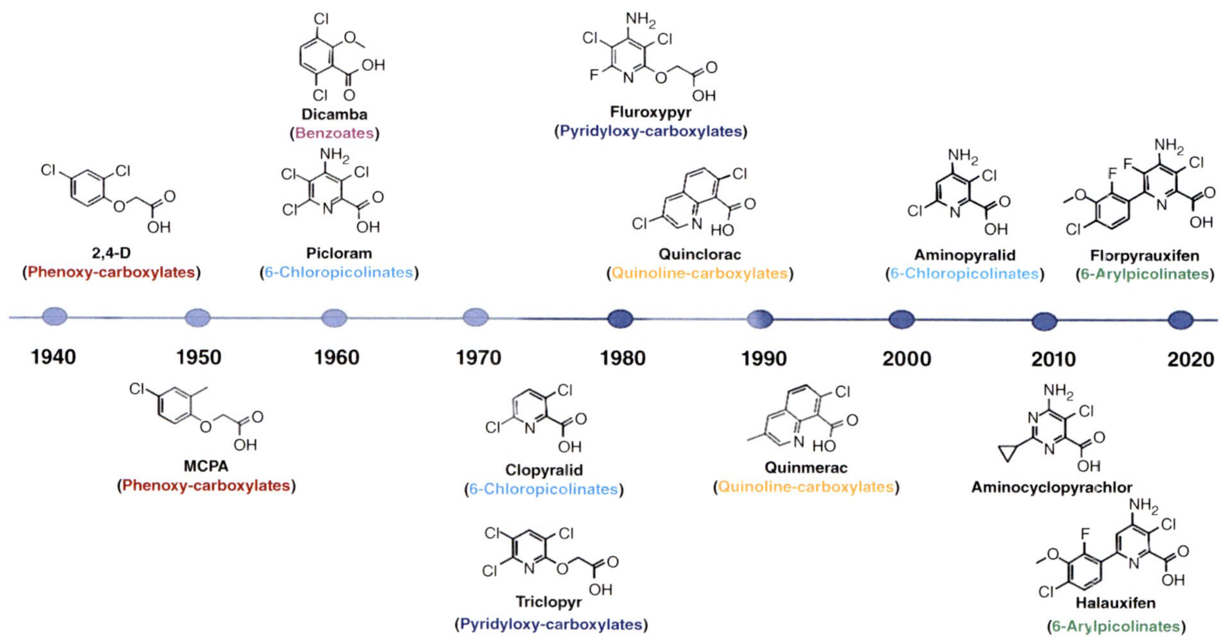

Figure 2.2 Chronology of discovery of representative commercial synthetic auxin herbicides (active moiety is shown with chemical family name in parenthesis).

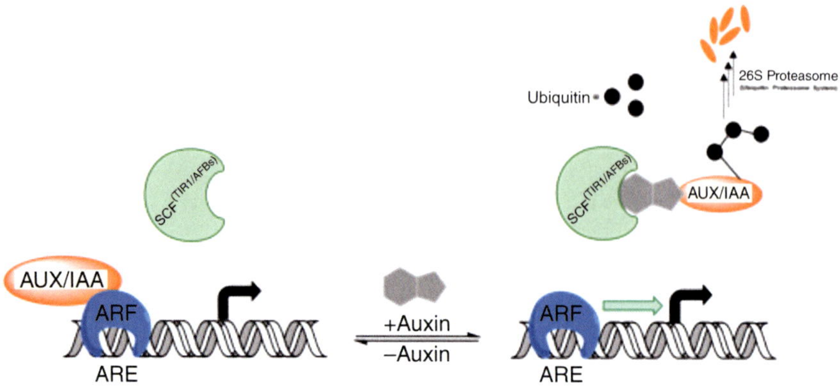

Figure 2.3 Canonical auxin signaling pathway.

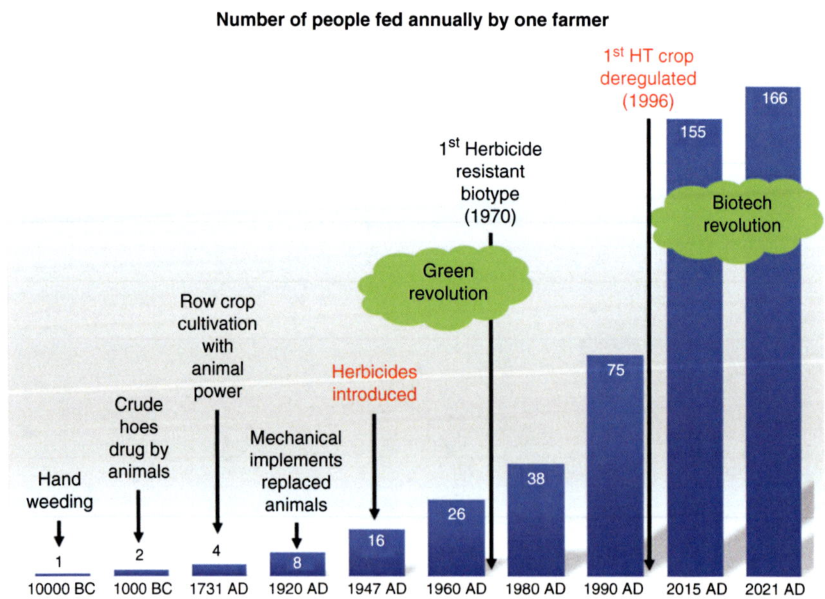

Figure 2.4 Chronology of agricultural production improvement in the United States and major discoveries associated with that growth. Source: Figure adapted from Monaco TJ, Weller SC, Ashton FM. 2002. Weed Science Principles and Practices, 4th Ed, John Wiley & Sons, Inc., NY and Fast Facts About Agriculture & Food, American Farm Bureau Federation, https://www.fb.org/newsroom.

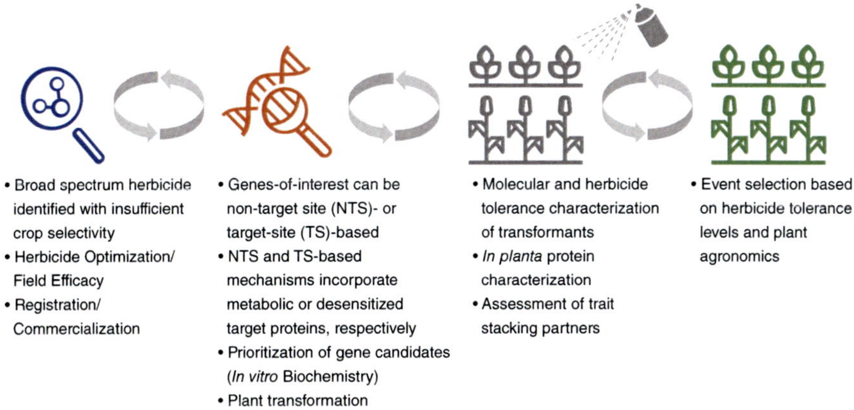

Figure 2.5 General processes for herbicide tolerance trait discovery.

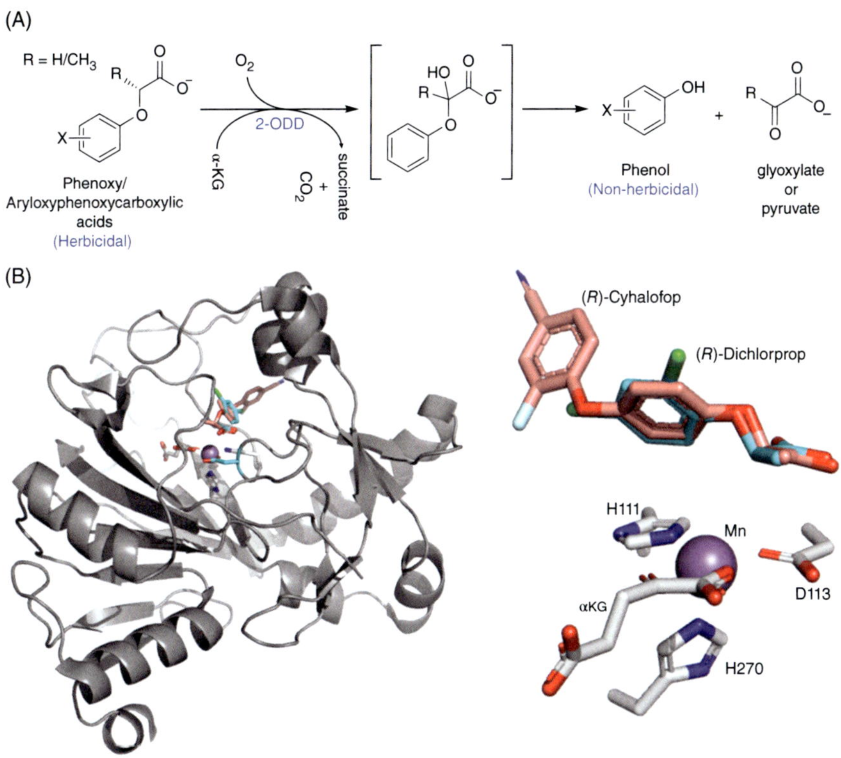

Figure 2.8 (A) General reaction catalyzed by 2-ODD during the initial step of herbicide degradation. (B) AAD-1 co-crystal structures demonstrating binding interactions various herbicide substrates (left, overall structure showing the catalytic site and right, overlay of bound AOPP and auxinic herbicides).

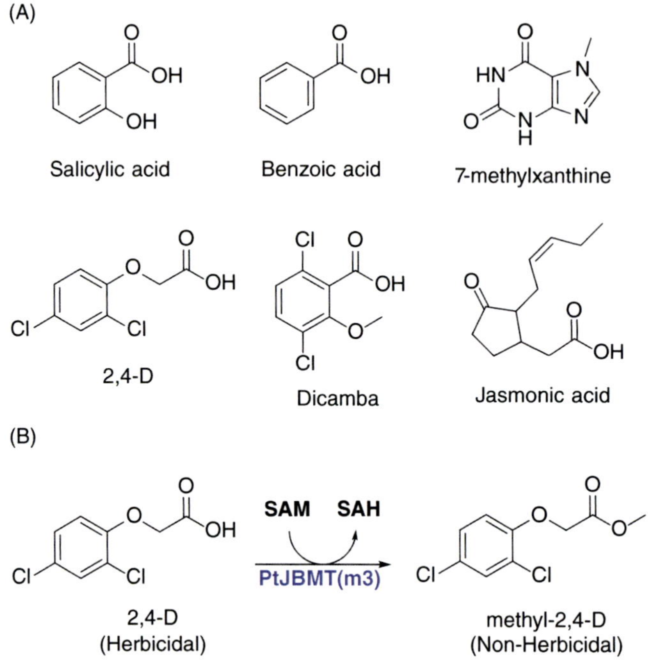

Figure 2.10 (A) Common endogenous substrates and synthetic auxins utilized by the SABATH methyltransferase family of enzymes. (B) PtJBMT (m3) catalyzed esterification of 2,4-D.

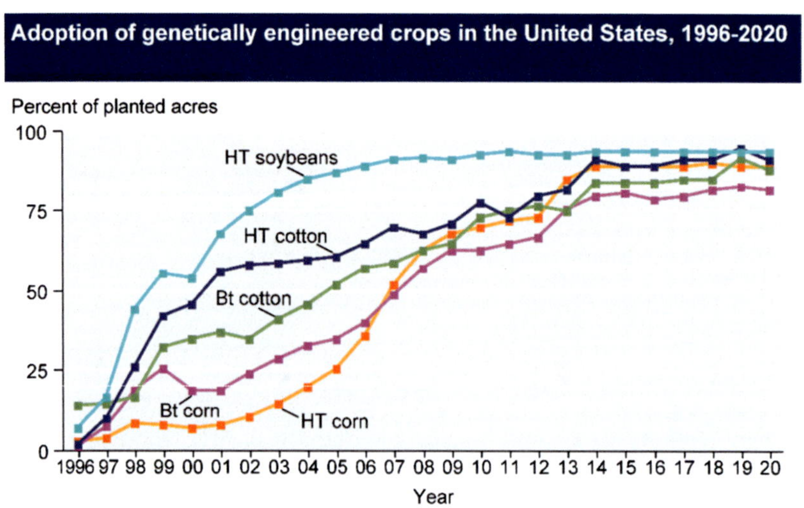

Note: HT indicates herbicide-tolerant varieties; Bt indicates insect-resistant varieties (containing genes from the soil bacterium Bacillus thuringiensis). Data for HT/Bt corn and cotton are not mututally exclusive, as HT and Bt categories include those varieties with overlapping (stacked) HT and Bt traits.
Source: USDA, Economic Research Service using data from the 2002 ERS report, *Adoption of Bioengineered Crops* (AER-810) for the years 1996-99 and National Agricultural Statistics Service, (annual) June Agricultural Survey for the years 2000-20.

Figure 2.12 Adoption of genetically engineered crops in the United States. Source: U.S. Department of Agriculture/CC BY 4.0.

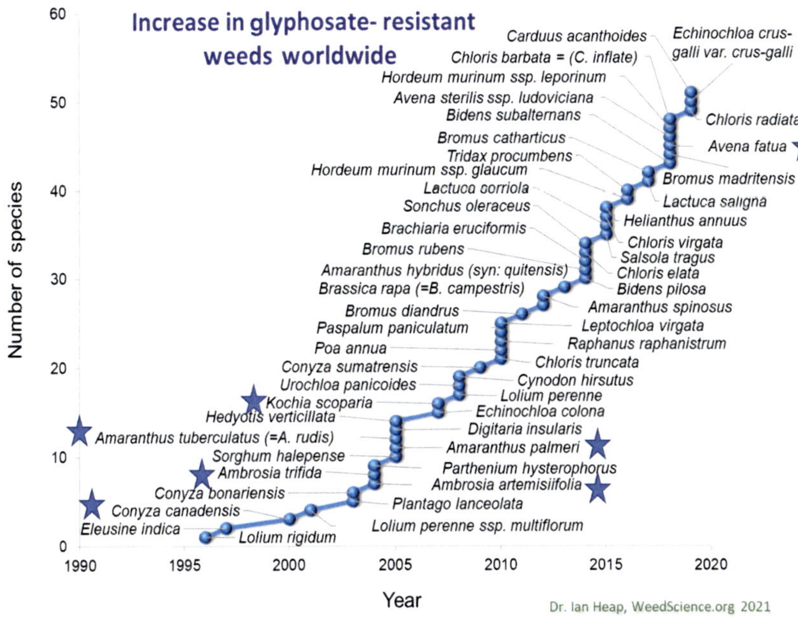

Figure 2.13 Species developing resistance to glyphosate over time. Species indicated with a star are those determined troublesome or most common by poll of growers (2017 WSSA Survey, Canadian Weed Scientists, Farm Journal Pulse Poll).

(A)

(B)

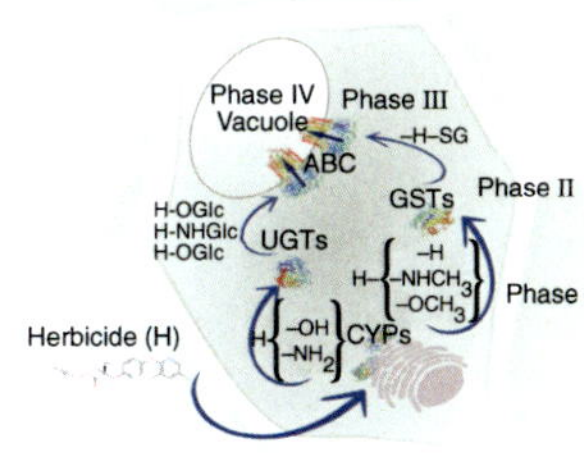

Figure 4.1 (A) Black grass and (B) herbicide detoxification in four phases by the xenome. ABCs=ATP-binding cassette transporters, CYPs=cytochrome P450s, GSTs=glutathione-*S*-transferases, UGTs=UDP-glucose-dependent glycosyltransferases.

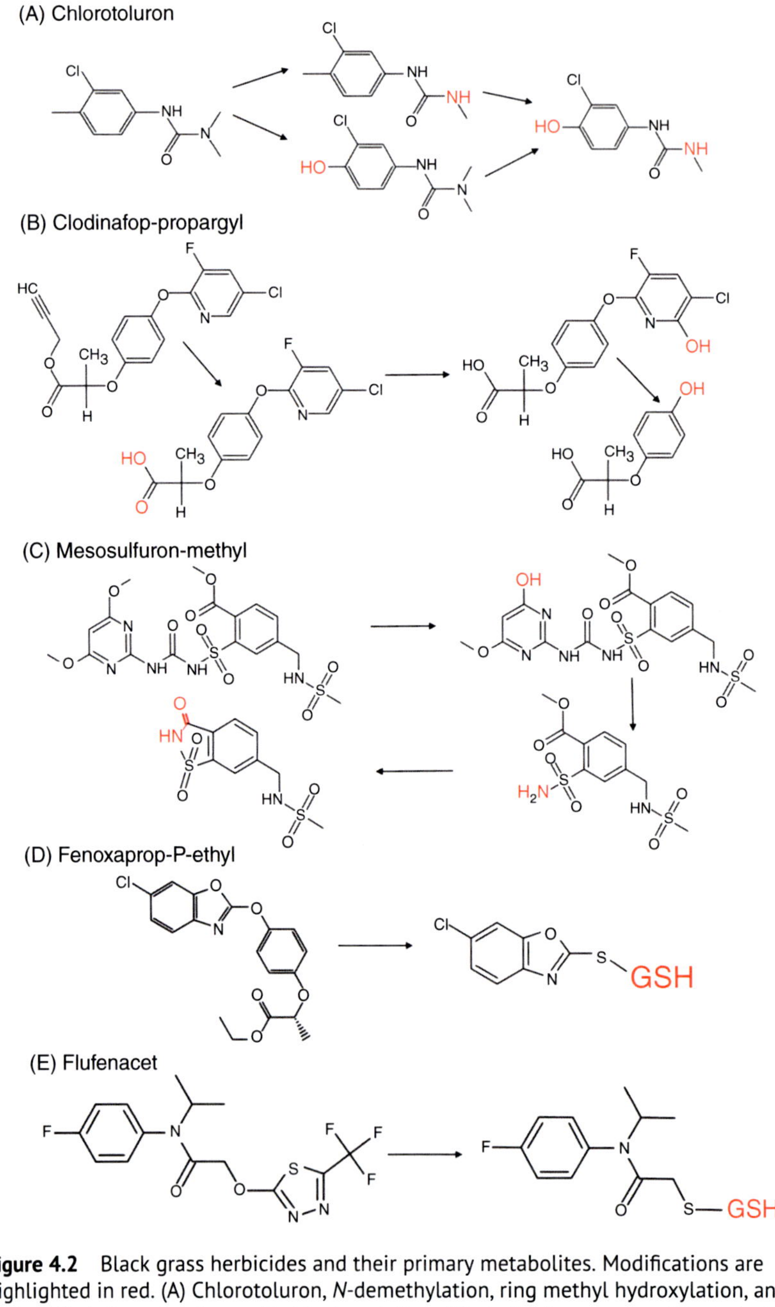

Figure 4.2 Black grass herbicides and their primary metabolites. Modifications are highlighted in red. (A) Chlorotoluron, *N*-demethylation, ring methyl hydroxylation, and *N*-demethylation plus ring methyl hydroxylation; (B) clodinafop-propargyl, hydrolysis, ring hydroxylation, and ether cleavage; (C) mesosulfuron-methyl, demethylation, cleavage, and cyclization; (D) fenoxaprop-*P*-ethyl and glutathione conjugation; and (E) flufenacet and glutathione conjugation.

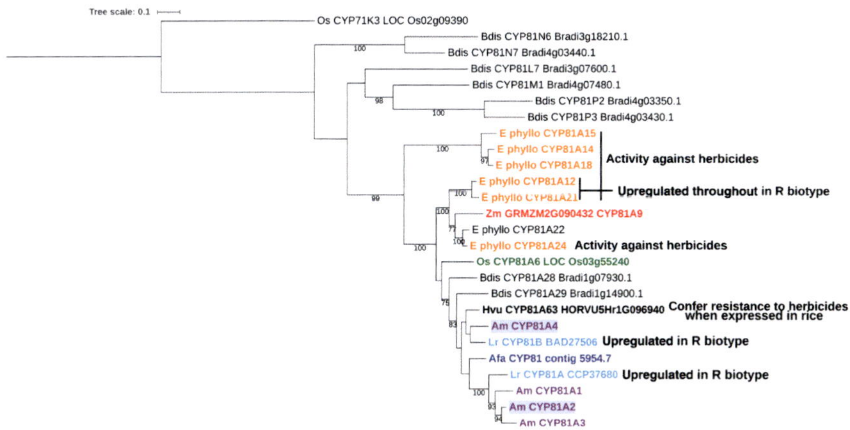

Figure 4.3 Phylogenetic analysis of CYP proteins from family 81. Amino acid sequences of CYPs from black grass (Am, *Alopecurus myosuroides*), brachypodium (Bdis, *Brachypodium distachyon*), wild oat (Afa, *Avena fatua*), maize (Zm, *Zea mays*), annual ryegrass (Lr, *Lolium rigidum*), rice (Os, *Oryza sativa*), later water grass (E phyllo, *Echinochloa phyllogon*), and barley (Hv, *Hordeum vulgare*) were used for maximum likelihood alignment for phylogenic analysis. The number on the branch represents the bootstrap support values above 70%. The scale bar indicates the inferred number of substitutions per site.

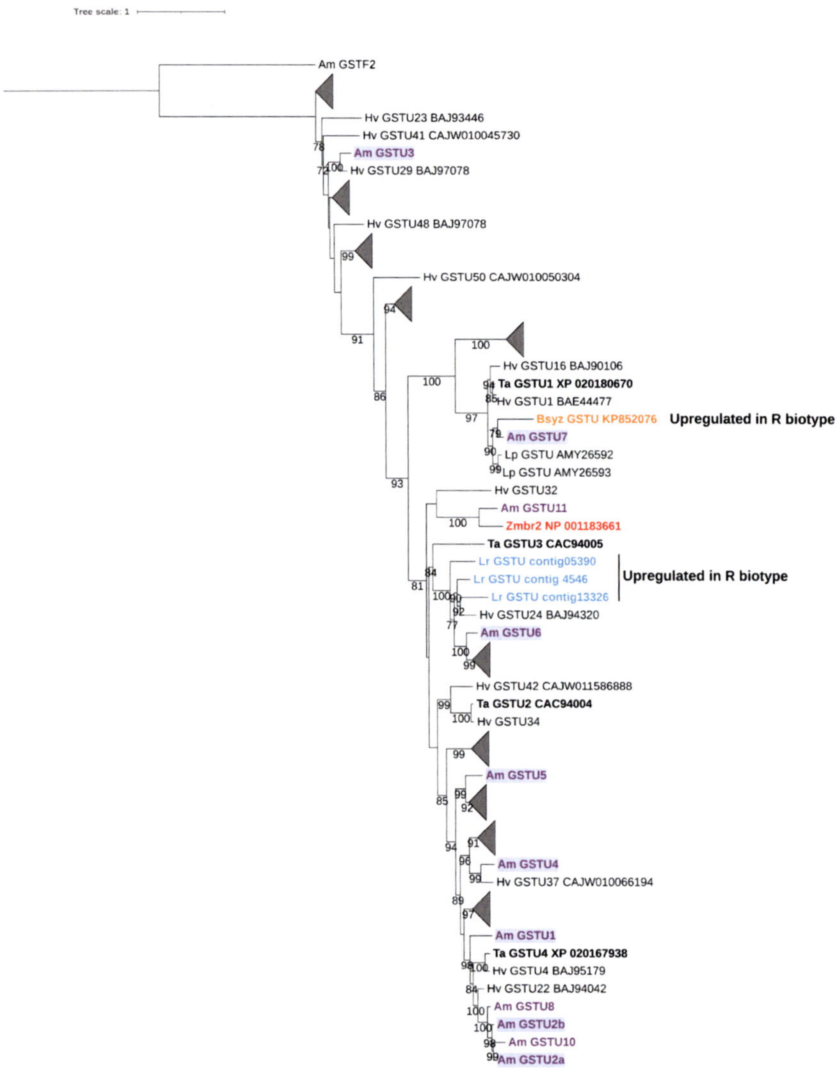

Figure 4.4 Phylogenetic analysis of GSTUs proteins in grass weeds and crop plants. Amino acid sequences of GSTUs from black grass (Am, *Alopecurus myosuroides*), barley (Hv, *Hordeum vulgare*), wheat (Ta, *Triticum aestivum*), maize (Zm, *Zea mays*), annual ryegrass (Lr, *Lolium rigidum*), perennial ryegrass (Lp, *Lolium perenne*), and American slough grass (Bsyz, *Beckmannia syzigachne*) were used for maximum likelihood alignment for phylogenic analysis. The number on the branch represents the bootstrap support values above 70%. The scale bar indicates the inferred number of substitutions per site. Clades comprising exclusively barley sequences were collapsed into triangles.

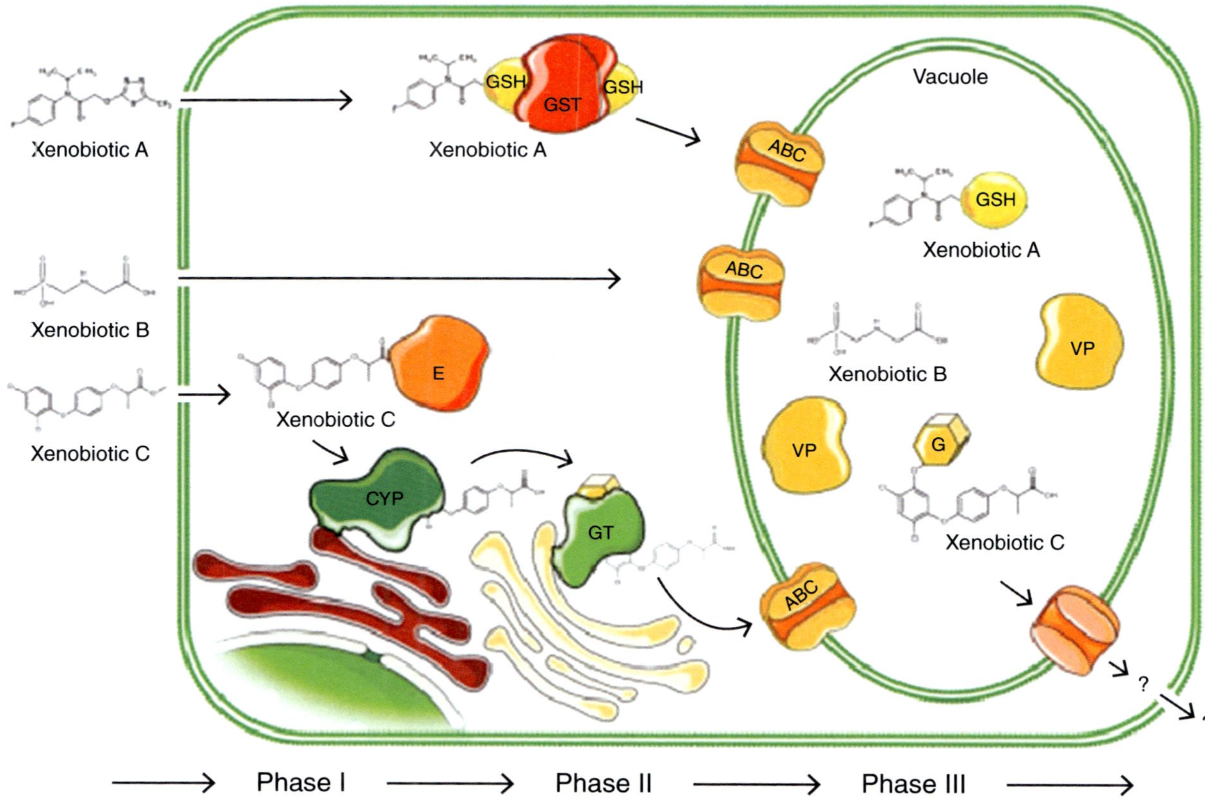

Figure 5.1 Summary of the main resistance mechanisms. Detoxification of xenobiotics in plants using the examples of flufenacet (xenobiotic A), glyphosate (xenobiotic B), and diclofop-methyl (xenobiotic C). The pathways include hydrolysis and hydroxylation catalyzed by esterases (E) and cytochrome P450 (CYP) monooxygenases in phase I; conjugation with glutathione (GSH) and glycosyl residues catalyzed by glutathione-*S*-transferases (GSTs) and glycosyl-transferases (GTs), respectively, in phase II. Phase III includes the transport into the vacuole, e.g., via ATP-binding cassette (ABC) transporters and degradation by vacuolar enzymes (VPs) as well as catabolism and compartmentalization. Source: Adapted from Refs.[35–39]

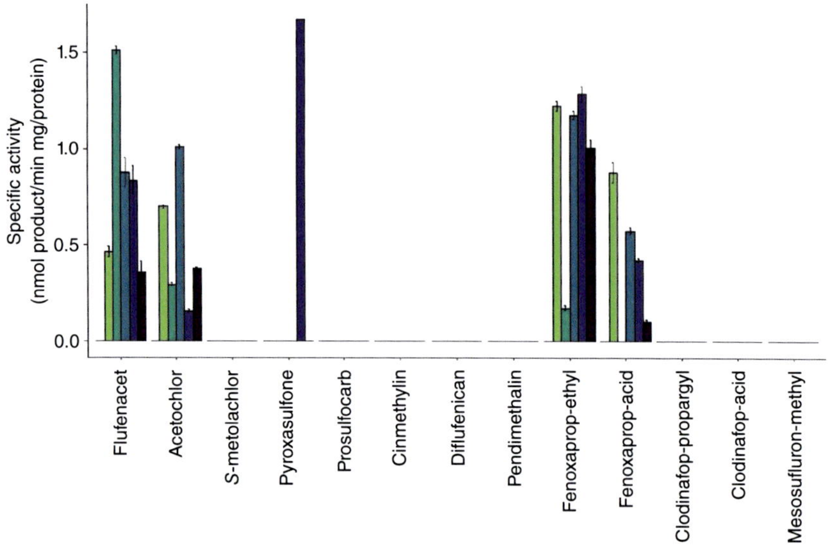

Figure 5.2 Enzymatic activity of recombinant EmGFP (■), ALOMY3G13667 (■), ALOMY3G13668 (■), ALOMY3G13670 (■), ALOMY5G35766 (■), and AmGSTF1/ALOMY3G11300 (■) on different substrates (flufenacet, acetochlor, *S*-metolachlor, pyroxasulfone, prosulfocarb, cinmethylin, diflufenican, pendimethalin, fenoxaprop-*P*-ethyl, fenoxaprop-acid, clodinafop-propargyl, clodinafop-acid, and mesosulfuron-methyl). The error bars represent the standard error of the mean ($n = 4$).

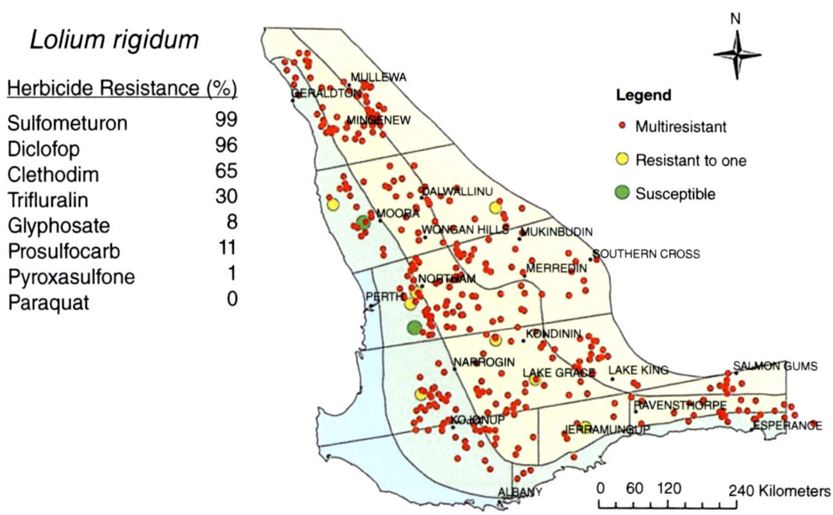

Figure 6.1 Distribution and abundance of multiple-resistant *Lolium rigidum* across Western Australia in 2015. Source: Adapted from Owen and Beckie.[11]

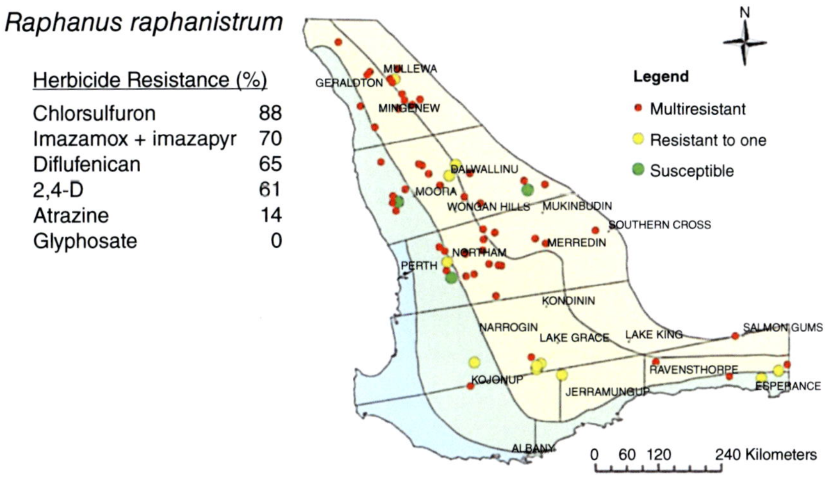

Figure 6.2 Distribution and abundance of multiple-resistant *Raphanus raphanistrum* across Western Australia in 2015. Source: Adapted from Owen and Beckie.[11]

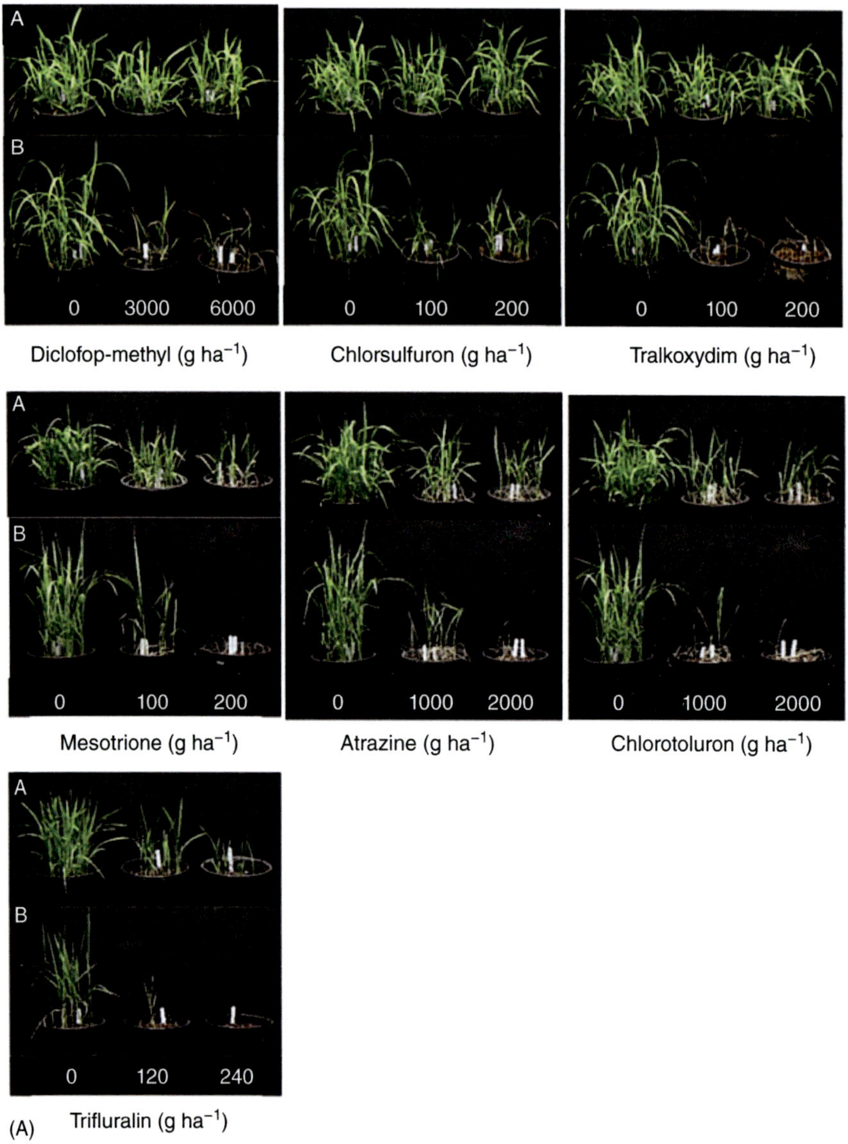

Figure 6.4 (A) Sensitivity of transgenic rice seedlings expressing CYP81A10v7 and green fluorescent protein (GFP) (negative control). The four leaf stage rice seedlings were treated with the respective herbicide, and photos were taken three weeks after treatment. A: CYP81A10v7, B: GFP. Source: Adapted from Han et al.[43]

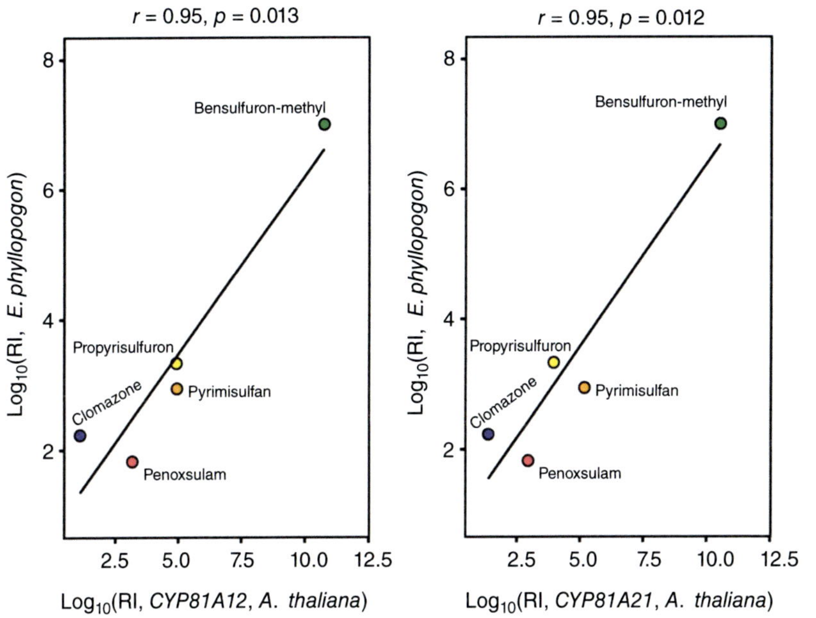

Figure 8.1 Correlation analysis of the resistance indices (RI) of *CYP81A12* and *CYP81A21* in transgenic *Arabidopsis thaliana* and in the multiple herbicide-resistant (MHR) line of *Echinochloa phyllopogon* to bensulfuron-methyl, penoxsulam, clomazone, pyrimisulfan, and propyrisulfuron. Source: Adapted from Dimaano et al.[2]

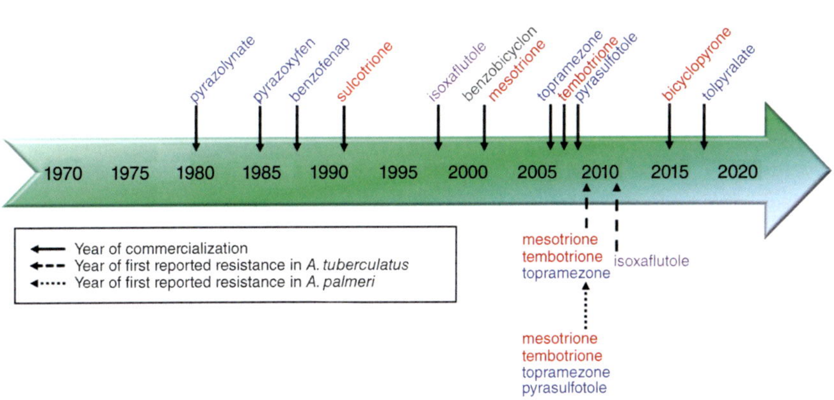

Figure 10.1 Timeline of commercialization of HPPD-inhibiting herbicides and the first year of reported resistance in waterhemp (*Amaranthus tuberculatus*) and Palmer amaranth (*A. palmeri*) populations. Year of commercialization, year of first reported resistance in waterhemp, and year of first reported resistance in Palmer amaranth are indicated by solid arrows, dashed arrows, and square-dotted arrows, respectively.

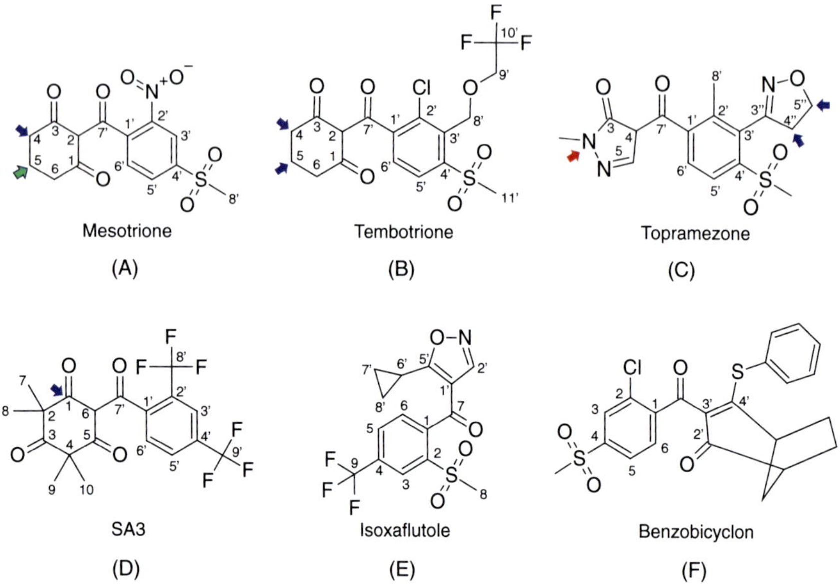

Figure 10.2 Structures of HPPD-inhibiting herbicides and reported primary sites of metabolism in weedy *Amaranthus* spp., maize, and rice.

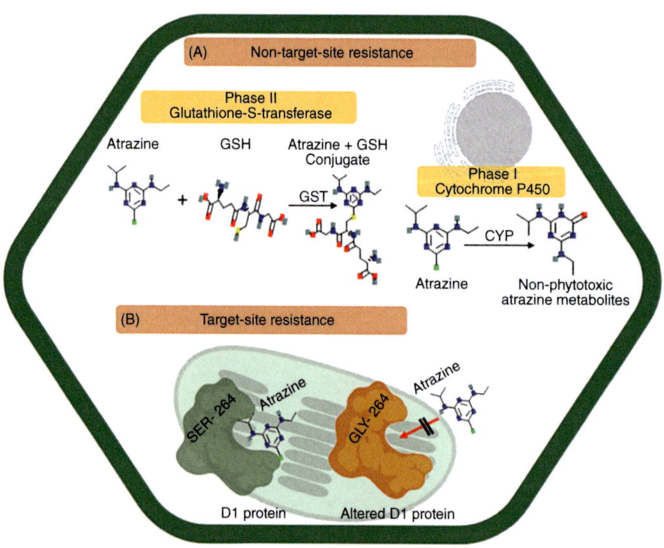

Figure 11.1 Illustration of known mechanisms conferring photosystem II (PS-II) inhibitor resistance in plants, using atrazine as a model herbicide.

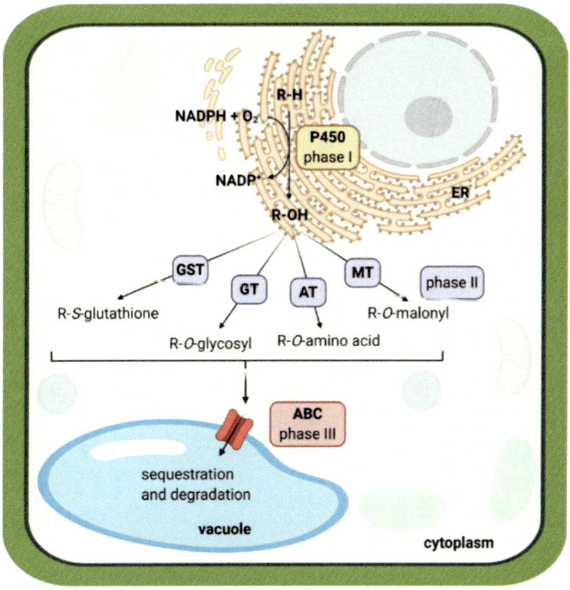

Figure 13.1 Phases of herbicide metabolic process, which herbicides can undergo when inside the plant cells. Source: Rigon et al.[3]/Ingenta.

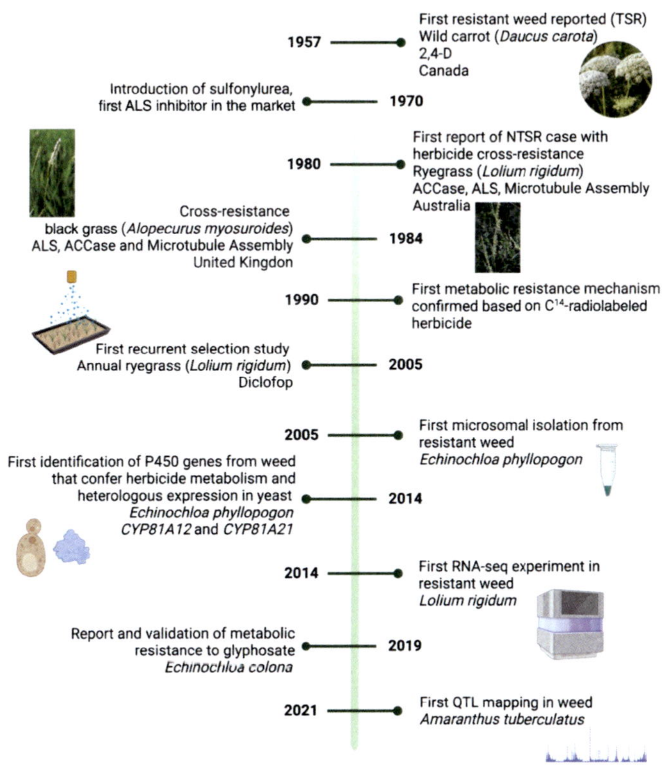

Figure 13.2 Timeline of key milestones in understanding metabolic resistance mechanisms in weeds over the past decades. Source: Patrik Stedrak/Adobe Stock Photos.

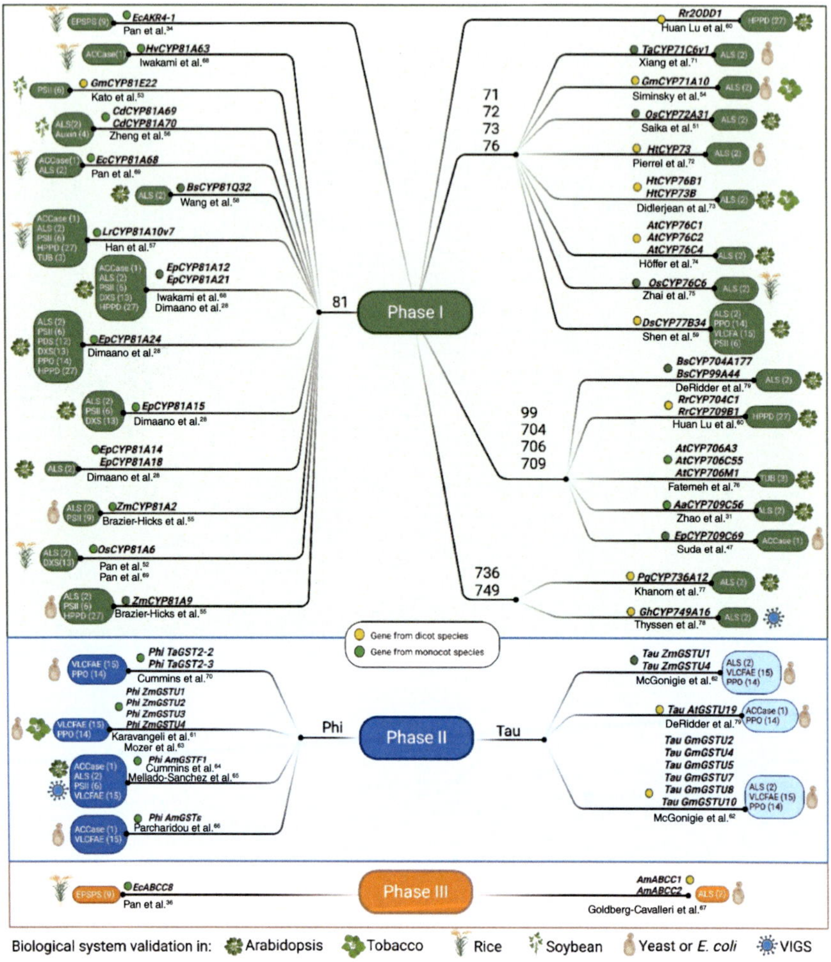

Figure 13.3 Illustration of herbicide metabolic resistance genes in weeds and their corresponding herbicide mechanisms of action (HRAC) for metabolism. Genes are separated by their function in the metabolic process as phase I (activation), phase II (conjugation), and phase III (compartmentalization and incorporation). Dark gray and light gray circles represent genes found in grass and broadleaf species, respectively. Note: VIGS – virus-induced gene silencing, Aa – *Alopecurus aequalis*, Am – *Alopecurus myosuroides*, At – *Arabidopsis thaliana*, Bs – *Beckmannia syzigachne*, Cd – *Cynodon dactylon*, Ds – *Descurainia sophia*, Ep – *Echinochloa phyllopogon*, Ec – *Echinochloa colona*, Gh – *Gossypium hirsutum*, Gm – *Glycine max*, Ht – *Helianthus tuberosus*, Hv – *Hordeum vulgare*, Lr – *Lolium rigidum*, Os – *Oryza sativa*, Pg – *Panax ginseng*, Rr – *Raphanus raphanistrum*, Ta – *Triticum aestivum*, Zm – *Zea mays*. Genes: *AKR* – aldo-keto reductase, *CYP* – cytochrome P450 monooxygenase, *GST* – glutathione-*S*-transferase, *ABC* – ATP-binding cassette.